On Quanta, Mind and Matter

Fundamental Theories of Physics

An International Book Series on The Fundamental Theories of Physics:
Their Clarification, Development and Application

On Quanta, Mind and Matter

Hans Primas in Context

edited by

H. Atmanspacher

*Institut für Grenzgebiete der Psychologie,
Freiburg, Germany
and
Max-Planck-Institut für extraterrestrische Physik,
Garching, Germany*

A. Amann

*Universitätsklinik für Anästhesie,
Leopold-Franzens-Universität,
Innsbruck, Austria*

and

U. Müller-Herold

*Departement für Umweltnaturwissenschaften,
ETH Zürich, Switzerland*

SPRINGER-SCIENCE+BUSINESS MEDIA, B.V.

A C.I.P. Catalogue record for this book is available from the Library of Congress.

ISBN 978-0-7923-5696-7   ISBN 978-94-011-4581-7 (eBook)
DOI 10.1007/978-94-011-4581-7

Printed on acid-free paper

Hans Primas

TABLE OF CONTENTS

Mind

Appendices

PREFACE
INSTEAD OF A "FESTSCHRIFT"

In June 1998 Hans Primas turned 70 years old. Although he himself is not fond of jubilees and although he likes to play the decimal system of numbers down as contingent, this is nevertheless a suitable occasion to reflect on the professional work of one of the rare distinguished contemporary scientists who attach equal importance to experimental *and* theoretical *and* conceptual lines of research. Hans Primas' interests have covered an enormous range: methods and instruments for nuclear magnetic resonance, theoretical chemistry, C*- and W*-algebraic formulations of quantum mechanics, the measurement problem and its various implications, holism and realism in quantum theory, theory reduction, the work and personality of Wolfgang Pauli, as well as Jungian psychology.

In many of these fields he provided important and original food for thought, in some cases going far beyond the everyday business in the scientific world. As is the case with other scientists who are conceptually innovative, Hans Primas is read more than he is quoted. His influence is due to his writings. Even with the current flood of publications, he still performs the miracle of having scientists eagerly awaiting his next publication.

His external life, by way of contrast, is not very spectacular. With the exception of a brief period as a guest professor at Washington University at St. Louis, he has never been away from Zurich for any length of time. He has never been awarded any prizes, never organized a congress, never done any organizational work in a scientific society. He deliberately distanced himself from the hustle and bustle of national and international scientific business.

Hans Primas' professional career started in 1945 with an apprenticeship as a chemical laboratory assistant for an industrial company in Oerlikon, close to his home city of Zurich. In 1948 he began to study chemistry at the Technikum in Winterthur. In 1953 he became a research chemist at the Laboratory for Organic Chemistry, which was headed at the time by the

Nobel laureate Leopold Ruzicka. In addition to taking courses in mathematics, he also attended lectures by Wolfgang Pauli during this period.

Under aegis of Hans-Heinrich Günthard, Primas became one of the founding members of the Laboratory for Physical Chemistry at the ETH in Zurich, where he stayed until his retirement. In 1960 he received his habilitation in the Chemistry Department and became an associate professor for Physical Chemistry one year later. 1966 saw him appointed as full professor for theoretical chemistry. He was Dean of the Chemistry Department for the periods 1967–68 and 1976–78. In 1991 his former doctoral student and co-worker Richard R. Ernst was awarded the Nobel Prize in Chemistry for his work on nuclear magnetic resonance.

The present volume takes up some of the discussions Primas has initiated or inspired. It deals with fundamental problems in the theory of matter and related philosophical questions. The individual articles have not been professionally reviewed and any editing has been restricted to matters of form. As a special feature of this book, and as a special service for the reader, each contribution is preceded by a brief introduction ("contextual background"), written by the editors, which places it in its scientific context in general or in the context of Primas' work in particular. The first three articles are an exception in this respect; to some extent they can be understood as background material themselves. In addition to the subjects they address, they also show aspects of Hans Primas' stature as a scientist.

The main body of the volume is organized according to the keywords "matter" and "mind", the first category focusing on physically oriented topics and the second on those of a philosophical nature. At the end of the volume, the reader will find a list of publications by Hans Primas until 1998, the addresses of all contributors, and an index.

After his retirement in 1995, the life of Hans Primas keeps being dedicated to scientific and general cultural matters. His workload is still enormous. And he is still no friend of pomp and circumstance – in fact he strongly dislikes such ceremonies. What he does like – and above all respect – is competent and constructive criticism. Habent sua fata libelli: this book is intended to provide valuable ideas and give rise to perceptive criticism or innovative research projects. This is the spirit from which it emerged, and this is the purpose to which it is dedicated.

October 1998

Innsbruck   *Anton Amann*
Freiburg/Garching   *Harald Atmanspacher*
Zürich   *Ulrich Müller-Herold*

THE PRIMAS EFFECT: ON QUANTA AND BEYOND

ULRICH MÜLLER-HEROLD
Departement für Umweltnaturwissenschaften,
ETH Zürich, CH-8093 Zürich-Hönggerberg, Switzerland

A university committee meeting. The discussion getting nowhere very fast. Old, familiar facts are being trotted out, with nothing more than slight variations, now and then. Progress has come to a halt; the air hangs heavier and heavier. Sometime, at long last, Hans Primas takes the floor. Suddenly . the problem appears to be transformed. There is a breath of fresh air, and the whole discussion takes a totally different turn. Such is the Primas effect. The term was coined by Konrad Osterwalder, the quantum field theorist who, having observed the Primas effect, officially coined this term on 9th November 1995.

What had struck Osterwalder is something that had also been noticed by many others who had encounters with Primas over the years: his virtually uncanny ability to break an object down into quasi-atomic components, to examine each of these components, to accept or reject them and then to start piecing them together again in his own unique way. For Primas himself, it was simply his way of learning, of tackling matters that were new to him. The amazing thing was that the result of his reconstructions looked unmistakably different from the original.

For his scientific colleagues it was often very disconcerting to be presented with such a radically different approach to things with which they were familiar from the literature or from practice, and which had become virtually second-nature to them. For science it was ideal: whenever Primas took a closer look at anything, it generally resulted in a new question, a new approach, a new frame of reference. And as Primas took a close look at a wide variety of things (within the mathematical physics community, he used to be referred to ironically as "the world's greatest pre-print collector"), the result was that, in the course of forty-five diligent years, an unusually large number of approaches, proposals and reformulations emerged. Several of his ideas Primas implemented himself. They form the bulk of his published scientific writings and yet, at the same time, they are the smaller part of the whole. Much simply fell by the wayside, either because of time, because of reviewers who lacked the necessary insight, or because he himself did not believe that there were a sufficient number of suitable addresses worth trying.

Actually, the core of his approach was always the same: he simply took seriously anything that was said or written. Given that science is anything else than consistent and is full of scarcely concealed jargon, he kept coming up against contradictions. In order to resolve them, he would go right back to the basics and check the underlying assumptions and their subsequent synthesis into a theory. In this respect he was ruthless.

The biggest undertaking of this nature – and also the one that was to have the most far-reaching consequences – concerns the untidy relationship between chemistry and quantum mechanics. It all started with the over-hasty announcement by P.A.M. Dirac, in the first flush of enthusiasm in 1929, that all the necessary knowledge was there for basing chemistry on quantum mechanics (Dirac 1929). This message became the credo of physical chemistry. But how could that be? Quantum mechanics states that there are no facts without measurements or, to be more precise, that facts do not exist in their own right, but only in relationship to a given measuring arrangement. If the method of measurement is altered, then the facts previously established lose their currency. As a chemist, on the other hand, Primas knew well that handbooks contain molecular properties, calculated by quantum mechanics, without, however, referring to any particular measuring device. How then was it possible for quantum mechanics to be a theory of chemistry if the way in which chemistry states its results contradicts quantum mechanics?

Starting in 1967, he set about getting to the roots of this problem and, in the years that were to follow, dedicated his term research lectures to it. On account of the fundamental nature of the question, his analysis included the whole foundation of quantum mechanics, in particular the various interpretations placed on it, the problem of measurements, and the status of formalism. The essence of these investigations was recorded in a series of ten sets of lecture notes entitled "The Quantum Mechanics of Large Molecules". These texts were studied throughout the world, with many a reader painstakingly decoding the German writings with the assistance of a dictionary. It was these writings that made Primas known and led to correspondence with such leading lights as Fock and Jauch.

Inquiring into the problem led to a revision of the fundamentals of "pioneer quantum mechanics" – as he liked to put it. After lengthy discussions, Josef Maria Jauch finally convinced Primas in May 1970 that the solution to the problem might lie in the introduction of so-called classical observables (Jauch actually preferred the term "essential observables") along the lines of what can be derived from the superselection rules in quantum field theory. The postulate of irreducibility in von Neumann's codification of quantum mechanics (von Neumann 1932) had first to be abandoned before quantum mechanics could be given fundamental competence for chemistry.

The postulate of irreducibility (and thus the impossibility of introducing classical observables and superselection rules into pioneer quantum mechanics) is characteristic for systems with a finite number of degrees of freedom, such as isolated molecules with a Coulomb Hamiltonian. Quantum field theory, on the other hand, which is precisely where the superselection rules were established, deals with systems with an infinite number of degrees of freedom. But, from a technical point of view, where could the superselection rules come from when dealing with a finite number of molecules having a finite number of degrees of freedom? At the time Primas saw the solution to this conundrum in a weak coupling between the molecules and their environment, as a result of the electromagnetic field which they themselves inevitably create.

However, his corresponding paper "Classical Observables in Molecular Quantum Mechanics" was never published. After a brief, heated exchange of correspondence with the publisher of the International Journal of Quantum Chemistry, Primas decided not to pursue the idea of publication. At the time, theoretical chemistry, in the first flush of its infatuation with the steadily increasing number of computational possibilities, was not interested in such questions. It was not until many years later that the outcome of this battle of strength was published in Primas' major monograph "Quantum Mechanics, Chemistry and Reductionism" (Primas 1983).

The hallmark of Primas' work is not just his personal, highly individual manner of proceeding but also its distinctive background in the history of science. His style of thought and research is pretty much pre-1940 and is thus rooted in a time before chemistry and physics became "big science" – an epoch with which today's scientists feel little more than vague notional ties. (It is a style to which German tradition has given the term "Geheimratswissenschaft", privy-council science.) Primas' research questions, on the other hand, are influenced by the fundamental crises of mathematics and physics in the early years of the twentieth century, which had led to a particular opening-up vis-à-vis philosophy.

For anyone living in the present day and age, it is far from easy to envisage a real-life image of "privy-council science". It was individual and elitist and, in its best representatives, reached standards of quality that have not been matched since. They often had the title of "privy councilor" bestowed on them, giving them a particularly distinguished position within society. It was based on the achievements of men whose education had been directed toward the development of personality and who had a firm system of values implanted in them. Such people were guided by an inner compass enabling them to tread the right path, regardless of the prevailing circumstances. This type of being is gifted with a clear sense of direction and tenacity, as well as self-discipline and self-sacrifice. The nineteenth century, with its cult of genius and its unshakeable belief in progress, provided the basis for the recognition – not to say veneration – of leading scientists of this stature. The dynamism of scientific excellence, recognition by society, and a heightened sense of personal responsibility led to a development spiral, which elevated scientists to lofty heights.

Primas is the very embodiment of this type of scientist. It did, however, drive him further and further away from the mainstream of science. The 1940s witnessed the emergence of a very different type of scientist. The reasons for this shift were the introduction of leveling research instruments, evolutions in society and large-scale military research programs, calling for huge sums of money and intricate coordination. This type of scientist no longer has an in-built compass but a sort of radar device, continuously scanning the environment in search of signals. Such scientists are highly flexible, have a sense of team spirit, can deploy marketing skills when the need arises and have a creative ability to seek out new opportunities. They possess organizational talent, secure their surroundings through research-policy networking and are successful in fighting for a share of the scant resources available for increasingly expensive research.

Taking a positive view of scientists of this type, we can see them as the product of successful adaptation to ever-accelerating changes in science and society. However, this development does have its price. The question as to the academic significance of a research project has, to a large extent, been superseded by an evaluation of its fundability and its chances of practical implementation. What were once major problems in science have now receded behind projects whose scientific importance is not always apparent.

Even more striking than its style is the philosophical thrust of Primas' research. At the start of the twentieth century, mathematics and physics were ejected from research routine on account of internal difficulties and were forced to debate basic questions that could not be solved with the everyday instruments of their disciplines. In mathematics, it was the collapse of the Hilbert program – the experience that there are more things that are true than can be proven – that was linked to the central question of what, conceptually and logically, "proof" actually is. The questioning of this, its most important, tool has shaken mathematics right up to the present. In quantum mechanics, it was the existence of incompatible observables and the so far unresolved measurement problem that led inevitably to the question: what sort of reality is it that quantum mechanics actually describes?

In one way or the other, latter-day natural scientists have always felt that they were moving nearer to a predetermined given reality, existing independently of them, and that to each element of a good theory there corresponds an element of reality. Philosophically speaking, this is the position of classical realism. All at once developments within physics had the effect of pushing a philosophically acceptable view into the "offside" position: quantum mechanics is not compatible with realism – at least not with its historical varieties. In a certain sense, quantum mechanics itself has turned out to be more clever than its founding fathers. Its very formalism disproved the philosophical prejudices of its inventors.

Establishing that formalized theories in the natural sciences could render philosophically acceptable basic positions obsolete represented a momentous turning-point. It made no difference that since Kant philosophy itself had opened classical realism up to question. The decisive point lay somewhere else; whereas philosophers had fought with words, quantum mechanics showed that a philosophically tenable position can be refuted by the facts of natural science, by means of the formalism of an empirically unchallenged theory. If classical realism is true, then quantum mechanics must be empirically wrong. Period. That was philosophizing with a hammer; it was an Archimedian point – like an age-old dream coming true. In this way, quantum mechanics took on a titanic aura.

Within the scientific community this led to a split: there were those who regarded quantum mechanics simply as a technique for calculating expectation values of physical parameters, who clung to a home-spun realism and chose not to bother too much about contradictions. This is the most common position. The alternative is to take seriously the difficulties arising between quantum mechanics and realism. Quantum mechanics does, of course, have some sort of bearing on reality. (How could it otherwise be confirmed so well empirically?) The question is: What precisely is this relationship? This is Primas' starting point, and it was precisely this position that transported his research projects into domains of fundamental significance.

Young scientists in particular have repeatedly been fascinated by the fact of there being someone engaged in research that casts such broad shadows. Research predominantly dependent on external funding – the more or less inevitable fate of today's young science graduate – is largely permeated with a dullness which is only thinly disguised by the pretended benefits and the flashiness of high-tech research apparatus. By contrast, Primas' dissertation themes always involved an intellectual venture. They were always couched in such a way that they could either culminate in a Nobel prize or in mental despair. Both these cases – the Prize and needing specialist medical care – are found among his PhD students.

His own credibility was never questioned – in contrast to other scientific borderliners. His extraordinary mathematical erudition, his computational skills and his outstanding knowledge of the literature ensured that any research projects would never appear doomed to failure from the outset. The factors that contributed decisively to the credit that was given to him included his reputation as a designer of nuclear magnetic resonance measuring devices, his numerous patents and his solid grounding in electronic engineering. It is also a well known fact that some of his former PhD students became professors of mathematics, chemistry, or physics.

The philosophical element in Primas' own research projects is not a manifestation of the maturity of his years. It is not an arm-chair philosophy, but one that emerged early in his career, due to the fact that he took note of the fundamental unanswered questions of quantum mechanics with his characteristic seriousness. Although he applied quantum mechanics throughout his scientific life, he viewed it less and less through the eyes of a user. The perception that the formalism of a good theory may lead further than the short-sightedness of its beginnings was a decisive lesson to be learnt.

Primas used to cherish the hope that the development of quantum mechanics would finally lead to results creating an impact on culture in general, beyond the bounds of natural science, and confronting philosophy, in particular, with a new situation. Such a hope is not unrealistic, as is shown by the quantum mechanical marvel that, starting from a theory of mechanics, formalism led to the correction that the material world is not a mechanism, i.e., not a mechanical system made up of individual components. Couldn't the same formalism – in a similarly fundamental step – also bring about the downfall of the Cartesian cut between mind and matter? That, however, would be a Primas effect of a novel and far-reaching kind.

References

Dirac P.A.M. (1929): Quantum mechanics of many-electron systems. *Proc. Roy. Soc. London A* **123**, 714–733.

Neumann, J. von (1932): *Mathematische Grundlagen der Quantenmechanik* (Springer, Berlin). English translation: *Mathematical Foundations of Quantum Mechanics* (Princeton University Press, Princeton, 1955).

Primas H. (1983): *Chemistry, Quantum Mechanics and Reductionism* (Springer, Berlin, 2nd edition).

HANS PRIMAS AND NUCLEAR MAGNETIC RESONANCE

RICHARD R. ERNST
Laboratorium für Physikalische Chemie,
ETH Zürich, CH-8092 Zürich, Switzerland

1. Preliminaries

Hans Primas in front of a cathode ray oscilloscope (Fig. 1), designing electronic circuitry? Hans Primas supervising an electronic design laboratory in order to build from scratch a nuclear magnetic resonance (NMR) spectrometer (Fig. 2)? – Inconceivable! We might ask whether we look at a stuntman. – No, indeed, it is himself personally. And those who met Hans Primas forty years ago will not be astonished at all.

More than any other scientist I know, Hans Primas covered the entire spectrum from practical industrial chemistry to theoretical physics and philosophy of science. He has set his landmarks on extraordinarily wide grounds. There is hardly a biographer who could adequately appreciate the superb quality of all his widespread achievements. Correspondingly, I will concentrate in this essay on just one aspect, his early contributions to NMR. Without exaggeration, one may claim that his approach, his particular line of thoughts contributed to the well-known breakthroughs in NMR in the 1960s and 1970s and to the unprecedented growth of the applicability of NMR from physics to clinical medicine.

How did Hans Primas come to work on NMR? His negative experience with the high school in Zürich, which he visited from 1941 to 1944, does not fill a particularly honorable page in the annuals of the Swiss schooling system of that time, reflecting inflexibility toward extraordinary students. What he experienced to be mindless exercises in Latin turned his interests from the subjects taught by teachers, who overestimated the importance of their own fields of knowledge, toward chemistry whose magic attraction fascinated him from the beginning. A typhoidal disease, finally, added the rest, and Hans Primas chose to start, instead of a humanistic education, an apprenticeship as a chemical laboratory assistant with the Werkzeugmaschinenfabrik Oerlikon, Bührle & Co.

Figure 1.   Hans Primas on 20 September 1958 in front of an oscilloscope testing an electronic circuit for a newly designed NMR spectrometer.

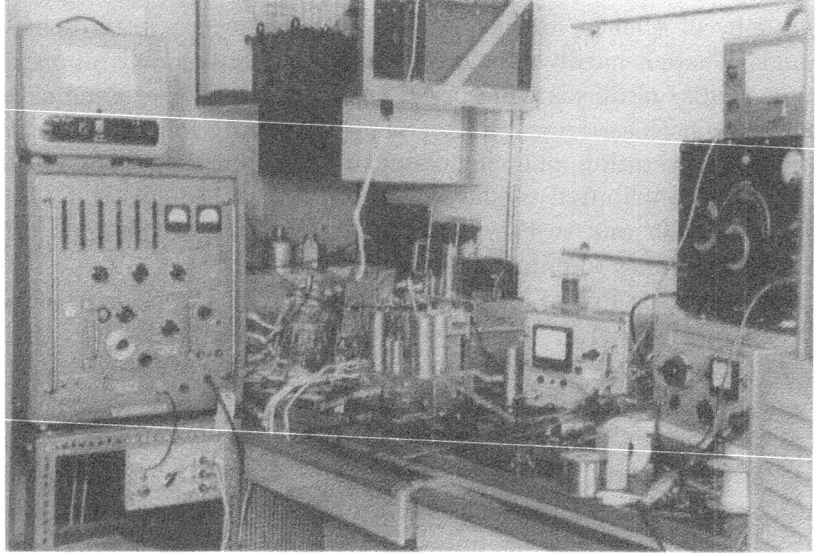

Figure 2.   Working place in Hans Primas' electronic design laboratory on 20 September 1958. In the center a home-built cascode preamplifier in the process of being tested.

Schweiz. Laboranten-Zeitung

Offizielles Organ der Schweizerischen Laborantenvereinigung

Basel, Januar 1950   Nr. 1   7. Jahrgang

Tüpfelreaktionen*)

von F. Osimitz und H. Primas

Unter Tüpfelreaktionen versteht man Semimikro-Reaktionen zur Identifikation von Anionen, Kationen und org. Radikalen. Im Allgemeinen werden sie auf der Tüpfelplatte oder auf Tüpfelpapier ausgeführt. Im weitern Sinne zählt man auch Reaktionen im Mikroreagenzglas und unter dem Mikroskop dazu.

Figure 3.   The first publication of Hans Primas which emerged from the work during his studies at the Technikum Winterthur.

During his apprenticeship, he seems to have felt deep satisfaction with his job. He eagerly learned whatever he could find on chemistry and its physical foundations. With his practical experience and his interests for basic aspects, he was properly prepared for the studies at the Technikum Winterthur from 1948 to 1951. His love for experiments and his analytical interests led also to his first two publications. They describe procedures for spot reactions [P1,P2], see Fig. 3. Especially, the second one gives a comprehensive survey on suitable reactions for the qualitative analysis of all major cations with a detailed description of the laboratory procedures. During this time he became fascinated by the work of Wolfgang Pauli who remained an idol for Hans Primas ever since then. He must have been a very extraordinary student with an unusual interest for the theoretical basis of chemistry. He rapidly caught the attention of Professor Rudolf Stieger who was the first to discover the exceptional talent of Hans Primas, and he motivated him to deepen his knowledge by continuing studies at ETH or at the University of Zürich.

It was also Professor Rudolf Stieger who established the first contacts with Privatdozent Hans Heinrich Günthard, working at the Laboratorium für Organische Chemie at ETH Zürich (Fig. 4), who was himself a brilliant absolvee of the Technikum Winterthur. Günthard suggested to Hans Primas to make up for the missing maturity examination, and Primas indeed

Figure 4. Professor Hs.H. Günthard on the occasion of his birthday 19 November 1958, inspiring his research group, including Hans Primas, by Campari and innovative concepts.

entered the private school Minerva just to find out, after one week, that this type of swotting was not according to his taste. He started then, 1951, as a special curricula student at ETH and Zürich University, registering for selected lectures in mathematics and theoretical physics, naturally without the possibility and also without the desire to graduate with a regular degree.

In 1953, he was employed by Günthard at the Laboratorium für Organische Chemie, ETH Zürich, as a scientific collaborator. And he remained at ETH Zürich for 42 years until his retirement in 1995. The first project assigned to Hans Primas was selected from the main field of activity of Günthard, the analysis of infrared spectra of organic molecules [P3–P9]. The mathematical aspects of normal mode analysis, combined with symmetry considerations, caught the analytical interests of Hans Primas. He explored the normal modes of chain molecules and extended the computational procedure by the introduction of circulant matrices which led to elegant solutions of the problem of diagonalization. Already in this early work one can discover his love, strongly supported by the scientific approach of Günthard, for formal calculations using operators and matrix representations. His incessant desire of focusing on the essential general aspects becomes apparent already at this early stage. He never had to face the danger of getting lost in irrelevant details.

After a few months of intensive work in optical spectroscopy, Günthard motivated Hans Primas to switch to nuclear magnetic resonance and to build up an independent research group in this promising new field. It kept Hans Primas busy for the following ten years.

2. Günthard's Spectroscopic Universe

It took some time, after the revolutionary discovery of quantum mechanics in the twenties by Schrödinger, Heisenberg, Pauli, and others, before molecular spectroscopy entered chemistry as a most powerful tool for accessing the molecular nature of matter. Without the development of electronics and optics during the second world war, practical routine applications of spectroscopy would hardly have been conceivable. It is thus not astonishing that chemical spectroscopy had a booming development during the late 1940s and the 1950s, especially in the United States and in England.

During the 1950s, physical chemistry in Switzerland was in a rather deplorable, antiquated state. A little bit of classical thermodynamics and some electro-chemistry formed the limited activities of a self-satisfied domain with little international radiance. Hans-Heinrich Günthard, vigorously supported by the head of the Laboratorium für Organische Chemie of ETH Zürich, Professor Leopold Ruzicka, recognized the need of the moment and decided to develop and assemble a comprehensive set of modern spectroscopic tools for chemistry, since 1954 as an associate professor within the Laboratorium für Organische Chemie and after his appointment 1958 as Professor of Physical Chemistry within his brand new Laboratorium für Physikalische Chemie. One after the other, like building blocks, nearly all known spectroscopic techniques were added to the arsenal. Obviously, infrared spectroscopy made the start, followed by Raman spectroscopy, ultraviolet spectroscopy, and optical circular dichroism. The first non-optical technique that was adopted was rotational microwave spectroscopy. Already in 1953, Günthard decided to add nuclear magnetic resonance, and finally, during the early 1960s, also electron paramagnetic resonance (EPR) was adopted.

Indeed, the Laboratorium für Physikalische Chemie soon became one of the world's centers of physico-chemical spectroscopy. Most of the instruments were designed and built in-house. A superb technical infrastructure was established by hiring some of the best mechanical and electronic engineers. It allowed Günthard and his research groups to solve even the most demanding technological problems. Usually, each PhD student had to build or modify an instrument before starting his own measurements. It is not astonishing that instrument design became a principal goal of the research activities during the early years. Numerous papers on the design and optimization of spectrometers were published, and signal-to-noise calculations and optimizations belonged to the daily routine. Many students, like myself, hardly ever recorded a relevant spectrum during their graduate studies. Nevertheless, they received a superb training in practical instrumentation, electronics, and signal processing and were well prepared for their later career. The practical work was accompanied by a very thorough training in the basic mathematical aspects of spectroscopy. Working sessions in small groups, led by Günthard himself, served the purpose of going through subjects such as group theory, based on the book by Wigner (1959); through quantum mechanics, in the matrix formulation by Born and Jordan (1930); through radiation theory, following the monograph by Heitler (1957); through optical spectroscopy, reading the three monumental volumes by Herzberg (1945, 1950, 1966); and treating many more subjects. In addition, mathematical treatises like those by Courant and Hilbert (1953) and Margenau and Murphy (1943) provided the necessary mathematical support of the theoretical activities.

3. The Emergence of NMR

The early historical development of NMR is well known by now. In 1944, Felix Bloch and Edward M. Purcell both attended a party honoring Isidor Rabi's Nobel prize in physics "for his resonance method for recording the magnetic properties of atomic nuclei". Rabi had observed magnetic resonance in a molecular beam experiment. Both Bloch and Purcell seem to have been stimulated to try to observe NMR also in condensed phase, and indeed they achieved this goal independently at the end of 1945 (Purcell et al. 1946, Bloch et al. 1946). At that time, nobody had the slightest idea that NMR could revolutionize analytical chemistry and become important even for clinical medicine. NMR was a tool for nuclear physicists who were interested in measuring nuclear magnetic moments in an attempt to obtain insight into the secrets of nuclear structure. In fact when a director of Dupont Chemical Company asked Purcell what the usage of NMR in chemistry could be, he answered: *"Absolutely none!"*.

Figure 5.   Hans Primas at work on 20 September 1958.

The first hints of a potential usefulness of NMR in chemistry were obtained when Proctor and Yu (1950) and Dickinson (1950) surprisingly found nuclear magnetic shielding effects caused by the electronic shell. The discovery of the chemical shifts stimulated a few brave and imaginative chemists to study the potential of NMR for distinguishing chemical components. In particular Herbert S. Gutowsky (Gutowsky and McCall 1951, Gutowsky et al. 1952) and, a little bit later, James N. Shoolery (1953) recognized the extraordinary power of NMR. Varian Associates started to commercialize NMR, producing and selling the first commercial NMR spectrometers with a proton resonance frequency of 40 MHz based on an electromagnet with a field strength of about 1 Tesla (10 kG).

It must have been these exciting developments, together with the Nobel prize in physics 1952 for Felix Bloch and Edward M. Purcell, which motivated Professor Günthard in 1953 to ask Hans Primas (Fig. 5) to start a major activity in NMR.

4.   The First High-Resolution NMR Spectrometer in Switzerland

In 1953, Hans Primas knew virtually nothing about nuclear magnetic resonance, he had very little experience with electronics, and he had never before designed a complex electronic device. It was concluded that a rapid start in the new field required some hands-on experience. Fortunately, Hans

Staub was appointed professor of physics at the nearby University of Zürich in 1949. He had spent the time from 1938 to 1949, finally as full professor, at Stanford University and had a close collaboration with Felix Bloch. He was mostly interested in accurate measurements of nuclear magnetic moments, for example of the neutron, proton, and various other nuclei. After his return to Zürich, Professor Staub also continued activities in solid-state NMR, and Hans Primas was admitted to spend several months in Staub's research group to learn how to construct a working NMR spectrometer. And Primas learned very fast!

In the meantime, Günthard organized the finances and could on 31 December, 1953, write from Glarus to Hans Primas: "Lieber Herr Primas, vor einigen Augenblicken erhielt ich Nachricht, dass der grösste Teil der Mittel, die für die Radiospektroskopie nötig sind, bewilligt wurden. Dies ist zwar ein unpersönliches Weihnachtsgeschenk, aber es ist nicht weniger erfreulich als andere. Vor allem hoffe ich, dass Sie mir helfen werden, dies Geld mit Erfolg zu verbrauchen. Mit herzlichen Grüssen, Ihr Hs.H. Günthard." In 1954, Günthard spent some sabbatical months in Cambridge, Massachusetts, and collected ideas, designs and electronic components that could be useful for the future NMR spectrometer. In a series of detailed letters he reported on his impressions and discoveries. In the meantime, Primas was working intensively on the design of the new instrument.

The design was done in a very professional manner. It was preceded by numerous calculations and optimizations. A trial and error approach was out of question. Primas and Günthard wanted to be on the right track from the beginning and hopefully, at the same time, to contribute novel design principles. For this reason, the construction took some time. The first real spectra could be recorded in 1955 (Fig. 6) and finally, 1957, a series of important papers on the design and the performance of the NMR spectrometer were published [P10–P13].

Primas and Günthard decided to use a permanent magnet for the indispensable stable and homogeneous magnetic field. They claimed that the principal advantage of permanent magnets would be their geometric field stability in contrast to the spatially drifting domains in electromagnets [P11], a belief obviously based on a lack of experience with electromagnets. Permanent magnets had been used by several authors before for NMR experiments (see Andrew 1956, Lösche 1957). The use of a permanent magnet limited the proton resonance frequency to 25 MHz. The magnet was designed and built in-house after a careful check of the homogeneity of the ferromagnetic core and of the non-magnetic pole pieces. Aesthetically a rather pleasing design was chosen (Figs. 7–10). The homogeneity achieved by the careful design was in the order of 2×10^{-7} over a sample volume of approximately 5 mm^3.

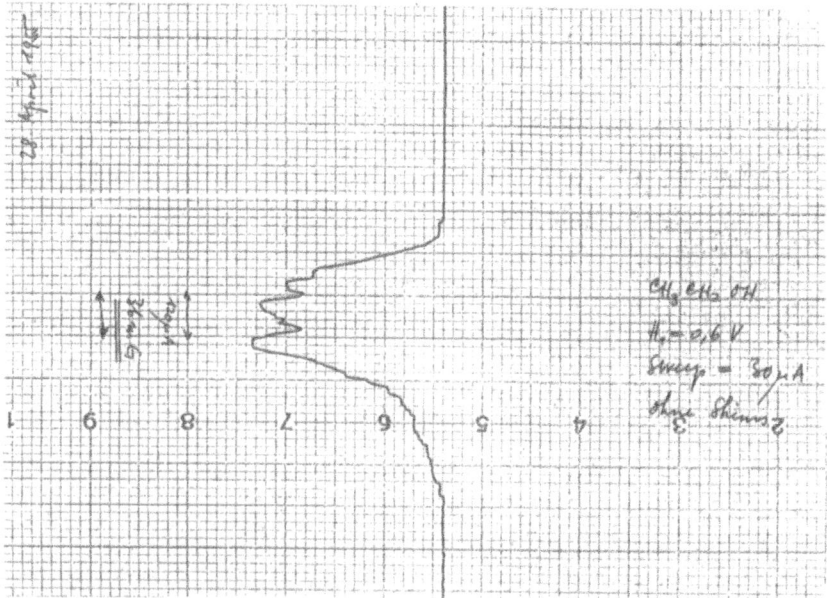

Figure 6. First documented "low-resolution" spectrum of ethanol recorded on 28 April 1955 by Hans Primas at 25 MHz proton resonance. Although no current shims were used, the three chemical shifts are visible (unpublished).

The inherent difficulties of achieving optimal homogeneity and minimal line width motivated Primas to perform a thorough study of lineshapes in inhomogeneous and time-dependent magnetic fields [P10]. This work is an early demonstration of his basic approach using linear response theory and the description of stochastic processes by the Wiener-Khintchine theorem. In some way, the response of a nuclear spin system to a stochastic perturbation appeared here for the first time. Some earlier references containing related lines of thought, e.g. by Gabillard (1951a,b), were not yet known [P11].

The achieved homogeneity of the magnetic field was then further improved by the development of systematically optimized current shims [P12]. It was found that a single pair of coils was sufficient to fully correct second-order gradients without the introduction of fourth-order gradients.

It was soon found that even permanent magnets are not sufficiently stable for demanding high-resolution NMR experiments. The temperature dependence of the magnetic field strength and the perturbations by the nearby streetcar line were disturbing. This led to one of the very first magnetic flux stabilizers applied to NMR spectroscopy (Fig. 9) [P13]. It used a pair of induction coils sensing magnetic field variations. The induced voltage was amplified by a rather delicate galvanometer amplifier, causing a lot

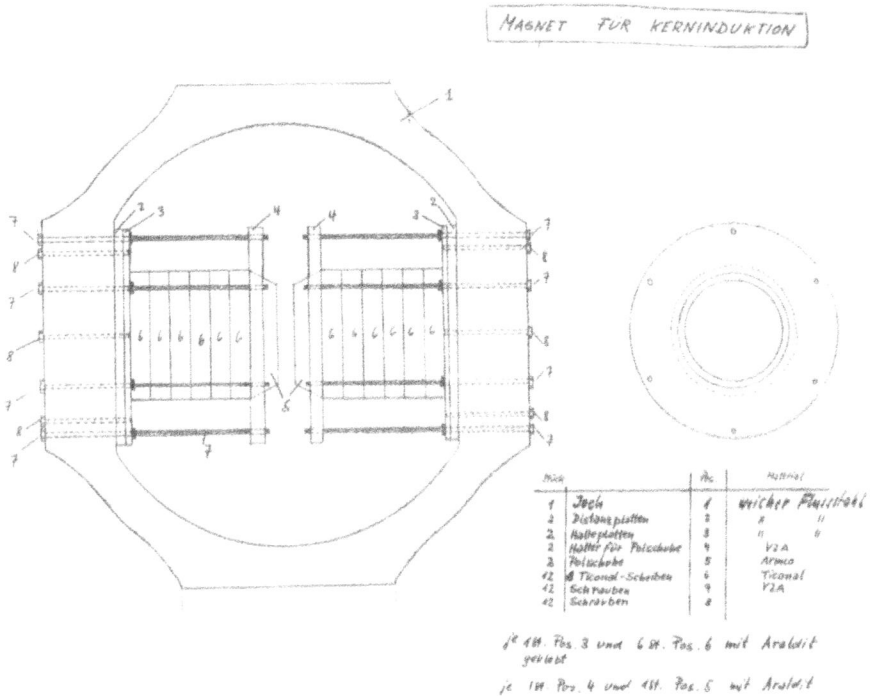

Figure 7. Hand drawing (1954) by Primas for the design of the 6 kG permanent magnet for NMR spectroscopy.

of headache in operation, and the integrated correction signal was fed back into a pair of correction coils. The performance was excellent and allowed for long-time sweep experiments (Fig. 10). A commercial superstabilizer for electromagnets was developed about the same time by Varian Associates (V-K 3506, 1956).

Later, in May 1961, Hans Primas described at the 5th European Congress on Molecular Spectroscopy in Amsterdam for the first time an NMR stabilizer for improving the magnetic field stability by locking the magnetic field strength on an internal reference line of the sample itself. This new stabilizer, using the tetramethylsilane (TMS) signal, gave a stability of better than 0.1 Hz over a month. Before, only field-frequency locks using a second, external sample with significantly inferior stability, have been described in the literature (Baker and Burd 1957).

Of particular originality was the heart of the spectrometer, the probe assembly which was designed with nearly complete rotational symmetry to minimize magnetic field gradients (Fig. 11). For the same reason, the

Figure 8. 25 MHz high resolution NMR spectrometer designed and built by Hans Primas and Hs.H. Günthard [P11].

sample volume was chosen to have spherical symmetry by blowing spherical cavities in cylindrical sample holders which were rotated in sapphire bearings, driven, as usual, by an air turbine. This design did not survive because of difficulties in manufacturing accurately spherical sample containers. The means for orthogonalizing receiver and transmitter coils was also quite novel.

Some irony is contained in a footnote in [P11]: *"Die in der Literatur angegebenen Relationen sind oft inkorrekt"*. The given relation for the filling factor, to which this footnote referred, contains also a sign error! But except for such minor deficiencies, the four papers [P10–P13] document a truly

Figure 9.   6 kG permanent magnet "in operation". The switch on the right is for the magnetic flux stabilzer, the bottles in top for the NMR samples and the ocasionally successful operators.

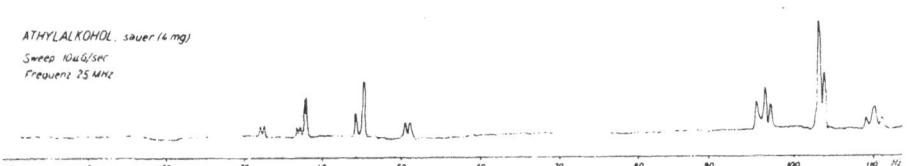

Figure 10.   High-resolution spectrum of ethanol recorded with a flux-stabilized magnetic field at 25 MHz proton resonance frequency with a total sweep time of 50 min. The single line on the left originates from the OH proton, the center quartet from the CH_2 group and the right triplet from the CH_3 group [P16].

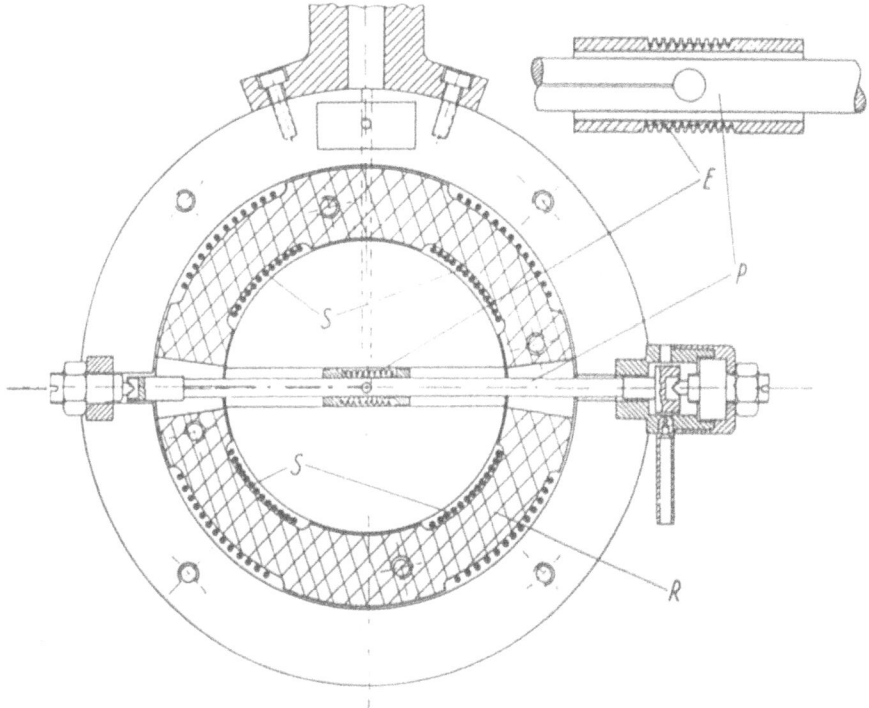

Figure 11. Drawing of the probe assembly by H. Primas for the 25 MHz proton magnetic resonance spectrometer [P11]. The receiver coil (E) is wound tightly around the sample tube (P). The transmitter coils (S) are mounted on a plexiglass ring (R) that can be rotated to adjust the orthogonality of the two coils.

heroic and highly successful effort to build from scratch a high-performance NMR spectrometer.

Professor Günthard was very well acquainted with the efficiency of modulation schemes for eliminating low-frequency noise and base-line instabilities. It is therefore not astonishing that also in this early NMR spectrometer a modulation scheme of the static magnetic field was used in order to eliminate leakage due to not perfectly orthogonal transmitter and receiver coils. Hans Primas prepared the necessary mathematical formalism for the treatment of modulation effects in NMR [P16]. He followed an earlier treatise by K. Halbach who developed a "model-free description of modulation effects in NMR" (Halbach 1954). Primas found, in addition, suitable solutions of Bloch's equations for arbitrary radio-frequency field strength and arbitrary modulation amplitude. He also thought about the equivalence of field and frequency modulation. His paper [P16] represents one of the first thorough discussions of modulation effects in NMR.

The decision of Professor Günthard to commercialize the brand new

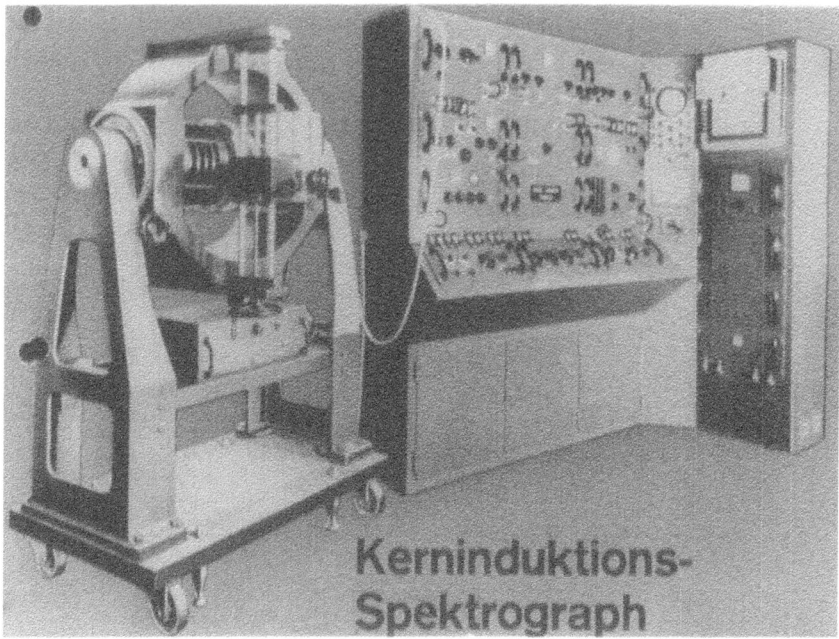

Figure 12. KIS I NMR spectrometer built by Trüb, Täuber & Company closely following the prototype developed by Hans Primas and Hs.H. Günthard.

spectrometer turned out to be an enormously beneficial move. It had a lasting effect on the development of analytical instrumentation in Europe. Günthard approached the Swiss company Trüb-Täuber which manufactured electronic measurement instruments and had also a department producing electron microscopes. And Trüb-Täuber, particularly Dr. L. Wegmann, became interested in diversifying the production. The prototype NMR spectrometer designed by Hans Primas was copied almost without change (Fig. 12), except for an attractive dark red color which was selected by the first customer contemplating in front of a half empty glass of red wine. The instrument was exhibited for the first time in 1958 at the instrument fair ACHEMA in Frankfurt under the name KIS I (Kerninduktions-Spektrometer I). KIS I became a reasonable commercial success, and Trüb-Täuber NMR instruments were installed in many European laboratories.

With an excellently performing NMR spectrometer at hand, the outside pressure increased to perform also some useful applications. Organic chemists, in particular Prof. Vladimir Prelog and Dr. Albert Eschenmoser, frequently knocked at the door, asking for spectra, not always with success as the remark of Professor Prelog upon a visit of the NMR laboratory on 8 September 1958 documents: "Nach diesem sehr guten und bewährten Rezept ist es bisher noch niemandem gelungen, Gold herzustellen." Nevertheless a few spectra were produced, some of them published [P15,P19], but the output remained low, partially because of the limited interest of Hans Primas in down-to-earth chemistry problems. He never developed a particular attachment to a specific class of chemical compounds or reactions and felt no urgent desire to record extended series of spectra. His primary motivation was and remained the exploration of the physical concepts underlying chemistry and nature in general.

5. Working with Graduate Students

While so far Hans Primas worked mainly by himself, supported by skilled mechanical and electronic engineers, and by the enthusiasm of Professor Günthard, it became time to expand the research group and to get involved also in educational activities by supervising graduate students. Officially, it appeared to be an impossible situation: a researcher without any university degree should foster and educate students toward a PhD! In practice, nobody objected, and we as his students respected him as our teacher at least as much as any "real" professor. He knew so much more than we and we could learn immensely from him as a scientist and as a very kind person who was serious and playful at the same time. It was an enormous pleasure to collaborate with him.

The atmosphere in the NMR laboratory of Hans Primas was very relaxed and friendly, as is exemplified by many notes in the Lab journal, such as the "Satz von Arndt" of 9 April 1958: *"Jede Schraube hat ein Rechtsgewinde mit Ausnahme der Linksschrauben"*; or the "Maximes and Definitions" of 28 March 1958: *"(1) Recht ist was dem Primas nützt.* (E. Billeter) *(1bis) Recht ist was der Quantenmechanik nützt.* (H. Primas)." Risky predictions were usually underpinned by a bet, mostly involving a bottle of Campari or Whisky. Correspondingly, well-performing sample tubes were called "Campari-Röhrchen". Hans Primas always had time for jokes and discussions. The real work he performed apparently during night hours and on weekends.

His first graduate student was Rolf Arndt, who graduated as a natural scientist (Abteilung X für Naturwissenschaften). He started 1957 to work on an extension of the existing 25 MHz instrument to a solid state spectrometer. He designed a low-temperature probe assembly and contributed, in addition, to the general instrument development. He tested his solid-state NMR equipment by measurements of the temperature-dependent second moment of cycloalkanes which show interesting ring-puckering motions (Arndt 1962). These measurements were motivated by some earlier work by Prof. Vladimir Prelog.

I began my PhD work in 1958 after one and a half years of military involvement. The subject of my thesis was defined by Prof. Günthard: "Von zu irreduziblen Darstellungen zugehörigen Linearkombinationen von Spineigenfunktionen". I had asked him for a theoretical piece of work as theory interested me particularly. What he really had in mind by the posed subject, I never found out, and I was not particularly inspired by it either. The start of my thesis work was correspondingly diffuse. Coming back alive from my military commitment, I found Hans Primas busy with a soldering-iron at the work bench (Fig. 2). So I got my own iron and started to solder randomly within electronic circuits without much success. After numerous 300 V shocks, I learned slowly, very slowly about the secrets of electronic vacuum tubes from books and from Hans Primas. "Vacuum Tube Amplifiers" by Valley and Wallman (1951) was our electronics bible at that time.

If I remember properly, my first assignment was to build a room thermostat to minimize the temperature-induced drift of the magnetic field. I learned a lot about positive and negative feedback. Our standard experience was that a stabilizer at first oscillates and an oscillator functions initially as a stabilizer. The regular occurrence of sign errors in experiments and calculations appeared to be an unfailing natural law. To compute and measure the frequency response of devices were daily routine, and we became capable of predicting the behavior of a device as an oscillator or stabilizer. Nyquist diagrams, Laplace and Fourier transforms were used regularly, and we learned how powerful the concepts of linear response theory are.

Not knowing yet what my final thesis subject would be, I started to concentrate on radio-frequency electronics, as Primas decided to build a new state-of-the-art "high-frequency" NMR spectrometer operating at 75 MHz proton resonance. Obviously, this required an electromagnet which again was designed from scratch by Hans Primas in two versions, the first one with a high voltage power supply (Fig. 13) and the second one with a high current supply. My task was to design a 75 MHz probe assembly and 75 MHz low noise preamplifiers. Although they formed part of my thesis (Ernst 1962), I do not recall that they ever worked properly for recording real spectra. Other parts of my thesis will be addressed in section 7.

At that moment, we felt it was time to summarize and document our experience building NMR equipment in the form of a more extensive publication. This finally resulted in a series of four papers [P20–P23] which describe all conceivable aspects we felt to be essential for spectrometer designers. Due to the publishing in a somewhat obscure German journal, the impact of these papers was relatively limited.

The 75 MHz spectrometer design was again adopted by Trüb, Täuber & Co. AG, this time incorporating more modifications and upgrading it into a 90 MHz spectrometer which was then sold under the designation KIS II. It was already a powerful NMR spectrometer for multinuclear applications

Figure 13. Electromagnet designed by Hans Primas for a 75 MHz proton magnetic resonance spectrometer with parts of the spectrometer.

in liquids and solids and was equipped with an NMR field-frequency lock system for field stabilization.

Again came a period where the treatment of real chemical problems was needed to justify the past instrumental investments. In particular, we urgently needed some chemically relevant results for a presentation at the International Meeting of Molecular Spectroscopy in Bologna 1959. As a curiosity within Primas' list of publications, we determined additive group contributions to the proton chemical shifts of organic compounds [P27]. Although we had a rather limited data set of chemical shifts, the additivity relations worked surprisingly well. Later this type of approach became quite popular for the prediction of chemical shifts.

In order to further enhance the output of chemically relevant results, Peter Bommer, a devoted chemist, was accepted by Primas as his third graduate student in the fall of 1958. Peter Bommer concentrated on experimental spectroscopy of chemical compounds in collaboration with users from industry and other university laboratories. During 1959, one day per week was reserved for service spectra. Peter Bommer recorded in this time about 200 spectra at 25 MHz proton resonance, some of which were published (TCC Information 10-60, P. Bommer and H. Primas, "Einige Beispiele zur Anwendung der Magnetischen Kernresonanz-(KR-)Spektroskopie für die Strukturaufklärung in der organischen Chemie", Trüb, Täuber, Zürich, 1960). For some time, he had even a commercial Varian A60 Spectrometer available. I recall just one occasion when a customer sample produced a spectrum that could in no way be interpreted as originating from a single molecular species. It indicated the presence of a mixture. Indeed, an inquiry led to the customer's confession that in order to save half of the spectroscopy fee he mixed two unrelated samples together! Peter Bommer seemed not to be too happy with the missing support and lacking interest of Primas for such down-to-earth problems.

For his PhD thesis, Peter Bommer extended the concept of incremental group contributions to the chemical shift using a much larger set of chemical shift data. To pay his compulsory duties in instrument design, he extended the 25 MHz NMR spectrometer for the performance of double resonance experiments. He also contributed toward a generalized theory of double resonance for strongly coupled spin systems. He graduated 1963 with a thesis on the subject: "Beiträge zur Bestimmung von Chemical Shifts: A. Additivität des Chemical Shifts, B. Doppelresonanz" (Bommer 1963).

The fourth graduate student, Hans Kummer, came 1959 and chose a theoretical subject for his thesis work: the question whether a given NMR spectrum uniquely determines the underlying spectral parameters, such as chemical shifts and spin-spin coupling constants. This is a question of some practical relevance. The outcome of the study was rather comforting as no

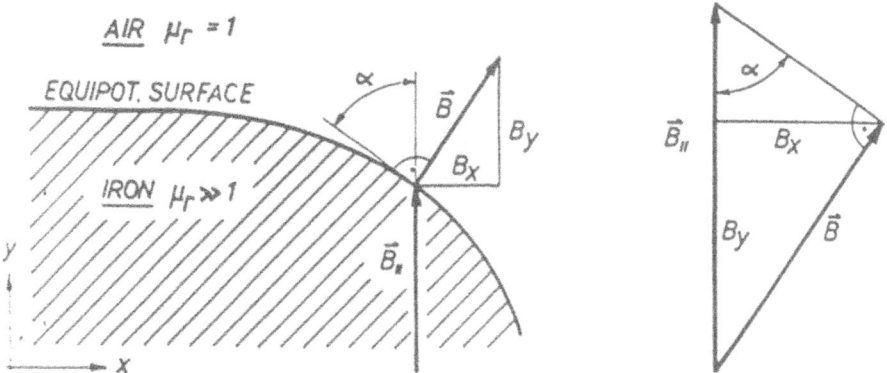

Figure 14.   Design principle for magnet pole caps shaped according to a magnetic equipotential surface [P34].

critical cases of importance could be found – in agreement with the basic theorems proven by Kummer (1963a,b).

The fifth and last student who had the unique chance to collaborate with Hans Primas on NMR was Adalbert Huber, a physicist who in 1962 started his thesis on "Eine Apparatur zur Präzisionsmessung von Resonanzverschiebungen in der hochauflösenden magnetischen Kernresonanz-Spektroskopie". Huber completed the construction of the yet unfinished 75 MHz proton-resonance spectrometer. In addition, he developed a procedure for the very accurate measurement of differences of resonance frequencies. By locking the magnetic field to a fixed resonance frequency of a first line and locking a second frequency oscillator to a second line, it was possible to measure the difference frequency between the two lines with an accuracy of 1% of their widths (Huber 1969).

Adalbert Huber was also involved in the last major contribution to NMR instrumentation and in the last paper written by Primas on an NMR-related subject [P34]. Extending ideas expressed for the first time by Kumagai (1960), Primas found that in order to obtain homogeneous magnetic fields at any field strength the surface of the pole pieces of the electromagnet has to form an equipotential surface $U =$ const. This implies that the magnetic field $\underline{B} = -\operatorname{grad} U$ is invariably perpendicular to the surface and the magnetic field inside of the pole cap constant for a sufficiently large permeability μ. Possible saturation effects are thus homogeneous throughout the entire pole cap and do not distort the magnetic field configuration (Figs. 14,15). Shaped pole caps following equipotential surfaces have been used successfully in many commercial magnet systems, including those of Trüb, Täuber & Co. AG in the KIS II spectrometer.

In 1964, Hans Primas presented the last of his lectures at NMR confer-

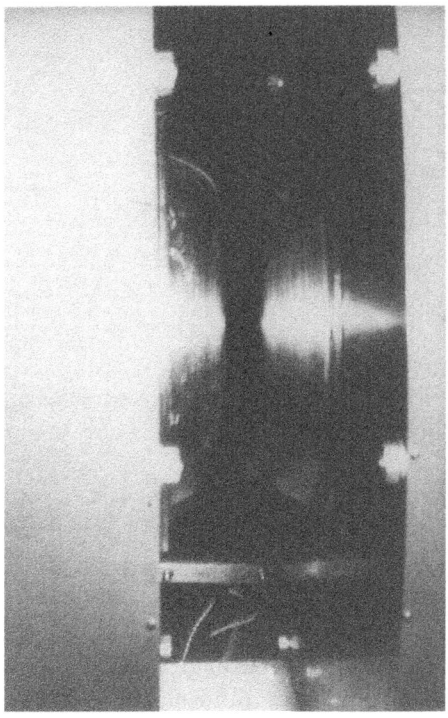

Figure 15. Shaped pole caps mounted within the 75 MHz proton resonance magnet designed by Hans Primas.

ences. He spoke at the Fifth Experimental NMR Conference in Pittsburgh on "A New, Universal, High-Homogeneity Magnet for Zero to Twenty-Five Kilogauss and its Precise Stabilization". The main subject concerned his invention of the shaped pole caps, and he started his lecture with the following question: *"Why build a magnet? The best way to get a good magnet is without any question to buy one. But there is a type of 'complementarity' between what you can buy just now and what you would like to have just now. We needed for some special experiments a magnet of the highest degree of homogeneity combined with a field-geometry that is independent of the field strength. By definition, a physical chemist is a mathematician who can blow glass; why should he not be able to build his own magnet, we thought. It has taken us several years to learn that magnet building is a highly developed art."* Then he continued with his "Basic Commandments" regarding magnet design.

In 1964, when the important paper on wide range electromagnets of high homogeneity [P34] was written, the thoughts of Hans Primas (who in the meantime, 1961, had become an Associate Professor of Physical Chemistry) had left NMR and concentrated on quantum chemistry, its foundations and implications. About the same time, the company Trüb, Täuber & Co. AG closed its operations in 1965 and parts of it were sold to other companies, the electron microscopy operation to Balzers Instruments, and the NMR operation was regrouped and formed the kernel of a new company, named Spectrospin AG, which was founded 1965. It became part of the Bruker group, headed by Professor Günther Laukien. Initially, the KIS II spectrometer design was adopted for the Spectrospin spectrometers, and the original ideas of Hans Primas inspired indirectly also the instrument development at the highly successful Spectrospin AG. In this way, Hans Primas had a very significant and lasting impact on NMR instrumentation.

6. Theoretical Work of Hans Primas Related to NMR

The theoretical work of Hans Primas, which accompanied the instrumental developments in NMR, seemed to have as a common goal the development of a universal operational description of spectroscopy, a description which does not require, from the beginning, the discussion of special cases, a description which allows one to perform formal calculations, and which reveals the common aspects of many spectroscopic methods. I remember how much Hans Primas disliked the almost weekly appearance of new spectroscopic techniques or of minor improvements of them in the literature. Many times he mentioned that it would be desirable to find a general procedure that would allow one to logically predict once and for ever the optimum way of performing an experiment or deducing the required information. He dreamed, so to say, of a "world formula" of experimental design. I am sure that he would have felt this desire even stronger had he been faced, 20 years later, with the multitude of multi-dimensional experiments that were proposed with all their crazy acronyms. For many experimentalists, on the other hand, these nearly unlimited possibilities of creativity form a major attraction of NMR; it has the fascination of a playground for juvenile spectroscopists who do not like to become "mature".

Hans Primas' attitude was different. Although he was often playful during the daily work, although he liked experimental surprises as well as jokes and small talk, his approach to science did not allow for any compromises. He wanted to grasp the absolute truths in as simple terms as possible. Special cases and occasional divergences of expansions did not interest him as long as they were not more than an occasional nuisance. Perhaps this is one of the reasons why he later concentrated on the general aspects of quantum theory and on the philosophy of science. NMR was too restricted and too mundane for his profound mind.

The theoretical work of his early period is mostly formal in nature. General expressions are being deduced, but seldom applications are worked out in detail nor are experimental verifications attempted. A sentence springs to my mind which he often used in an ironic mood: *"Never perform experiments and develop theories on the same problem. They might lead to nasty contradictions!"* Indeed the theory of NMR is so accurate that experiments are often superfluous and can be replaced by computer simulations. Logically, a theoretician must conclude that NMR represents a solved problem and can be left to the engineers. This might have been a motivation for Hans Primas around 1964 to abandon the field forever.

The first of the four major theoretical contributions to NMR consisted of "a new method for the direct calculation of the spectrum of the radiation absorbed or emitted by a quantum mechanical system" [P17,P18]. Although Primas mentions the work by Kubo and Tomita (1954) and Kubo (1957a,b), it seems that he developed largely independently a linear response formalism where instead of the spectrum its Fourier transform $K(t)$, called correlation function, is computed directly:

$$K(t) = \text{Tr} \{ X(t) X(0) \} \tag{1}$$

with

$$X(t) = \text{e}^{-\text{i}Ht} P \, \text{e}^{\text{i}Ht}, \tag{2}$$

where P is an arbitrary Hermitian (observable) operator. He generalized then Eq. (1) by introducing a complex "correlation function"

$$\xi(t) = \frac{1}{2} \text{Tr} \{ \text{e}^{-\text{i}Ht} P^- \, \text{e}^{\text{i}Ht} P^+ \}. \tag{3}$$

Obviously, at this point, the idea of Fourier spectroscopy is just around the corner! He then developed sum relations for the intensities and an approximation method for the computation of a Hamiltonian of the form

$$H = H_o + \epsilon H_1, \tag{4}$$

where he obtained signal intensities and frequencies as power series in ϵ.

The second contribution [P25] extends the perturbation treatment used above by applying a systematic operator notation without ever having to use a specific matrix representation. He starts again from a "perturbed" Hamiltonian

$$\mathcal{H} = K + V \tag{5}$$

and attempts to determine a unitary operator e^G to transform $\mathcal{H}$ into the "level-shift representation"

$$\text{e}^G (K + V) \text{e}^{-G} = K + W, \tag{6}$$

where it is required that $[K, W] = 0$ so that K and $K + W$ can simultaneously be diagonalized. With $V = \sum_{n=1}^{\infty} \epsilon^n V_n$, W and G are then obtained as two power series in ϵ:

$$W = \sum_{n=1}^{\infty} \epsilon^n W_n, \quad G = \sum_{n=1}^{\infty} \epsilon^n G_n. \tag{7}$$

The terms W_n and G_n can conveniently be expressed by the "resolvent superoperator" $1/\underline{k}$ where

$$\underline{k}(A) = [K, A] \tag{8}$$

is a "derivation superoperator":

$$W_1 = \langle V_1 \rangle, \qquad W_2 = \langle V_2 \rangle + \tfrac{1}{2} \langle [\tfrac{1}{\underline{k}} V_1, V_1] \rangle, \quad \cdots$$

$$G_1 = \frac{1}{\underline{k}} V_1, \qquad G_2 = \tfrac{1}{\underline{k}} V_2 + \tfrac{1}{2\underline{k}} [\tfrac{1}{\underline{k}} V_1, V_1 + W_1], \quad \cdots \tag{9}$$

Here $\langle V_1 \rangle$ denotes the diagonal part of V_1 with respect to the Hamilton operator K. An explicit evaluation of $\tfrac{1}{\underline{k}}(X)$ can take advantage of Lie function relationships.

In this work, Primas for the first time utilized superoperators in a systematic way. Indeed there is an extensive appendix to [P25] which defines and characterizes the superoperators used in this paper. Of course, Primas did not invent the concept of superoperators, and he heavily refers to authors such as Fano (1957) and Crawford (1958). But he was most likely the first to recognize the importance of this concept for formal calculations in spin dynamics. Today, the usage of superoperators is very widespread in NMR (Ernst et al. 1987, Jeener 1982) and is also implemented in general computer software tools (Smith et al. 1994).

Hans Primas did not like the idea of popularizing superoperators too much, but the outside pressure on him was appreciable. And when a guest arrived for being introduced to NMR and no suitable research project with chemical relevance was available for him, Primas agreed to let Dr. C.N. Banwell collect and systematize the superoperator relations contained in his notebooks and seminar protocols. For the formal training of the graduate students, a density operator and superoperator seminar was organized, and Primas presented weekly his new discoveries found in the literature and his own developments. The result was an unpublished, but very useful collection of operator equations and, finally, the well known paper by Banwell and Primas [P31] which, together with [P17], still today is a classical reference for superoperator relations. That the algebraic relations of operators, acting on Hilbert space, could be transferred without any difficulties to superoperators acting on Liouville space and spanning themselves a super-Liouville space was very pleasing and clearly revealed the common properties of formalized algebras. This matched the lines of thoughts of Primas who desired, in general, to discover common rather than distinctive properties.

7. Stochastic Resonance

The last contribution of Hans Primas to be mentioned in this context has been essential for the further development of modern NMR. Primas has been intrigued for a long time by the work of Norbert Wiener, Claude E. Shannon, Andrei N. Kolmogorov, A.I. Khinchin, and others on stochastic processes in the context of electronic measurements and data transmission. Books such as "Random Processes in Automatic Control" by Laning and Battin (1956) and "An Introduction to the Theory of Random Signals and Noise" by Davenport and Root (1958) belonged to the daily reading. They were later supplemented by Norbert Wiener's "Extrapolation, Interpolation and Smoothing of Stationary Time Series with Engineering Applications" (Wiener 1949) and "Nonlinear Transformation of Random Processes" (Wiener 1958). The NMR signals were invariably weak and noise was everywhere. Optimized data processing was an absolute must, and filtering theory was in daily use for the suppression of random noise.

In this situation it was a true revelation that noise could itself be used in a beneficial manner for the characterization of physical systems. The application of noise to linear time-invariant systems, white noise in and colored noise out, was considered as trivial. But the effects of nonlinear systems on random noise or the effects of random noise on nonlinear systems were truly exciting. Testing of nonlinear nuclear spin systems by the response to random noise seemed to open up new dimensions. Here the concept of broadband excitation and detection of nuclear spin resonances was articulated. It was at that time (about 1959) unknown that Russel Varian had already put similar ideas into a patent application (Varian 1956).

The project of using random noise for testing quantum mechanical systems led to the Habilitationsschrift of Hans Primas, published as [P24], and to the second part of my own thesis, published as [P29]. Although the sensitivity and information advantage of using broadband excitation and detection was never explicitly mentioned, it formed at least subconsciously a strong motivation of pursuing investigations in this direction. It was also characteristic that never any experiments were performed nor planned. None of us believed in the practical usefulness. We were just interested in general concepts.

The motivation in [P24] for exploring quantum mechanical systems with a stochastic Hamiltonian was kept brief and general: (a) description of systems with certain external parameters varied purposely in a stochastic manner, and (b) description of dissipative systems by the coupling of a quantum mechanical system to a macroscopic environment. The main result of this paper refers to a stochastic Hamiltonian of the form

$$H(t) = H_o + \lambda r(t) V \,, \tag{10}$$

where $r(t)$ represents a normalized Gaussian white noise process. The solution for the expectation value $a(t)$ of an observable A,

$$a(t) = \mathrm{Tr}\left\{A\rho(t)\right\}, \tag{11}$$

is expressed in the form of a series expansion

$$a(t) = K_o + \sum_{n=1}^{\infty} \int_{-\infty}^{\infty} d\tau_1 ... \int_{-\infty}^{\infty} d\tau_n K_n(\tau_1, ..., \tau_n) H_n(t - \tau_1, ..., t - \tau_n) \,, \tag{12}$$

where $H_n(t_j, ..., t_m)$ is a member of a set of orthonormalized stochastic Hermite polynomials, defined by

$$
\begin{aligned}
H_o &= 1 \\
H_1(t_j) &= r(t_j) \\
H_2(t_j, t_k) &= r(t_j)r(t_k) - \delta(t_j - t_k) \\
H_3(t_j, t_k, t_m) &= r(t_j)r(t_k)r(t_m) - r(t_j)\delta(t_k - t_m) \\
&\quad - r(t_k)\delta(t_j - t_m) - r(t_m)\delta(t_j - t_k) \\
.... &= ....
\end{aligned} \tag{13}
$$

and $K_n(\tau_1, ..., \tau_n)$ are operator kernels of the form

$$
\begin{aligned}
K_o &= \mathrm{Tr}\left\{A\,\Pi\right\}, \\
K_n(\tau_1, ..., \tau_n) &= \lambda^n(-i)^n U(\tau_1)U(\tau_2 - \tau_1) ... U(\tau_n - \tau_{n-1}) \\
&\quad \times \mathrm{Tr}\left\{[\overline{V}(-\tau_1), ...[\overline{V}(-\tau_n), \overline{\Pi}(-\tau_n)]...]A\right\} e^{-\omega|\tau_n|} \,, \tag{14}
\end{aligned}
$$

with the unit step function $U(t)$ and with the stationary solution Π of the second-order Karplus-Schwinger equation

$$i[H_o, \Pi] + \lambda^2[V, [V, \Pi]] = \omega\{\xi(1 - \beta H_o) - \Pi\}. \tag{15}$$

The bar indicates the transformation to the interaction representation

$$\overline{V}(t) = e^{iH_o t}\, V\, e^{-iH_o t}, \tag{16}$$

and $\xi = 1/\mathrm{Tr}\left\{1 - \beta H_o\right\}$ in the high-temperature approximation. Although some remarks are made in [P24] on the convergence of the series expansion, no detailed investigation of these aspects was undertaken.

It was finally my own task to work out an example to demonstrate, on paper, the application of the theory of systems with a stochastic Hamiltonian [P32]. Broadband homonuclear decoupling by a stochastic radio-frequency field was selected. The goal was the design of a technique by which all spin-spin splittings in a homonuclear spin system could be eliminated by strong irradiation of all resonance lines except for those under investigation. Although this is a highly desirable goal for applications, it was not attempted to develop a new practical method, but rather to explore the physical mechanism of decoupling phenomena. The hypothetical experiment employed white noise with a narrow empty frequency gap within which the lines to be observed were positioned. The final results showed phenomena similar to line broadening and line narrowing under chemical exchange, well-known in dynamical NMR (Ernst et al. 1987, Jackman and Cotton 1975, Kaplan and Fraenkel 1980). In this way, a link between the usually deterministic phenomena of spin decoupling and the random phenomena of chemical exchange was established, a fact which we realized later (Ernst 1966).

Unfortunately, a rather ugly computational error sneaked into the final results, represented by Fig. 4 in [P32] and Table 13 in Ernst (1962). The obtained asymmetry of peak shapes, apparent in Fig. 4 of [P32], is due to an incorrect evaluation of the (correct) Eq. (4322-8) in Ernst (1962), neglecting a portion of the imaginary part of γ_3^- and the resulting dispersive part of the line shape. On physical grounds, it is plainly obvious that the peak shapes must be symmetric with respect to the center of the doublet (see also Ernst 1966). That both myself and Hans Primas overlooked this error may be symptomatic for our attitude at that time. The formalism and its elegance were more important than the numerical results. Obviously, we did not think much about possible implications. Nevertheless, this work became the germ of the later development of noise decoupling (Ernst 1966) which was relevant for the development of carbon-13 spectroscopy and had, for some time, a sizeable commercial significance, becoming part of analytical high-resolution spectrometers.

Stochastic resonance was proposed later (Ernst 1970, Kaiser 1970, Blümich and Ziessow 1983a,b) also for the testing of the linear and nonlinear response of spin systems as an alternative method to one-dimensional and multi-dimensional pulse-Fourier spectroscopy (Ernst et al. 1987, Ernst and Anderson 1966). Again conceptually attractive work, but so far without much practical relevance, resulted. In an indirect way, these efforts of using broadband excitation and detection have certainly influenced the development of pulse-Fourier spectroscopy where similar concepts are applied. In general, the system-oriented approach, apparent in this work and in the thinking of Hans Primas, had a major impact on modern time-domain spectroscopy, more than I can demonstrate in this brief paper and probably also more than Hans Primas himself is aware of.

8. Concluding Remarks

Indeed an exceedingly rich harvest was produced by Hans Primas between 1953 and 1963. He achieved in ten years more than other successful scientists create during a life time. He has fertilized NMR instrumentation in an essential time of its development and he has laid some of the founding cornerstones of modern NMR methodology.

Sometimes, it is difficult for an outsider to understand why a scientist leaves a field of research in which he has been exceptionally successful. Although it seems that Primas has truly enjoyed electronic design and to build sturdy instruments, his profound interest was, from the beginning, more on the fundamental side, and it is not without inner logic that he got more and more involved with the true foundations of science.

It is gratifying to know that one of the great theoreticians experienced himself such a wide range of aspects, from the recipe-executing laboratory technician to the practical chemist, to the electronic instrument designer, to the inventor mastering engineering mathematics, developing fundamental theories, and finally leading to the activities for which Hans Primas is mostly known today.

I am sure that my personal interpretation of the ten years NMR spectroscopy will not do justice to Hans Primas and his oeuvre. It represents just an inept attempt to express my gratitude for his lasting contributions to the field of NMR, to physical chemistry in general, and to my own personal scientific formation in particular.

References

Andrew E.R. (1956): *Nuclear Magnetic Resonance* (Cambridge University Press, Cambridge).

Arndt R. (1962): Kernresonanzspektroskopische Untersuchungen einiger Cyclane im festen Zustand. Dissertation No. 3158, ETH Zürich.

Baker E.B. and Burd L.W. (1957): High stability nuclear magnetic resonance spectrograph. *Rev. Sci. Instr.* **28**, 313–321.

Bloch F., Hansen W.W., and Packard M. (1946): Nuclear induction. *Phys. Rev.* **69**, 127.

Blümich B. and Ziessow D. (1983a): Nonlinear noise analysis in nuclear magnetic resonance spectroscopy. 1D, 2D, and 3D spectra. *J. Chem. Phys.* **78**, 1059–1076.

Blümich B. and Ziessow D. (1983b): Multidimensional spectroscopy I. Perturbation theory. *Mol. Phys.* **48**, 955–968.

Bommer P. (1963): Beiträge zur Bestimmung von Chemical Shifts, Additivität des Chemical Shifts, Doppelresonanz. Dissertation No. 3354, ETH Zürich.

Born M. and Jordan P. (1930): *Elementare Quantenmechanik* (Springer, Berlin).

Courant R. and Hilbert D. (1953): *Methods of Mathematical Physics* (Interscience, New York).

Crawford J.A. (1958): An alternative method of quantization: the existence of classical fields. *Nuovo Cim.* **10**, 698–713.

Davenport W.B. and Root W.L. (1958): *An Introduction to the Theory of Random Signals and Noise* (McGraw-Hill, New York).

Dickinson W.C. (1950): Dependence of the F^{19} nuclear resonance position on chemical compound. *Phys. Rev.* **77**, 736–737.

Ernst R.R. (1962): I. Kernresonanz-Spektroskopie mit stochastischen Hochfrequenzfeldern. II. Zur Konstruktion eines optimalen Kernresonanz-Messkopfes. Dissertation No. 3300, ETH Zürich.

Ernst R.R. (1966): Nuclear magnetic double resonance with an incoherent radio-frequency field. *J. Chem. Phys.* **45**, 3845–3861.

Ernst R.R. (1970): Magnetic resonance with stochastic excitation. *J. Magn. Reson.* **3**, 10–27.

Ernst R.R. and Anderson W.A. (1966): Application of Fourier transform spectroscopy to magnetic resonance. *Rev. Sci. Instr.* **37**, 93–102.

Ernst R.R., Bodenhausen G., and Wokaun A. (1987): *Principles of NMR in One and Two Dimensions* (Clarendon Press, Oxford).

Fano U. (1957): Description of states in quantum mechanics by density matrix and operator techniques. *Rev. Mod. Phys.* **29**, 74–93.

Gabillard R. (1951a): *Comtes Rendues* **232**, 1477–1479.

Gabillard R. (1951b): *Comtes Rendues* **233**, 39–41.

Gutowsky H.S. and McCall D.W. (1951): Nuclear magnetic resonance fine structure in liquids. *Phys. Rev.* **82**, 748–749.

Gutowsky M.S., McCall D.W., McGarvey B.R., and Meyer L.H. (1952): Electron distribution in benzene derivatives. *J. Am. Chem. Soc.* **74**, 4809–4817.

Halbach K. (1954): Über eine neue Methode zur Messung von Relaxationszeiten und über den Spin von Cr53. *Helv. Phys. Acta* **27**, 259–282.

Heitler W. (1957): *The Quantum Theory of Radiation* (Clarendon Press, Oxford).

Herzberg G. (1945): *Infrared and Raman Spectra of Polyatomic Molecules* (Van Nostrand, New York).

Herzberg G. (1950): *Spectra of Diatomic Molecules* (Van Nostrand, New York).

Herzberg G. (1966): *Electronic Spectra and Electronic Structure of Polyatomic Molecules* (Van Nostrand, New York).

Huber A. (1969): Eine Apparatur zur Präzisionsmessung von Resonanzverschiebungen in der hochauflösenden magnetischen Kernresonanz-Spektroskopie. Dissertation No. 4232, ETH Zürich.

Jackman L.M. and Cotton F.A. (1975): *Dynamic NMR Spectroscopy* (Academic Press, New York).

Jeener J. (1982): Superoperators in magnetic resonance. *Adv. Magn. Reson.* **10**, 1–51.

Kaiser R. (1970): Coherent spectrometry with noise signals. *J. Magn. Reson.* **3**, 28–43.

Kaplan J.I. and Fraenkel G. (1980): *NMR of Chemically Exchanging Systems* (Academic Press, New York).

Kubo R. (1957a): Statistical-mechanical theory of irreversible processes. I. General theory and simple applications to magnetic and conduction problems. *J. Phys. Soc. Japan* **12**, 570–586.

Kubo R. (1957b): Stochastic theory of magnetic resonance. *Nuovo Cim. Suppl.* **X6**, 1063–1080.

Kubo R. and Tomita K. (1954): A general theory of magnetic resonance absorption. *J. Phys. Soc. Japan* **9**, 888–919.

Kumagai H. (1960): On a design of wide range magnet for cyclotron. *Nucl. Instr. and Meth.* **6**, 213–216.

Kummer H. (1963a): Beitrag zur Analyse komplizierter Protonenresonanzspektren. Dissertation No. 3378, ETH Zürich.

Kummer H. (1963b): Das Eindeutigkeitsproblem in der hochauflösenden Protonenreso-nanzspektroskopie. *Helv. Phys. Acta* **36**, 901–936.

Laning J.H. and Battin R.H. (1956): *Random Processes in Automatic Control* (McGraw-Hill, New York).

Lösche A. (1957): *Kerninduktion* (VEB Deutscher Verlag der Wissenschaften, Berlin).

Margenau H. and Murphy G.M. (1943): *The Mathematics of Physics and Chemistry* (Van Nostrand, New York).

Proctor W.G. and Yu F.C. (1950): The dependence of a nuclear magnetic resonance frequency upon chemical compound. *Phys. Rev.* **77**, 717.

Purcell E.M., Torrey H.C., and Pound R.V. (1946): Resonance absorption by nuclear magnetic moments. *Phys. Rev.* **69**, 37–38.

Shoolery J.N. (1953): Correlation of proton magnetic resonance chemical shifts with electronegativities of substituents. *J. Chem. Phys.* **21**, 1899–1900.

Smith S.A., Levante T.O., Meier B.H., and Ernst R.R. (1994): Computer simulations in magnetic resonance. An object-oriented programming approach. *J. Magn. Reson.* **A106**, 75–105.

Valley G.E. and Wallman H. (1951): *Vacuum Tube Amplifiers* (McGraw-Hill, New York).

Varian R. (1956): US Patent 3287629, filed Aug 29, 1956, issued Nov 22, 1966.

Wiener N. (1949): *Extrapolation, Interpolation and Smoothing of Stationary Time Series with Engineering Applications* (MIT Press, Cambridge).

Wiener N. (1958): *Nonlinear Problems in Random Theory* (Wiley, New York).

Wigner E.P. (1959): *Group Theory and its Application to the Quantum Mechanics of Atomic Spectra* (Academic Press, New York).

Quoted articles by Primas, arranged in the sequence of their publication:

[P1] Osimitz F. and Primas H. (1950): Tüpfelreaktionen. *Schweiz. Laboranten-Zeitung* **7**, 2–7.

[P2] Primas H., Lasman H., and Osimitz F. (1950): Moderne Vorschriften zur qualitativen Kationenanalyse. *Schweiz. Laboranten-Zeitung* **7**, 98–114.

[P3] Primas H. and Günthard Hs.H. (1953): Die Infrarotspektren von Ketten-molekülen der Formel R'CO(CH"CH")$_n$COR". I. Rocking- und Twisting-Grundtöne. *Helv. Chim. Acta* **36**, 1659–1670.

[P4] Primas H. and Günthard Hs.H. (1953): Die Infrarotspektren von Kettenmolekülen der Formel R'CO(CH"CH")$_n$COR". II. Die Normalschwingungen des Symmetriety-pus B_u. *Helv. Chim. Acta* **36**, 1791–1803.

[P5] Primas H. and Günthard Hs.H. (1954): Spectres infrarouges de dérivés carbonyliques du type R'CO(CH"CH")$_n$COR" contenant plus de dix groupes méthyléniques. *J. de Physique et le Radium* **15**, 209–211.

[P6] Primas H. and Günthard Hs.H. (1954): Theorie der Form von Absorptionsbanden suspendierter Substanzen und deren Anwendung auf die Nujolmethode in der In-frarotspektroskopie. *Helv. Chim. Acta* **37**, 360–374.

[P7] Primas H. and Günthard Hs.H. (1955): Theorie der Intensitäten der Schwingungsspektren von Kettenmolekeln. I. Allgemeine Theorie der Berechnung von Intensitäten der Infrarotspektren von grossen Molekeln. *Helv. Chim. Acta* **38**, 1254–1262.

[P8] Primas H. and Günthard Hs.H. (1956): Theorie der Intensitäten der Schwingungsspektren von Kettenmolekeln. II. Zur Berechnung der Intensitäten der Infrarotspektren von freien Kettenmolekeln der Symmetrie C_{2h}. *Helv. Chim. Acta* **39**, 1182–1192.

[P9] Günthard Hs.H. and Primas H. (1956): Zusammenhang von Graphentheorie und MO-Theorie von Molekeln mit Systemen konjugierter Bindungen. *Helv. Chim. Acta* **39**, 1645–1653.

[P10] Primas H. (1957): Ein Kernresonanzspektrograph mit hoher Auflösung. I. The-orie der Liniendeformation in der hochauflösenden Kernresonanzspektroskopie. *Helv. Phys. Acta* **30**, 297–314.

[P11] Primas H. and Günthard Hs.H. (1957): Ein Kernresonanzspektrograph mit hoher Auflösung. II. Beschreibung der Apparatur. *Helv. Phys. Acta* **30**, 315–330.
[P12] Primas H. and Günthard Hs.H. (1957): Herstellung sehr homogener axialsymmetrischer Magnetfelder. *Helv. Phys. Acta* **30**, 331–346.
[P13] Primas H. and Günthard Hs.H. (1957): Field Stabilizer for High Resolution Nuclear Magnetic Resonance. *Rev. Sci. Instrum.* **28**, 510–514.
[P14] Primas H. and Günthard Hs.H. (1957): Hochauflösender Kernresonanzspektrograph. *Chimia* **11**, 130–132.
[P15] Primas H., Frei K., and Günthard Hs.H. (1958): Protonenresonanzspektren einfacher cyclischer Aether und Ketone I. *Helv. Chim. Acta* **41**, 35–38.
[P16] Primas H. (1958): Ein Modulationsverfahren für die Kernresonanzspektroskopie hoher Auflösung. *Helv. Phys. Acta* **31**, 17–24.
[P17] Primas H. and Günthard Hs.H. (1958): Eine Methode zur direkten Berechnung des Spektrums der von quantenmechanischen Systemen absorbierten bzw. emittierten elektromagnetischen Strahlung. *Helv. Phys. Acta* **31**, 413–434.
[P18] Primas H. (1959): A new method for analyzing spectra in high resolution NMR spectroscopy. In *Proceedings of the Conference of Molecular Spectroscopy*, ed. by R. Thornton and H.W. Thompson (Pergamon Press, London), pp. 19–25.
[P19] Primas H. (1959): Anwendungen der magnetischen Kernresonanz in der Chemie. *Chimia* **13**, 15–23.
[P20] Primas H., Arndt R., and Ernst R. (1959): Die Konstruktion von Kernresonanz-Spektrographen hoher Auflösung Ia. *Z. für Instrumentenkunde* **67**, 293–300.
[P21] Primas H., Arndt R., and Ernst R. (1960): Die Konstruktion von Kernresonanz-Spektrographen hoher Auflösung Ib. *Z. für Instrumentenkunde* **68**, 8–13.
[P22] Primas H., Arndt R., and Ernst R. (1960): Die Konstruktion von Kernresonanz-Spektrographen hoher Auflösung. II. Die Konstruktion des Hochfrequenzteiles von Kernresonanz-Spektrographen hoher Auflösung. *Z. für Instrumentenkunde* **68**, 21–29.
[P23] Primas H., Arndt R., and Ernst R. (1960): Die Konstruktion von Kernresonanz-Spektrographen hoher Auflösung. III. Einige aktuelle Probleme der Kernresonanz-Instrumentierung. *Z. für Instrumentenkunde* **68**, 55–62.
[P24] Primas H. (1961): Ueber quantenmechanische Systeme mit einem stochastischen Hamiltonoperator. *Helv. Phys. Acta* **34**, 36–57.
[P25] Primas H. (1961): Eine verallgemeinerte Störungstheorie für quantenmechanische Mehrteilchenprobleme. *Helv. Phys. Acta* **34**, 331–351.
[P26] Primas H. (1962): 35 Jahre Quantenchemie. *Chimia* **16**, 281–289.
[P27] Primas H., Arndt R., and Ernst R. (1962): Group contributions to the chemical shift in proton magnetic resonance of organic compounds. In *Advances in Molecular Spectroscopy (Proceedings of the International Meeting of Molecular Spectroscopy, Bologna 1959)*, ed. by A. Mangini (Pergamon Press, Oxford), pp. 1246–1252.
[P28] Ernst R. and Primas H. (1962): High resolution NMR-instrumentation: Recent advances and prospects. *Disc. Faraday Soc.* **34**, 43–51.
[P29] Ernst R. and Primas H. (1963): Nuclear magnetic resonance with stochastic high-frequency fields. *Helv. Phys. Acta* **36**, 583–600.
[P30] Primas H. (1963): Generalized perturbation theory in operator form. *Rev. Mod. Phys.* **35**, 710–712.
[P31] Banwell C.N. and Primas H. (1963): On the analysis of high-resolution nuclear magnetic resonance spectra. I. Methods of calculating NMR spectra. *Mol. Phys.* **6**, 225–256.
[P32] Ernst R. and Primas H. (1963): Gegenwärtiger Stand und Entwicklungstendenzen in der Instrumentierung hochauflösender Kernresonanz-Spektrometer. *Ber. Bunsengesellschaft phys. Chemie* **67**, 261–267.
[P33] Primas H. (1964): Was sind Elektronen? *Helv. Chim. Acta* **47**, 1840–1851.
[P34] Huber A. and Primas H. (1965): On the design of wide range electromagnets of high homogeneity. *Nucl. Instruments and Methods* **33**, 125–130.

FRAGMENTS OF AN ALGEBRAIC QUANTUM MECHANICS PROGRAM FOR THEORETICAL CHEMISTRY

ANTON AMANN
Universitätsklinik für Anästhesie, Leopold-Franzens-Universität,
Anichstr. 35, A-6020 Innsbruck, Austria

AND

ULRICH MÜLLER-HEROLD
Departement für Umweltnaturwissenschaften,
ETH Zürich, CH-8093 Zürich-Hönggerberg, Switzerland

A major cause of poverty in the sciences
is mostly imagined wealth.

(B. Brecht 1994)

1.

In the winter semester 1978/79, Hans Primas organized a remarkable weekly seminar entitled "Great Unsolved Problems in Chemistry". He invited along all his colleagues from the renowned chemistry department of the ETH in Zürich. The issue under discussion was: What are the basic problems – from a contemporary point of view – in chemistry as an academic science, i.e., not directed towards practical application? For however successful chemistry may seem to the outside world, and however much it can point to glorious successes, chemists themselves fall silent when the subject arises of the genuinely scientific open questions in their field. This is in sharp contrast to biology and physics, where even the students are aware of unsolved basic questions.

Although the search lasted for the whole semester, at the end there emerged no single problem that was deemed worthy of the epithet "great". It is true that every chemist had several of his own scientific problems that seemed to him difficult or even unsolvable. But none of the examples presented was of such a general and fundamental nature that anyone was inclined to describe it as "great". Is it really the case that chemistry has developed into a sort of arts and crafts subject, finding its justification in dealing with practical matters but having abandoned all ambition to extend the horizons of knowledge? And is it really the case, as was claimed – amid great applause – by the Nobel Prize winner Vladimir Prelog in Primas' seminar, that "in chemistry there are no 'great' problems, in chemistry there are just 'great' solutions"?

Where are the "great" questions in chemistry? Do they really not exist or have they merely been repressed – repressed perhaps because dealing with them seems such a futile undertaking that it is better to let sleeping dogs lie? This second possibility would seem to gain support from the irritated manner in which chemists tend to react whenever such questions arise, and also from the ease with which they can be found. They are the questions put to a chemistry teacher in school, which he has to talk himself out ·of as best as he can, whereas at university level one simply turns a deaf ear: What are molecules? What are chemical substances? What is the relationship between molecules and chemical substances?

2. Quantum Mechanics and the Existence of Molecules

Under the influence of quantum mechanics, there has been a strong tendency in chemistry to redefine many of its pre-quantum mechanical, chemical concepts in terms of quantum mechanics. However, this turned out to be quite a tricky matter, and up to today there is only partial success. In the light of this development, the opposition between molecule and substance, for example, seems to be a special case of the polarity between a quantum theoretical molecule mechanics and classical thermodynamics. The statistical mechanics of molecular partition functions works as a sort of provisional link between both modes of description. In the meantime, everyday practice, which – with a certain amount of success – sees substances as Boltzmann ensembles of identical molecules, thus more or less identifying both approaches, has not stood up to closer examination. Even inveterate chemical reductionists can hardly claim that liquid water consists of identical H_2O molecules.

Physical chemistry usually describes molecules by means of quantum mechanics and a "recipe." The recipe claims that the nuclei in a molecule "cannot" be delocalized beyond a given minimum in the Born-Oppenheimer potential. The Born-Oppenheimer picture can be traced back to the traditional ideas about chemical structure and does not admit a rigorous mathematical formulation from a fundamental point of view. Nor is it completely correct from a phenomenological point of view. The ammonia maser transition, for example, is a transition between states which are delocalized over two minima of the Born-Oppenheimer potential of ammonia. On the other hand, quantum mechanics predicts molecular spectra with surprising accuracy. Might it be possible to bring the quantum mechanical formalism to a level where it predicts the above-mentioned recipe from its internal mathematical structure, instead of ad hoc superimposing traditional chemical ideas upon the basic quantum mechanical formalism?

Primas was puzzled by this "riddle of chemical structure" and set out to attack it in an unconventional manner. He wanted to test whether the intrinsic coupling between molecules and the surrounding radiation field would destabilize certain delocalized states in such a way that the "usual" chemical states would survive. Even today, it is far from clear whether Primas' proposal is correct or not. Almost all numerical quantum chemists regard it as absurd and utter nonsense. Primas' "Einige Fragen eines theoretischen Chemikers an den rechnenden Quantenchemiker" usually lead to an immediate increase in adrenaline level. Why is it so difficult to arrive at a reasonable conclusion with regard to Primas' suggestion?

From a conceptual and technical point of view, the problem is that the radiation field brings in infinitely many degrees of freedom, one for each possible frequency. Consequently, even a single molecule should not be treated as consisting of finitely many nuclei and electrons, but as a joint system consisting of some nuclei and electrons and infinitely many modes of the radiation field. Every mode of the radiation field is described by a pair of momentum and position operators fulfilling the Heisenberg commutation relations. Instead of finitely many momentum and position operators (for the nuclei and electrons), one therefore has to deal with infinitely many momentum and position operators.

It has been clear since the 1940s that two sets of infinitely many momentum and position operators acting on a Hilbert space can have very different properties. It fascinated Primas that in systems with infinitely many degrees of freedom classical properties may appear by the formalism itself. "Classical observable" means that the pure states of the respective quantum system can be classified by expectation values of the respective observable in such a way that pure states with different such expectation values "cannot be superposed". In the formalism, such pure states can be represented by vectors, and these vectors can be formally superposed, just as in the quantum mechanics of finitely many degrees of freedom. However, it turns out that this formal superposition is a mixture and no longer pure, something which can never occur in the quantum mechanics of finitely many degrees of freedom.

3. Algebraic Quantum Mechanics and Thermal States of Matter

The structure of quantum mechanics with infinitely many degrees of freedom was mathematically developed into an elaborate theory of C*-algebras
and W*-algebras. In algebraic quantum mechanics, one has three possible
types of W*-algebra: type I, II, and III. This classification was invented
by John von Neumann in the 1930s. All the matrices acting on a complex
vector space form a type I W*-algebra, and the same holds true for all
bounded operators on an arbitrary Hilbert space. Type II W*-algebras are
relevant for infinite-temperature states. As far as type III W*-algebras are
concerned, von Neumann could prove that there exists at least one example. Some thirty years later, in the 1960s, it turned out that there exist lots
of examples of type III W*-algebras, and shortly after Tomita's theory of
modular Hilbert algebras and KMS-states was popularized by Masamichi
Takesaki around 1970, Alain Connes developed a classification of type III
W*-algebras. Originally, type III W*-algebras were considered to be "monsters," particularly in group theory, but the development of KMS-theory –
referring to thermal states – showed that "all reasonable" thermodynamical
systems with infinitely many degrees of freedom should be described by a
type III algebra. (For more details, see Bratteli and Robinson 1981, 1987.)

In quantum systems which reside on a Hilbert space of finite dimension (e.g., spin systems), thermal states D_β can be introduced by density
operators defined as

$$D_\beta = \frac{\exp\{-\beta H\}}{\text{trace}\,(\exp\{-\beta H\})} \,. \tag{1}$$

Hence, for every inverse temperature β there exists a unique thermal state.
This, of course, is not appropriate for a chemical substance, where one may
have different phases at some particular temperature which are characterized by different thermal states with the same temperature.

The concept of a β-KMS state fulfilling the Kubo-Martin-Schwinger
(KMS) boundary condition generalizes thermal stability to systems with
infinitely many degrees of freedom, as well as systems of finitely many
degrees of freedom where the trace $(\exp\{-\beta H\})$ does not exist. For systems
with infinitely many degrees of freedom, several different β-KMS states
may exist at the same temperature, and furthermore, such states refer to
different values of some classical observable. Hence, in KMS-theory, classical
observables may arise from the formalism as such, without using additional
recipes.

The concepts of quantum mechanics for systems with infinitely many
degrees of freedom are not changed by this mathematical development,
but their practicability is improved. Nevertheless, even the most ingenious

mathematical theorems did not bring the theory of infinite quantum systems to a point where it could be handled so easily that concrete results about specific molecules could be derived. Infinite quantum systems were – and unfortunately still are – difficult to deal with.

Though this remark also applies to the KMS-theory of thermodynamical equilibrium states, the theory of equilibrium states brought some progress and illustrated the generation of classical observables in quantum systems with infinitely many degrees of freedom. Primas was puzzled by the possibilities of KMS-theory, so much so that he proposed informally that substances and substance classes in chemistry might actually be describable by KMS-states on appropriate W*-algebras of type III.

4. On the Relation between Molecules and Substances

In his book "How To Solve It", the Hungarian mathematician Georg Polya presents a recipe for how one can attempt to proceed when there is a mathematical problem that one cannot solve. Polya says there is often another problem, a simpler one, that one cannot solve either. He then suggests: find it! (Polya 1985, p. 114). The basic idea is obvious: with a problem that is simpler but structurally similar, one may perhaps move forward a little and thus get some sort of clue as to how to tackle the original problem.

One of Primas' most ingenious ideas was about the relation between molecules and substances. Even though we do not yet have the technical aids to work out these ideas, let us nevertheless take a closer look at this example to see the boldness with which he tackled difficult questions, and how he was able to seek inspiration from his great fund of theoretical knowledge.

The simpler problem, in Polya's terms, was here seen by Primas in the relation between a classical particle and a quantum mechanical elementary system. This relation was historically established in Bohr's famous correspondence principle, which played an important role in the development of quantum mechanics in the 1920s. Nevertheless, the correspondence principle is just a recipe, which needs to be applied with a great deal of common sense. It is by no means a clearly and unequivocally formulated principle. How can it be that such a vague concept actually works so well in practice?

From a modern point of view the correspondence principle is nothing more than an attempt – a somewhat clumsy one – to transfer Galileian kinematics from classical to quantum mechanics. In mechanics one usually distinguishes between kinematics and dynamics. Originally, kinematics was the study of the geometrical motions of rigid bodies. Today, kinematics is understood as that part of mechanics reflecting the geometrical structure of four-dimensional space-time. Depending on the selected space-time model different kinematics are obtained. In non-relativistic mechanics, such as that of Aristotle, space and time are completely independent. In relativistic mechanics, space-time is defined by a kinematical group connecting space and time. In Galilei-relativistic Newtonian mechanics it is defined by the Galilei group, and in the Lorentz-relativistic special theory of relativity of Einstein it is defined by the Lorentz group.

The relevant kinematics for a theory of chemical phenomena is fundamentally Galilei-relativistic. In the sense of an (energetically) satisfactory approximation, Galileian kinematics is a possible choice for chemistry, provided that the relevant velocities are small in comparison with the speed of light. Furthermore, the adoption of Galileian kinematics as a possible choice for chemistry is also necessary ·for very different reasons! For only with this choice are there separate conservation laws for mass and energy, and only with this choice are there electrons at all. In a Lorentz-relativistic theory there is only one joint conservation law for mass and energy, and there is only one undivided electron-positron-field.

The kinematics of a classical (Newtonian) point particle is defined by a transitive representation of the Galilei group as a transformation group acting on the configuration space, in this case $\mathbf{R}^3$, whereas the corresponding quantum mechanical analog is generated by a unitary ergodic representation of the Galilei group in Hilbert space. As representatives of the physical state, the points in $\mathbf{R}^3$ and the rays in Hilbert space reflect all the fundamental differences between classical mechanics and quantum mechanics. From the covariance of both classical and quantum mechanical systems under Galilei transformations emerge certain family similarities. These make it possible, with a recipe and some luck, to use the knowledge of classical point particles to guess at analogous characteristics of quantum mechanical elementary systems (see Fig. 1).

Primas transferred this example heuristically to the relation of molecules to substances. In the case of n various types of molecules, the observables form a type I W*-algebra, and the time evolution is given by a one-parameter group of inner automorphisms generated by the Hamiltonian. In a chemical system of temperature $T > 0$, consisting of n different chemical components, the observables are given by a type III W*-algebra, and the time evolution is given by a one-parameter group of outer automorphisms, generated by the modular operator of the algebra fulfilling the KMS-condition for a fixed temperature T.

How are the various types of chemical components reflected in this system? In case of particles without internal degrees of freedom this can be seen easily in a Fock space representation. In case of n different species there are n different mutually commuting particle number operators N_j, $j = 1, \ldots, n$, bounded from below. Each of them generates a compact unitary

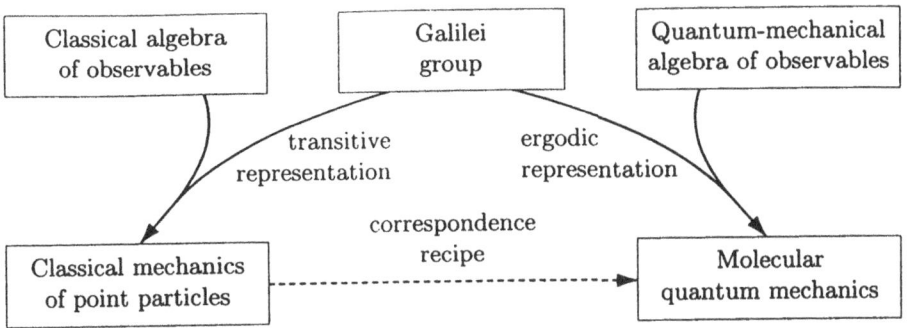

Figure 1.   On the relation between classical point particles and quantum mechanical elementary systems.

$1, \ldots, n$, bounded from below. Each of them generates a compact unitary one-parameter group acting as a gauge group. In the case of n different constituents, this gauge group is the n-torus (Primas 1981, p 189f). This situation can immediately be transferred to substance mixtures: in this case, the material variety is reflected in the same n-dimensional compact Abelian group, acting as gauge automorphisms on the type III algebra of the thermodynamic observables (Müller-Herold 1984).

So far the situation seems perfectly clear. The central question is what happens when the "internal", more chemical differences in the various molecules and substances come into play. In the statistical mechanics of partition functions, the center of mass of a structured particle corresponds to the unstructured particle. So for a structured particle there will be a sort of particle number operator together with the corresponding gauge group. The internal degrees of freedom, by contrast, must manifest themselves differently.

This brings us to the decisive point in Primas' speculative idea. It goes back to the Erlangen program of the mathematician Felix Klein who in his inauguration lecture in 1872 suggested characterizing the diversity of geometrical objects by means of a specific group of invariance (Klein 1893). In this spirit, Primas proposed an additional internal gauge symmetry, which he called pattern group (cf. Primas 1981, p. 329f), characterizing the specifically chemical features of the material components. This pattern group, if it exists, would imply that automorphic representations on type I algebras lead to molecules and automorphic representations on type III algebras lead to related substances. Whatever the differences between molecules and substances, there would be a common species-typical invariance group under-

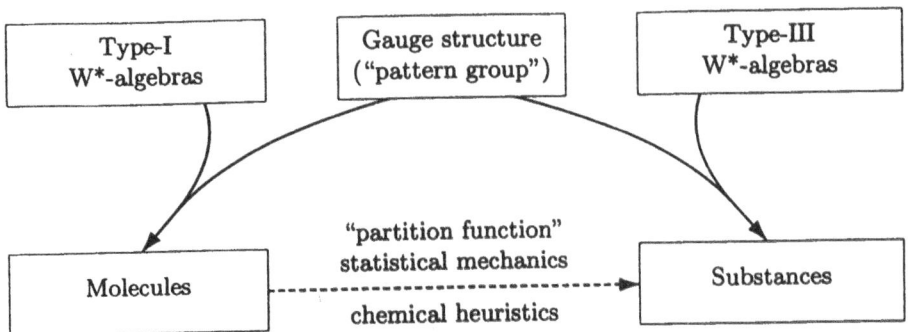

Figure 2. On the relation between molecules and substances according to the hypothetical "pattern group" as proposed informally by Primas.

lying the successes of synthetic chemistry in designing chemical reactions of substances through considerations about molecules (see Fig. 2).

This is as far as Primas got with this (unpublished) proposal. It is clear what has to be expected by anyone who would like to step in: the unsolved problems of the Erlangen program, the difficulties of the (probably) non-Abelian gauge groups, the question of how to find non-trivial examples, the anticipated incomprehension of the scientific community in the event of any solution being reached. On the positive side there are the powerful tools of a mathematically well-examined structure and the experimental indications from nuclear magnetic resonance spectroscopy that one is not completely on the wrong track.

5. Perspectives of the Algebraic Quantum Chemistry Research Program

In all these matters Primas' hope was to find a quantum-mechanical theory of chemistry which would get rid of external assumptions and traditional ideas about chemical structure. Such a theory would gain its relevance from its internal structure alone, and it would be able to describe macroscopic classical observables and microscopic structure. KMS-theory, in fact, can be nicely worked out for mean-field theories. However, chemical substances consisting of infinitely many molecules or consisting of finitely many molecules coupled to the radiation field involving infinitely many degrees of freedom were and are out of reach.

Assuming that the mathematical problems could be solved in such a way that quantum systems with infinitely many degrees of freedom could to illustrate these problems in more detail. There are some reasons to believe that a single quantum system, e.g., a single piece of solid or a single container of ethanol, should be treated as being in a pure state. Thermal states, on the other hand, are non-pure and do not change under the dynamics implemented by the Hamiltonian of the system in question. Should a thermal state be considered as the proper state describing an individual quantum system? Or should a thermal state be considered as an average over all different behaviors (pure states) the system can have (in this way thermal states are treated in classical statistical mechanics)? As long as only thermodynamical quantities (such as pressure, overall magnetization, etc.) are considered, these questions do not arise. But as soon as the expectation value of the position of some nucleus in a substance is considered, this problem comes up immediately.

A major problem with Primas' suggestions concerning either single molecules coupled to the radiation field or infinitely many molecules (modeling a substance coupled to the radiation field) is that the limit of infinitely many radiation field modes or the limit of infinitely many molecules is somewhat obscure. In some way, a large but finite number of field modes should approximate infinitely many modes of the field etc., but this is not reflected in the structure of the corresponding W*-algebra. For a quantum system with finitely many degrees of freedom, the algebra of observables is always of type I, whereas for infinitely many degrees of freedom, the algebra of observables is always of type III. In other words: there is no "approximate" type III algebra for many (but finitely many) degrees of freedom. Similarly, strict classical observables only appear with infinitely many degrees of freedom and never for many (but finitely many) degrees of freedom.

Primas wanted to understand such choices from a more fundamental point of view: as a proper choice but also with an understanding of its shortcomings when compared with the proper quantum mechanical structure and dynamics. In this program, the necessity of making a choice is taken seriously (for example by constructing its classical observables as operators in the center of the respective W*-algebra), but it is also viewed in a broader perspective. Chemistry and the chemical concepts offer important examples to develop the quantum-mechanical formalism in a way that evaluates different points of view together with the corresponding classical observables. At the same time the shortcomings of each of these points of view can be worked out "quantitatively".

This program turned out to be much more difficult than originally anticipated. What would the final goal be? Besides a proper understanding of molecular structure, substances, and substance classes, one should get a mathematically and conceptually rigorous description of all sorts of *single* individual quantum systems. Single-molecule spectroscopy, in particular, gives experimental results which are far from fundamentally understood. Also all sorts of stochastic dynamics for "approximately" pure states, the different aspects of the Hartree approximation, and the structure of molecular Hamiltonians (all derived by Hartree approximation, since the radiation field must disappear in a "molecular" description) belong to the class of problems which remain entirely unresolved.

6. Outlook: Approximately "Classical" States of Matter

The recipe of chemical structure, used by quantum chemistry for the description of molecules, was adapted from traditional chemistry to quantum mechanics and leads to a kind of classical observable. There are a number of major problems in this respect:

- Are structures, which are predicted on the basis of chemical wisdom, correct from a rigorous quantum mechanical point of view?
- From a quantum mechanical point of view, every molecule possesses states which do not conform to chemistry. How fast do such states decay and what is the role of the molecular environment, e.g., electromagnetic field, gravitational field, neighboring molecules, in this decay process?
- Can one find exceptional examples of chemical species, which do not conform to chemical wisdom? What role does ammonia play in this respect? In the gas phase? In liquid and solid phase?
- Can a rigorous quantum dynamics for such exceptional examples be formulated which incorporates the environmental effects and allows us to understand the chemical recipes?

Answers to these questions are difficult to provide. Nevertheless, one may sift out one specific point from quite a lot of confused evidence: The classical observable for the nuclear structure of a molecule – forbidding superpositions of states on different minima of the Born-Oppenheimer potential – is an "approximate" classical observable. We propose to take such approximately classical observables as the prototypes of chemically relevant observables, for, e.g., chemical structure, substance classes, etc. In other words: quantum mechanics can only be successful in explaining chemical features in a rigorous manner if strictly classical observables are replaced in the quantum-mechanical formalism by approximately classical observables. This is not a simple task at all, and therefore it has not been done or even been tried. In the following we try to explore some ideas in this direction.

The expectation value of pure states with respect to approximately classical observables is in "most cases" an expectation value of some particular classes of states – e.g., one of many Born-Oppenheimer sinks. But even if the number of degrees of freedom is large, some pure states (e.g., a superposition of states with different classical expectation values) still arise which do not show classical properties. It is only in the limit of infinitely many degrees of freedom that a strictly classical observable appears. Hence the question is important as to "how quickly" the quantum aspects of an approximately classical observable disappear with increasing number of degrees of freedom. In chemistry this means that it should become clear how approximately classical nuclear structures of some molecular species can be related to a strictly classical nuclear structure (arising in the limit of infinite nuclear masses). Similarly, strictly pure states should be replaced by approximately pure states.

This requirement is an immediate consequence of the use of approximately classical observables. Since certain superpositions of states lead to a mixed state in the limit of infinitely many degrees of freedom (if a strictly classical observable exists) but to a pure state for a large but finite number of degrees of freedom (because a classical observable cannot be strict for finitely many degrees of freedom), it follows that the latter pure state is "almost" mixed. Consequently, "purity" of a state is not stably defined and therefore one should not accept the usual mathematical definition of a pure state *as physically relevant* in situations such as for systems with a large but finite number of degrees of freedom. Actually, in systems with infinitely many degrees of freedom it is often (for certain algebras of observables $\mathcal{A}$) the case that the set $P(\mathcal{A})$ of pure states is dense in the set $E(\mathcal{A})$ of all possible (non-pure) states (Dixmier 1977). This does not only mean that the notion of a mathematically pure state must not be confused with an empirically accessible physical state. It means that infinitesimal variations of a pure state lead to states which are not even approximately pure. There is no way to "scale" purity as a function of distance from a pure state.

We get the following picture:

– For "small" isolated quantum systems, pure states accidentally make sense. Classical observables do not arise.
– For systems with infinitely many degrees of freedom, pure states should not be rendered prominent. Strictly classical observables exist.
– For systems with many degrees of freedom, pure states lose their prominence (with increasing number of degrees of freedom). Approximately classical observables exist which turn into strictly classical observables in the infinite limit.

It might seem strange that chemical concepts can only be incorporated with difficulty into quantum theory and that nevertheless quantum mechanics works so well in usual spectroscopy for "ensembles" of molecules. The reason is that pure states do not play a role in this context. "Averages" of states of many molecules are non-pure from the very beginning. The respective density-operator dynamics is relatively simple, quantum jumps are not necessary and even obsolete, and some conceptual problems disappear altogether. On the other hand, ensemble spectroscopy cannot make absolutely definite statements about chemical structure.

Considering, for example, NF_3 as a rigid rotator, the (microwave) rotation spectrum can be computed and compared with the experimental transition frequencies $\nu(^{14}NF_3) = 21.362$. GHz and $\nu(^{15}NF_3) = 21.258$ GHz, resulting in an N-F bond length of 1.37 Å. This argument is, of course, conditional: only under the condition that NF_3 has a molecular structure does the notion of the N-F bond length make sense. If NF_3 does not have a molecular structure at all, then 1.37 Å is just some length scale characterizing the microwave spectrum of NF_3 without any further significance.

The usual link between density-operator theory for averages of molecular states on the one hand and single molecules on the other is a decomposition of non-pure states into pure or approximately pure states. This decomposition is "not unique". Even more important, by a particular "choice" for the decomposition one imposes a certain heuristic idea on the correct density-operator dynamics. It was, for example, demonstrated by Primas in his lecture courses how Fermi's Golden Rule can be derived from a simple density operator-dynamics *under the assumption* that all the molecular states arising are energy eigenstates of the molecular Hamiltonian. Other pure-state dynamics are also compatible with one and the same density-operator dynamics.

Hence, different "interpretations" of one and the same average behavior of a collection of molecules (or another quantum system) in terms of pure states (or some other class of states) are possible. Such interpretations are, of course, never entirely correct, but incorporate only one particular point of view. In other words, such interpretations are based on bias and prejudice, and they are incompatible with any Schrödinger dynamics. Hence, a genuine quantum theory of single open quantum systems, not based exclusively on decompositions of density-operator dynamics, would be highly desirable. First steps have been indicated by Amann and Atmanspacher (1998).

The old problem: "How is it possible for quantum mechanics to be a theory of chemistry if the way in which chemistry states its results contradicts quantum mechanics?" together with the initial conjecture "that the solution to the problem might lie in the introduction of so-called classical observables" thus appears in a new form. Faced with the formal elements of mathematics, strictly classical observables have proved to be a too demanding goal, unattainable to achieve.

From the contemporary point of view, the search for "approximately" classical observables appears as a suitable variant of the program. The idea that such observables in the limit of infinitely many degrees of freedom could develop into strictly classical observables is a conjecture that Primas expressed back in the early 1970s. One of his "major" dissertation themes – handed out but never worked through – was "inequivalence in the finite". The basic idea is that in the case of very large, but finite systems classical observables are already pre-formed in a more or less "embryonic" way. They are only given birth to in the transition to infinitely many degrees of freedom. At that time it was not clear that the key to the problem lies in the relation between pure and mixed states. In this way a central problem of theoretical chemistry has been given a novel perspective.

References

Amann A. and Atmanspacher H. (1998): Fluctuations in the dynamics of single quantum systems. *Stud. Hist. Phil. Mod. Phys.* **29**, 151–182.

Bratteli O. and Robinson D.W. (1981): *Operator Algebras and Quantum Statistical Mechanics Vol. 2* (Springer, New York).

Bratteli O. and Robinson D.W. (1987): *Operator Algebras and Quantum Statistical Mechanics Vol. 1* (Springer, New York, 2nd revised edition).

Brecht B. (1994): *Life of Galileo* (Arcade, New York).

Dixmier J. (1977): C^*-*Algebras* (North Holland, Amsterdam).

Klein F. (1893): Vergleichende Betrachtungen über neuere geometrische Forschungen. *Math. Ann.* **43**, 63–100.

Polya G. (1985): *How To Solve It* (Princeton University Press, Princeton), 2nd edition.

Primas H. (1981): *Chemistry, Quantum Mechanics, and Reductionism* (Springer, Berlin).

Müller-Herold U. (1984): Algebraic theory of the chemical potential and the condition of reactive equilibrium. *Lett. Math. Phys.* **8**, 127–133.

CONTEXTUAL BACKGROUND

Anton Amann and Harald Atmanspacher:
C*- AND W*-ALGEBRAS OF OBSERVABLES

In his "Habilitationsschrift" of 1960, Hans Primas examined the influence of an external classical stochastic radiation field on finite spin systems (Primas 1961). One of the reviewers, the mathematical physicist Res Jost, made a marginal note in his copy: Here, wrote Jost, in relation to the classical radiation field, one should perhaps use the term "ring of macroscopic observables".

What Primas was unaware of in 1960 was already familiar to mathematical physicists such as Jost: the attempts to find an algebraic version of quantum mechanics, as prepared in some of Pascual Jordan's early writings (Jordan 1932, 1933, 1934) and in John von Neumann's famous papers on "Rings of Operators" (von Neumann 1961) in the late 1930s. A few years later, Rudolf Haag and Daniel Kastler published a seminal paper (Haag and Kastler 1964) paving the way for a breakthrough in algebraic quantum mechanics. They showed that the choice of a topology for algebras of observables has a physical significance which the ideas prevailing at the time had been rather imprecise about.

Haag's and Kastler's result needs to be seen in the context of a re-evaluation of the role of "technical assumptions" when proving theorems in mathematical physics. Besides assumptions based on a firm physical foundation, one usually needs a series of so-called technical assumptions in order to fulfill the mathematical prerequisites essential for a proof of a physical theorem. During the 1950s, however, it became increasingly clear that technical assumptions are by no means as "innocent" as one had been led to believe. On the contrary: it emerged that physics had been smuggled in via technical assumptions, the epitome of this being von Neumann's irreducibility postulate. Haag and Kastler demonstrated that topologies have a physical meaning and, accordingly, the choice of a topology has to be substantiated physically.

During the early years of quantum mechanics, individual observables were treated more or less as $n \times n$ matrices. Calculations were purely formal and no proof of their existence had to be furnished. For various reasons – unknown at that time – unbounded operators or operators with absolutely continuous spectra could be regarded as limiting cases of finite matrices without doing too much mathematical harm. In the case of whole algebras of operators, however, it turned out that the structure of the limiting algebras is crucially dependent on how the limit $n \to \infty$ is carried out.

In other words, the structure depends on the topology used in the closure operation. Von Neumann's choice was inspired by the practical quantum mechanics of those days: a sequence of operators converges if the corresponding sequences of expectation values converge. In modern terminology, this is called convergence in the weak *-topology (read: star-topology), which is precisely the relevant topology for von Neumann algebras or W^*-algebras. What was excitingly innovative in the work of Haag and Kastler was that it fostered the use of C^*-algebras (instead of W^*-algebras). C^*-algebras are closed in the norm-topology, as first proposed in 1947 by Irvine Ezra Segal for applications in quantum mechanics (Segal 1953).

The central question for Primas over a number of years was which type of algebras is "more fundamental" in quantum mechanics. It was and remains an indisputable fact that both W^*-algebras and C^*-algebras are indispensable tools in algebraic quantum mechanics. In addition, there are many technical connections between the two: each W^*-algebra is a C^*-algebra, and a given C^*-algebra can always be embedded in larger W^*-algebras by GNS-construction. Going back to Primas' question, however, there is a world of differences between the two algebras. C^*-algebras are the starting point of all existing models, and W^*-algebras are the exclusive choice in quantum logics.

In the late 1960s, European and Japanese scientists obtained important technical results enabling the use of *-algebras in advanced quantum mechanical research to become everyday routine. In his (unpublished) model of the measurement process of 1969, Primas used W^*-algebras as a matter of course, and after 1975, C^*- and W^*-algebras (the standard textbook is Bratteli and Robinson 1979, 1981) became the standard tools of his research group at the ETH in Zürich. From that time on, at the latest, Primas started to pose his crucial questions in the language of normed algebras: What is an observable in physical terms? What does it mean to define the state of a physical system? Do our experimental findings – as described by contextual observables und epistemic states – refer (only) to our observational knowledge or do they refer to systems not necessarily under observation, i.e., to intrinsic observables and ontic states?

In Primas' research these questions were discussed at great length, but not ultimately answered. The contribution by Amann and Atmanspacher reviews an interpretation of C^*- and W^*-algebras of intrinsic and contextual observables, together with an associated distinction of ontic and epistemic states, respectively. They then go on to introduce a third, intermediate level of description that links epistemic states with intrinsic observables. This formal possibility is tailored to a thorough examination of certain aspects of measurements, broadly understood in terms of a change of perspective from an ontic to an epistemic scheme of interpretation.

The two basic steps of the measurement problem are first of all the decomposition of the states of a system into (dressed) product states for the object to be measured and the environment ("Heisenberg cut"), and secondly the corresponding dynamical time evolution of some appropriate observable. A Heisenberg cut is physically reasonable if it minimizes the Einstein-Podolsky-Rosen-correlations between object and measuring apparatus, thus allowing for "approximative" objects. To this end, Amann and Atmanspacher propose novel approaches based on maximum entropy and stability considerations. They conclude with a brief presentation of the relations between the stochastic dynamics of measurements as K-flows and the concept of a non-commutative time observable.

References

Bratteli O. and Robinson D.W. (1979): *Operator Algebras and Quantum Statistical Mechanics 1* (Springer, Berlin), 2nd revised edition 1987.

Bratteli O. and Robinson D.W. (1981): *Operator Algebras and Quantum Statistical Mechanics 2* (Springer, Berlin), 2nd revised edition 1997.

Haag R. and Kastler D. (1964): An algebraic approach to quantum field theory. *J. Math. Phys.* **5**, 848–861.

Jordan P. (1932): Über eine Klasse nichtassoziativer hyperkomplexer Algebren. *Nachr. Ges. Wiss. Göttingen, Mathematisch-Physikalische Klasse* 1932, 569–575.

Jordan P. (1933): Über die Multiplikation quantenmechanischer Größen. *Zeitschrift für Physik* **80**, 285–291.

Jordan P. (1934): Über die Multiplikation quantenmechanischer Größen II. *Zeitschrift für Physik* **87**, 505–512.

Neumann J. von (1961): *Collected Works, Vol. III: Rings of Operators*, ed. by A.H. Taub (Pergamon Press, Oxford).

Primas H. (1961): Über quantenmechanische Systeme mit einem stochastischen Hamiltonoperator. *Helv. Phys. Acta* **34**, 36–57.

Segal I.E. (1953): Postulates for general quantum mechanics. *Ann. Math.* **48**, 401–457.

C*- AND W*-ALGEBRAS OF OBSERVABLES, THEIR INTERPRETATIONS, AND THE PROBLEM OF MEASUREMENT

ANTON AMANN
*Universitätsklinik für Anästhesie, Leopold-Franzens-Universität,
Anichstr. 35, A-6020 Innsbruck, Austria*

AND

HARALD ATMANSPACHER
*Institut für Grenzgebiete der Psychologie,
Wilhelmstr. 3a, D-79098 Freiburg, Germany*
and
*Max-Planck-Institut für extraterrestrische Physik,
D-85740 Garching, Germany*

1. Introduction

Algebraic formulations of quantum theory are an outstanding example of attempts to axiomatize a theoretical framework after the phase of its discovery. The main pioneering phase in the development of quantum mechanics was finished in the early 1930s and has been summarized excellently in von Neumann's monograph (1932) on the mathematical foundations of quantum mechanics. Algebraic quantum theory started in the mid 1930s essentially with Jordan's approach toward non-associative algebras of observables, so-called Jordan algebras. Mathematicians and mathematical physicists like von Neumann, Gel'fand, Naimark, Segal, Wightman, Emch, Haag, and many others have contributed to the further development of algebraic quantum theory. A compact account of the main features of this development has been given by Primas (1983).

The main conceptual difference between pioneer quantum mechanics and algebraic quantum mechanics consists in their different points of departure. The pioneers of quantum mechanics started with classical ideas and tried to find how these correspond to the typical features of quantum systems. Algebraic quantum theory starts with the most typical one of these features, the non-commutativity of quantum mechanical observables, and is basically characterized as a representation theory of the corresponding canonical commutation relations.

This different perspective allows classical systems to be represented within algebraic quantum theory. Methodologically, this representation does not make use of any quantization procedures of classical systems and it does not require any correspondence principle as in the understanding of Bohr. Algebraic quantum theory tries to derive (commutative) classical observables from a basic (non-commutative) quantum structure. In this sense, a basic idea behind algebraic quantum theory is the conviction that quantum theory is a fundamental description of the physical world.

The modern achievements of algebraic quantum theory make clear in what sense pioneer quantum mechanics as well as classical mechanics can be considered as limiting cases of the general theory (Primas 1990a). Compared to the framework of von Neumann's monograph (1932), important extensions are obtained by giving up the irreducibility of the algebra of observables (admitting only non-commuting observables) and the restriction to locally compact state spaces (admitting only finitely many degrees of freedom). As a consequence, modern quantum physics is able to deal with open systems in addition to isolated ones, it can involve infinitely many degrees of freedom such as the modes of a radiation field, it can properly consider interactions with the environment of a system, superselection rules can be formulated that would be impossible in an irreducible algebra of observables, there are in general infinitely many representations inequivalent to the Schrödinger representation, and irreversible (non-unitary) dynamical evolutions can be successfully incorporated.

In addition to this remarkable progress, it has become possible to address quite a number of unresolved conceptual and interpretational problems of pioneer quantum mechanics from a new perspective. This has been a major focus of the work of Hans Primas during the last two decades. First of all, his contributions to a clarification of a number of different concepts of states as well as observables provide a much better understanding of many confusing issues in earlier conceptions. Second, a clear-cut characterization of these concepts is a necessary precondition to explore new approaches, beyond von Neumann's projection postulate, toward the central problem that pervades all quantum theory since its very beginning: the measurement problem.

In our present contribution we intend to describe in detail the progress which has been achieved concerning three different categories of states, observables, and their associated dynamics together with their relevant interpretation. This categorization makes essential use of the distinction between C*- and W*-algebraic formulations; in this respect it is an original contribution even within algebraic quantum theory. Subsequently we shall discuss the measurement problem, basically in two steps: (i) the Heisenberg cut that serves as a metaphor for a proper dressing procedure, i.e., a proper tensorization of the pure state of an isolated system into pure states forming a product state, and (ii) the actual process of measurement, i.e., the dynamics of the product state (object plus environment) that can finally lead to the "material constitution" of an object in an operationally accessible sense. A number of indications will be given for the way in which the concept of a non-classical time operator (in several differing formulations) plays a major role for a proper dynamical description of measurement.

2. States and Observables

2.1. ONTIC STATES WITH INTRINSIC OBSERVABLES

A *strictly isolated* quantum system is in a *pure* state ϕ_t for any time t and evolves under unitary Schrödinger dynamics. The pure state ϕ is called an *ontic* state. It provides a complete specification of the system and is independent of any measurement- or observation-based information gained by external observers. For such a system, concepts like entanglement with its environment and dissipation of energy do not make sense. A pure state ϕ is often represented by a state vector Ψ in the sense that the expectation value $\phi(\hat{A})$ of an observable $\hat{A}$ is given by $\phi(\hat{A}) = \langle \Psi | \hat{A} \Psi \rangle$.

The observables of a strictly isolated quantum system generate a C*-algebra $\mathcal{A}$ (Primas 1994) and are called *intrinsic* observables. They are defined without any reference to empirical measurement or observation. Moreover, isolated quantum systems as such are not observable since any observation would destroy the property of being isolated. They are sometimes called *endo-systems* (Primas 1994), expressing the fact that they do not refer to any external perspective. They are precisely the way they are, and any limited knowledge or partial information of observers is irrelevant at this level of consideration.

From a conceptually rigorous point of view, traditional quantum theory in the framework of the Stone-von Neumann theorem (sometimes called "pioneer quantum mechanics") deals exclusively with ontic states since it deals exclusively with isolated systems. In this framework, the relevant algebra $\mathcal{A}$ of observables is always considered to be irreducible (von Neumann's irreducibility postulate; no commuting observables), and every representation of the canonical commutation rules is considered to be equivalent to a Schrödinger representation.

A strictly isolated quantum system described by a particular pure state ϕ has a *holistic* structure which is encoded in ϕ. In general, it is not legitimate to consider an endo-system to be constituted by sub-endosystems, since the restrictions of the pure state ϕ to the subsystems in general are not pure states any more. Therefore a system in its pure state is an *individual, single* quantum system which does not consist of subsystems (e.g., objects). Since such a system is *not* observable without breaking the isolation from its environment, there is *no* "reduction of the wave function"; hence all types of stochastic dynamics (leading to different eigenstates of any measured observable, with different probabilities) do not play any role. Similarly, transition probabilities between different pure states (as, e.g., specified by Fermi's Golden Rule) do *not* make sense.

2.2. EPISTEMIC STATES WITH CONTEXTUAL OBSERVABLES

For any *open* quantum system, sometimes called *exo-system*, the coupling to its environment is relevant and cannot be neglected. Even if the joint quantum system {exo-system & environment} is isolated and hence in a pure state ϕ, the restriction of ϕ to the exo-system is usually *not* pure any more. Such non-pure states η can be represented by a density operator D, in the sense that for observables $\hat{A}$ of the exo-system the expectation value with respect to the overall pure state ϕ is given as $\phi(\hat{A}) = \eta(\hat{A}) = \mathrm{Tr}(D\hat{A})$, where "Tr" denotes the trace of an operator (the state ϕ refers to all observables, the state η refers to the "relevant" observables of the exo-system only).

The density operator D represents only partial information and does not allow the overall pure state ϕ to be reconstructed. The non-pure state represented by the density operator D is called an *epistemic* state, since it refers to our *knowledge* about the state ϕ. The dynamics of an *open* quantum exo-system is not determined by the Schrödinger equation alone, additional "dissipative" effects usually play a role (Fick and Sauermann 1986, Kubo et al. 1985).

Which structure is appropriate for the observables of a quantum exo-system? Since the joint quantum system {exo-system & rest of the world} is an endo-system and its observables are described by a C*-algebra $\mathcal{A}$, the "relevant" observables of the exo-system generate a C*-subalgebra $\mathcal{B}$ of $\mathcal{A}$. For a given non-pure epistemic state η of the quantum exo-system, one can always construct an appropriate W*-algebra $\mathcal{M}_\eta$, namely the σ-weak closure of the C*-algebra $\pi_\eta(\mathcal{B})$ represented in the Gel'fand–Naimark–Segal Hilbert-space representation (briefly: GNS-representation) π_η of $\mathcal{B}$ with respect to the state η (Bratteli and Robinson 1987, Pedersen 1979, Dixmier 1977). This representation depends always on a context (e.g., a reference state), therefore the corresponding observables are called *contextual* observables (Primas 1994).

The advantage of this procedure is basically that the non-pure state η can now be represented by a density operator (maybe even a vector, also for non-pure states) in the corresponding GNS-Hilbert space. In an abstract C*-algebraic setting, density operators *cannot* be introduced (but non-pure states are well-defined). A further advantage of the W*-algebra is that it contains important classical observables such as operators for temperature (Takesaki 1970) and chemical potential (Müller-Herold 1980), which are not explicitly recognizable in the C*-algebraic formalism. It is often appropriate to construct a W*-algebra on a Hilbert space in such a way that *all* relevant non-pure states of the exo-system can be represented by density operators. This is a non-trivial task if the number of such states is uncountable (and if the Hilbert space should be separable nevertheless).

An exo-system in its non-pure epistemic state η may

- either describe an individual single system, if the state η arises by restriction of a pure state on a larger quantum object,
- or describe an ensemble of individual quantum systems, if the state η is actually a "mixture" of different pure states of the exo-system itself,
- or describe a single individual system, if the state η arises as the *average in time* of pure states ρ_t (though the actual average is not known).

Usually, these different possibilities are not clearly distinguished, because pure states are not considered in an appropriate exo-description.

For an open exo-system described by a particular non-pure state η, consideration of subsystems is admitted, and restriction of the states η_t to some subsystem produces other non-pure states. *Detailed* information about holistic entanglement is *not* encoded in the epistemic states η_t, since it is not clear whether these states arise as restrictions of overall pure states ϕ on a larger system or as averages of pure states of the exo-system itself. Even in the latter case holistic entanglement with respect to a tensorization $\mathcal{B} = \mathcal{B}_1 \otimes \mathcal{B}_2$ is not defined.

2.3. EPISTEMIC STATES WITH INTRINSIC OBSERVABLES

In this section we introduce a third, "hybrid" level of description in addition to those of ontic states with intrinisic observables and epistemic states with contextual observables. It is designed to address situations in which an exo-system is always in a *pure* state. Such an exo-system will be called an *exo-object*. The creation of an exo-object relies upon an appropriate *dressing procedure* (see Sect. 3) which guarantees that the restrictions of all relevant pure states on the overall joint quantum system {exo-system & environment} are pure states again. The dynamics of a quantum exo-object are not determined by the Schrödinger equation alone; additional *stochastic* elements come in which roughly correspond to the dissipative part of the density-operator dynamics for exo-systems as in Sect. 2.2 (see Primas 1990b). For an exo-object, single pure states and *probability distributions* μ *of pure states* are admitted. Averaging over a probability distribution μ of pure states ρ *of the exo-object* results in a *uniquely* determined non-pure state η_μ with density operator D_μ defined by

$$\eta_\mu(\hat{A}) = \mathrm{Tr}(D_\mu \hat{A}) = \int_{\text{all pure states } \rho} \rho(\hat{A})\mu(d\rho). \tag{1}$$

The converse is not true, i.e., to a given non-pure state η represented by a density operator D there exist many different probability distributions μ on the pure states of the exo-object such that Eq. (1) holds. Probability distributions of pure states on an exo-object are called *epistemic probability distribution* or simply *epistemic state* (Amann and Primas 1997), because they refer to our restricted knowledge concerning the "correct" pure state of the exo-object (see also Amann 1995, 1997).

While an *ontic* state refers always to an endo-system, an *epistemic* state may either be a non-pure state (for an exo-system) or a probability distribution of pure states (for an exo-object). Let us stress again that the two different types of epistemic states are not in one-to-one correspondence. To every epistemic state of a quantum exo-object there corresponds uniquely an epistemic state of a quantum exo-system, but the converse is not true.

Which structure is appropriate for the observables of an exo-object? It is the C*-algebra $\mathcal{B}$, without the σ-weak closure to be introduced for the W*-algebra of observables for exo-systems. Surprisingly, a W*-algebra often does *not* make sense for exo-objects! The reason for this is the following: Let us take a probability distribution μ of pure states on the C*-algebra $\mathcal{B}$ with associated non-pure state η (through averaging as in Eq. (1)). This non-pure state η can be *uniquely* extended to a σ-weakly continuous state on the W*-algebra $\mathcal{M}_\eta$ (generated by the C*-algebra $\mathcal{B}$ in the GNS-representation π_η with respect to η; a state on a W*-algebra is σ-weakly continuous if and only if it can be represented by a density operator). Nevertheless the pure states on $\mathcal{B}$ *cannot* be extended to σ-weakly continuous states on $\mathcal{M}_\eta$, and hence also the probability distribution μ (i.e., the epistemic state of the exo-object) does not make sense at the level of the W*-algebra $\mathcal{M}_\eta$. Even worse: Pure states on a typical W*-algebra of type III arising in thermal situations (Bratteli and Robinson 1981) are utterly pathological, because they do not even fulfill the (almost trivial) Jauch-Piron condition (Amann 1987). The reason for this is that the W*-algebra does not only contain interesting observables such as temperature or chemical potential, but also a lot of "garbage" which enters by taking the σ-weak closure of the C*-algebra $\mathcal{B}$. A remarkable exception are factor W*-algebras of type I, i.e., W*-algebras isomorphic to all bounded operators on a Hilbert space $\mathcal{H}$.

These are the algebras of traditional quantum theory, where pure states have particularly nice properties, and fulfill the Jauch-Piron condition.

An exo-object is described by probability distributions μ_t of pure states which – by averaging as in Eq. (1) – provide non-pure states η_t. This depends, of course, on the initial pure state or on the initial distribution μ_o of pure states. For a large class of stochastic dynamics of exo-objects, however, one ends up with *one* particular distribution μ_∞ of pure states in the limit $t \to \infty$ *independently of the initial conditions* (such dynamical exo-objects are called ergodic). Splitting of the underlying C*-algebra $\mathcal{B}$ into two subsystems with two C*-subalgebras $\mathcal{B}_1$ and $\mathcal{B}_2$, $\mathcal{B} = \mathcal{B}_1 \otimes \mathcal{B}_2$, is then admitted under certain conditions. In an ideal situation all those pure states onto which the probability measures μ_t extend are *product* states with respect to the tensor product $\mathcal{B} = \mathcal{B}_1 \otimes \mathcal{B}_2$. This situation never arises in practice, but "most" relevant pure states can be product states or almost product states, if the tensorization is chosen appropriately (see "dressing procedures" in the next section). One can introduce a quantitative criterion estimating a tensorization of $\mathcal{B}$ at time t by the positive number

$$I_{\mathcal{B}_1 \otimes \mathcal{B}_2} \overset{\text{def}}{=} \int_{\text{all pure states } \rho} p(\rho)\mu_\infty(d\rho), \qquad (2)$$

where $p(\rho)$ estimates the "degree" to which the pure state ρ is a product state (taking, e.g., the value $p(\rho) = 0$, iff ρ is a product state). Let us stress that one must *always accept approximations* when dealing with exo-objects. The exo-object itself may arise by a tensorization of some larger system in the way just described. Consequently one must always expect that the pure states used in the description of an exo-object are not exactly product states.

Remark: Consider the tensor product $\mathcal{B} = \mathcal{B}_1 \otimes \mathcal{B}_2$. Assume that a given probability measure μ_1 exists on product states $\rho_1 \otimes \rho_2$ only, and introduce the respective averaged non-pure state η, as defined in Eq. (1). Then the state η can be decomposed into pure states which are *not* of product form, i.e., there exists another probability distribution μ_2 of pure states on $\mathcal{B}$, which exists on non-product states but nevertheless results in η by averaging. Hence a non-pure state η does not encode the holistic entanglement between two subsystems with C*-algebras $\mathcal{B}_1$ and $\mathcal{B}_2$. For exo-objects one can find appropriate quantitative criteria to estimate the holistic entanglement for a given tensorization (which is impossible if only a non-pure state η is known).

An exo-object is always in a *pure* state, though this pure state is perhaps not precisely known and must be estimated (a probability distribution μ of pure states can be such an estimate). Therefore an exo-object is always an *individual* quantum object, described in a statistical way (namely by an epistemic probability distribution). The stochastic aspect of the time evolution (of pure states of the exo-object) stems from the fact that the (initial) state of the environment cannot be determined (and is therefore a stochastic variable). Starting from an initial pure state ρ_0, one gets time-evolved states $\rho_{t,\omega}$, where ω is the stochastic variable (i.e., the pure initial state of the environment). In an ergodic situation, the final distribution η_∞ of pure states (for large or infinite time) could – in principle – be regarded as an average in time *or* as a mixture of all the different pure states with respect to the stochastic variable ω. Since we consider an exo-object to be an individual quantum object (and not an ensemble of quantum objects), we shall interpret η_∞ as *average in time* of the pure states $\rho_{t,\omega}$, where the external (initial) state ω is kept fixed.

For a quantum-mechanical description of (mesoscopic or macroscopic) "common sense" objects, an exo-object formalism is appropriate. A cat, for example, is surely not an endo-system and obviously considered as an individual, single object. Hence a cat – when described quantum mechanically – is in a pure state, but this pure state is unstable under small external perturbations and can therefore only be estimated. As stressed above, *approximations are unavoidable* when exo-objects are discussed. An exo-object is described by a pure state only approximately. De facto, entanglement between an exo-object and its environment cannot be excluded completely since both have quantum aspects and since even an optimal dressing is not exact (see Sect. 3).

2.4. SUMMARY

According to the preceding detailed discussion, we have three categories of quantum systems: isolated endo-systems with pure states, open exo-systems with non-pure states, and exo-objects that are open systems with pure states. From a conceptual point of view, the pure states of endo-systems are ontic in the sense that they are independent of (epistemic) information or knowledge that observers may have about them. By contrast, the states of exo-systems and exo-objects are obviously epistemic, however in different ways. While states of exo-systems are described by density operators, states of exo-objects are described by probability distributions of pure states.

The intrinsic, operationally inaccessible observables of an endo-system are elements of a C*-algebra $\mathcal{A}$. If an endo-system is decomposed into an exo-object and its environment (by an appropriate dressing procedure), the C*-algebra of observables of this exo-object is given by a "relevant" subset $\mathcal{B} \subset \mathcal{A}$. The algebra of contextual observables of exo-systems is a W*-algebra $\mathcal{M}$, obtained by the σ-weak closure of $\mathcal{B}$ which can be obtained by a (contextual) GNS-construction. The corresponding three categories of dynamics are a Schrödinger dynamics of pure states for an endo-system, density-operator dynamics including Schrödinger dynamics and dissipative terms for exo-systems, and Schrödinger dynamics of pure states and stochastic terms for exo-objects.

These three categories of quantum systems need different interpretations. An individual and non-statistical interpretation is appropriate for endo-systems, a "density-operator statistical" interpretation is required for exo-systems, and an individual-statistical interpretation for exo-objects. There are two different interpretations for the different epistemic state concepts, expressing the notorious difference between a probability distribution (of pure states) as such and its moments (averages).

3. Heisenberg Cut and Dressing

By contrast to common-sense guided expectations, a "material object" is not at all defined in an obvious, a priori manner. Consider, for example, something like a molecule as such a material entity with its common molecular Hamiltonian

$$\hat{H} \overset{\text{def}}{=} \sum_j \frac{\hat{P}_j^2}{2m_j} + \frac{1}{4\pi\epsilon_\circ} \sum_{j<k} \sum \frac{e_j e_k}{|\hat{\vec{q}}_j - \hat{\vec{q}}_k|}, \qquad (3)$$

where the sums go over all the particles (electrons and nuclei) of the molecule, m_j are the respective masses and e_j are the respective charges. Such a Hamiltonian describes already a *dressed object*. "Originally" only the joint system {molecule & radiation field} is well-defined, since charged particles cannot be screened from the radiation field. The Hamiltonian of the joint system consists of three parts comprising, firstly, the kinetic energy of the nuclei and electrons, secondly, the energy of the radiation field and, thirdly, the interaction energy between the charged particles and the radiation field. By a clever choice of conventions one can single out from the third part an interaction between the charged particles, which is – by convention – *attributed* (Heitler 1954) to the molecule and appears now as the potential energy in Eq. (??). It could, of course, also be attributed to the radiation field, since it is the field which mediates the interaction. This choice/attribution corresponds to an appropriate "cut" of the joint endo-system

{molecule & radiation field},

described by a pure state ϕ, into a product of pure states according to

{dressed molecule & dressed radiation field}.

The lesson is that molecules as objects are not defined a priori. A clever choice is necessary and the resulting object is called a "dressed quantum object", e.g., a dressed molecule. The idea behind all this is that such a dressed object is in a pure (approximately pure) quantum state ρ_{mol}. Consequently, the joint system {dressed molecule & dressed radiation field} is a product state $\rho_{\text{mol}} \otimes \rho_{\text{rad}}$ of two pure states.

It is not clear from the very beginning whether such conventions are sensible. A choice or cut as mentioned above does not necessarily lead to a quantum *object*, but could result in a quantum *system* only (both exo, of course). In the latter situation, pure states would usually not play a role and should be replaced by non-pure states (density operators). In the particular case of molecules, this would mean that neither molecular structure, nor handedness, nor the monomer sequence of a macromolecule would make sense (Amann 1995, 1996, 1997), because non-pure states (such as thermal states) can be decomposed into pure states that may or may not admit molecular structure (handedness etc.).

Similarly, considering a cat from a quantum-mechanical point of view, we would hope that it is a quantum exo-object and not just a quantum exo-system. Being a quantum exo-system only would imply that holistic features (between different cats or between a cat and its environment) play a decisive role. Being a quantum exo-object means that only "internal" holistic effects persist rather than holistic entanglement with the environment. In the terminology of Sect. 2, molecules, cats, etc. are to be considered as exo-*objects*. Hence they are – as a rule – described by pure states, though non-pure states may play a certain role, e. g., for the spontaneous decay of molecular states.

Dressing procedures go by two steps:

(a) First one tries to reformulate the Hamiltonian, suggesting new particles or other structures. Second quantization (Guichardet 1970, Jost 1973), for example, is a procedure which starts with a direct-sum Hilbert space composed of subspaces referring to a fixed number n of "old" particles (arbitrary $n = 1, 2, \ldots$); then boson (or fermion) operators are introduced which give rise to "new particles", e.g., photons or phonons (whose observables – isomorphic to a factor W*-algebra of type I – are generated by a pair a_j and a_j^* of boson operators satisfying the canonical commutation relation $[a_j, a_m^*] = \delta_{jm}\mathbb{1}$, with commuting operators for different new particles). Nevertheless, non-product states between different "new particles" are admissible, i.e., the "new particles" are not necessarily in pure states (not necessarily exo-objects, but maybe only exo-systems).

(b) Secondly, one might insist that "new particles" are (approximate) exo-*objects*, either by declaration or by comparison with experiment or by some theoretical derivation. Often, this is indeed a mere *declaration*, which "forbids" certain superpositions of given basic pure states.

Usually, experiments describe ensembles of quantum systems/objects and therefore depend only on the averaged density-operator dynamics and not on the actual stochastic dynamics of the pure states. Such experiments cannot distinguish between different declarations (as long as both declarations are compatible with the density-operator dynamics). Consequently, such declarations are of practical use only and do not reproduce the reality of individual quantum objects.

Here we are mainly interested in a description of individual, possibly macroscopic quantum objects. Experiments then do *not* refer to an ensemble of such objects, but to individual systems or objects. In such a situation the distinction between exo-systems and exo-objects is vital and new structures cannot be introduced by mere declarations.

How can a proper dressing procedure be introduced in such an "individual" context? Let us stress again the dichotomy between dressed exo-objects on the one hand and quantum mechanical superpositions on the other:

- For proper dressed objects it is vital to have (approximate) pure product states $\rho_1 \otimes \rho_2$ for the joint system {dressed object & dressed environment}.
- The superposition principle is at the heart of quantum mechanics, but superpositions of product states are *not* product states any more. Their restriction to the dressed object or dressed environment is therefore a non-pure state.

An appealing way out of this conflict is a *stability analysis* of the pure states on the joint system {dressed object & dressed environment}. We consider this joint system to be described by a C*-algebra $\mathcal{B}$ from which the dressed exo-object is derived, with pure states under a stochastic dynamics. Though *all* pure states of this joint system "exist", some of them are unstable with respect to the stochastic dynamics and decay into more stable ones. As mentioned in Sect. 2.3, the final distribution μ_∞ of pure states for $t \to \infty$ is independent from the initial conditions if the dynamics is ergodic. Then different tensorizations $\mathcal{B} = \mathcal{B}_1 \otimes \mathcal{B}_2$ can be estimated quantitatively by the positive number $I_{\mathcal{B}_1 \otimes \mathcal{B}_2}$ defined in Eq. (2). If $I_{\mathcal{B}_1 \otimes \mathcal{B}_2} = 0$, then all relevant pure states are exact product states with respect to the chosen tensorization.

In all practical cases, one must expect that $I_{\mathcal{B}_1 \otimes \mathcal{B}_2}$ takes small but non-zero values and non-product states play some role. Even if $I_{\mathcal{B}_1 \otimes \mathcal{B}_2}$ is equal to zero, it is only guaranteed that the final distribution μ_∞ exists on product states. It is not guaranteed that an initial product state will be a product state for all finite times $t > 0$. Hence one may start with a product state and end with a product state (for large times), and nevertheless intermediate states can be of non-product form.

Usually, such non-product states are suppressed – either by accepting a Hartree approximation (Primas 1990b,c) from the very beginning (only product states are admitted) or by considering an appropriate *limit* with respect to which only product states appear (e.g., infinite volume, infinitely many particles, infinite nuclear masses in molecules, etc.). In limit situations, classical structures and classical observables may appear "automatically" through the mathematical formalism (Müller-Herold 1980, Bratteli and Robinson 1981, Primas 1983, Sewell 1986), partly suppressing the consequences of the superposition principle. In realistic situations one has, of course, finite volume, finitely many particles, or finite nuclear molecular masses, and so on, with unrestricted validity of the superposition principle. Hence, one is in the strange situation that for arbitrarily large but finite volume, finite nuclear masses, ..., the superposition principle holds unrestrictedly whereas in the limit of infinite volume, ..., its validity is lost. (More precisely: its validity is lost for classical observables in different superselection sectors, separated by superselection rules.)

From our point of view it is important to accept non-product states (with corresponding holistic entanglement between B_1 and B_2) and to quantitatively determine the probability that they appear. This means that

- we accept the Hartree approximation only as a first step toward a proper understanding of dressing procedures and non-linear stochastic time evolutions,
- and we are not only interested in limit considerations, but also in the discussion of systems/objects before the limit is reached, i.e., for large but finite volume etc.

In some relatively simple situations it is possible to discuss these problems explicitly: For a magnet consisting of N spins (each described by Pauli matrices σ_x, σ_y, σ_z) one may investigate the specific magnetization operator

$$\hat{m}_N \stackrel{\text{def}}{=} \frac{1}{N} \sum_{j=1}^{N} \sigma_{z,j} . \tag{4}$$

In the limit $N \to \infty$ one gets a classical magnetization $\hat{m} \stackrel{\text{def}}{=} \lim_{N \to \infty} \hat{m}_N$, i.e., $\hat{m}$ commutes with *all* other observables (constructed from local Pauli matrices). This fact implies that – in the limit – pure states with different expectation values of $\hat{m}$ are not superposed any more. It is no problem to verify these facts mathematically, but they are not easily understandable from a conceptual point of view, since such superpositions can be written down as pure states for any (whatever large but) *finite* number N of spins. A stability analysis in the case of a Curie–Weiss magnet (Amann 1995, Amann and Primas 1997) shows that these superpositions "die out" successively with increasing number N of spins, and the exponential rate of this process is given by an entropy in the sense of large deviations statistics.

4. Dynamics of Measurement

Any dynamical description of measurement has to start from a proper decomposition of a system into a dressed exo-object and its dressed environment. It is crucial to keep in mind that such a decomposition is a logical precondition for the dynamics of measurement insofar as the Hamiltonian of the composed system needs to be written as a sum

$$H = H_{\text{obj}} \otimes 1 + 1 \otimes H_{\text{env}} + H_{\text{int}}. \tag{5}$$

An illustrative example has been extensively discussed by Primas (1990b,c). Consider the simple case of a two-level quantum object (spin 1/2 system) with the Hamiltonian

$$H_{\text{obj}} = \frac{\hbar}{2} \sum_{\nu=1}^{3} \Omega_\nu \sigma_\nu, \tag{6}$$

a sufficiently nontrivial boson field environment

$$H_{\text{env}} = \sum_{\nu=1}^{3} \sum_k \omega_k \alpha_{k\nu}^* \alpha_{k\nu}, \tag{7}$$

and an interaction

$$H_{\text{int}} = \sum_{\nu=1}^{3} \sigma_\nu \otimes A_\nu, \tag{8}$$

where

$$A_\nu = \sum_k \lambda_{k\nu} \alpha_{k\nu} + c.c. \tag{9}$$

If such a decomposition has been properly carried out (cf. Sec. 3), then it is possible to derive the expectation values

$$M(t) = <\psi_t|\sigma|\psi_t> \tag{10}$$
$$\alpha(t) = <\chi_t|A|\chi_t> \tag{11}$$

with respect to the (approximate) product state

$$\Psi_t = \psi_t^{\text{obj}} \otimes \chi_t^{\text{env}}. \tag{12}$$

Corresponding to the product state Ψ_t, the C*-algebra of intrinsic observables in the composed system of dressed object and dressed environment is

$$\mathcal{A} = \mathcal{A}_{\text{obj}} \otimes \mathcal{A}_{\text{env}}. \tag{13}$$

$\mathcal{A}_{obj}$ is the C*-algebra of 2×2 matrices and $\mathcal{A}_{env}$ is the C*-algebra of intrinsic observables of an environment with infinitely many degrees of freedom.

The equations of motion for the expectation values $M(t)$ and $\alpha(t)$ are given by:

$$\dot{M}(t) = M(t) \times \Omega + M(t) \times \alpha(t), \tag{14}$$

$$\dot{\alpha}_{k\nu}(t) = -\omega_k \alpha_{k\nu} + \frac{i}{2}\lambda_{k\nu}M_\nu(t). \tag{15}$$

They describe the feedback between object and environment. More precisely, they describe the polarization M of the object under the influence of the environment and the motion of the environment observable α (boson operator) under the polarizing influence of the object. The solution of the second equation, referring to the observables of the environment (or the measuring system, respectively) splits into a retarded and an advanced part:

$$\alpha_{k\nu}^{ret} = \exp(-i\omega_k t)\alpha_{k\nu}(0)$$
$$- \frac{i}{2}\lambda_{k\nu} \int_0^t \exp(-i\omega_k(t-s))M_\nu(s)ds \quad (t \geq 0), \tag{16}$$

$$\alpha_{k\nu}^{adv} = \exp(-i\omega_k t)\alpha_{k\nu}(t)$$
$$+ \frac{i}{2}\lambda_{k\nu} \int_t^0 \exp(-i\omega_k(t-s))M_\nu(s)ds \quad (t \leq 0), \tag{17}$$

Selecting one of these solutions and disregarding the other requires the time inversion symmetry of the composed system to be broken. In other words: a bidirectionally deterministic system has to be described in terms of a superposition of a backward deterministic (forward non-deterministic) and a forward deterministic (backward non-deterministic) process which are equally relevant a priori. Then, one can apply the principle of causality (past-determinacy, error-free retrodiction, no anticipation) as a "heuristic" argument for the selection of one of the solutions. Causality is consistent with the retarded solution, whereas the advanced solution contradicts causality, i.e., it has the properties of future-determinacy, error-free prediction, and no memory (Primas 1992).

It has been argued that the retarded, i.e., the backward deterministic, forward non-deterministic, solution is a K-flow on a state space with infinitely many degrees of freedom (Primas 1990b,c, 1997a). Mixing alone is *not* sufficient to break the time reversal symmetry of the interaction between object and environment (Primas 1997b). In the simplest case, the relaxation time for this K-flow is the time constant τ_ν of an exponentially decaying correlation function (for details, see Primas 1990b)

$$K_\nu = \gamma_\nu \exp(-|t|/\tau_\nu). \tag{18}$$

At this point we are still at the level of description of intrinsic observables, needed for the specification of initial conditions of the K-flow. Conceptually, this K-flow represents a stochastic process for the exo-object (cf. Pearle 1976, Gisin 1984, Ghirardi et al. 1986) which corresponds to chaos in the sense of Wiener rather than chaos in the sense of Kolmogorov and Sinai (i.e., a dissipative exo-system dynamics). The transition from the former to the latter is made by averaging over a probability distribution of pure states and their trajectories, i.e., by introducing the level of description using density operators or, in classical systems, distribution functions and corresponding ensembles of trajectories. After such a procedure one deals with epistemic states and (GNS-constructed) contextual observables, where a context may have been introduced by a reference state with respect to which stability in a certain sense (hopefully more general than thermal equilibrium) can be checked. At this level of description, the evolution of the flow is an irreversible semigroup evolution with a finite number of degrees of freedom, whereas Wiener chaos is a stochastic process in an infinite-dimensional state space.

The fact that the dynamics of measurement can be described as a stochastic process in time suggests that it takes a certain amount of time until a measurement is finished. Of course, this statement applies to the act of measurement in the general sense of an interaction between an object and its environment, approximately decomposed from one another, and is not restricted to controlled laboratory measurements. The temporal cut determining a measurement as finished "by declaration" is intimately related to the Heisenberg cut approximately separating an object from its environment. It may be speculated that the relaxation time τ_ν of the K-flow is a function of the degree of stability of the pure states forming the dressed product state.

Since these deliberations apply to non-controlled types of "measurement" in a quite general sense, one may make use of them for the notorious discussion of "quantum jumps" or, more general, any relaxation (or excitation) of unstable quantum objects. Such relaxations do not occur for strictly isolated objects; for instance, an environment is needed to embed an atom at least in its own electromagnetic field. In this context, an interesting analog to the issue of the duration of measurements is the question as to whether quantum jumps happen in time (and, if yes: how much time they take). In the following we sketch how this question might be addressed from a novel perspective, providing an alternative to recent decoherence approaches as represented in Giulini et al. (1996).

An important implication of a K-flow is the fact that the corresponding system admits a time observable which is not in the center of the W*-algebra of contextual observables, i.e., does not represent a classical observable. There is a long tradition of approaches toward such a time observable, which is intimately connected with the history of energy-time uncertainty relations (cf. Atmanspacher 1994). A basic objection against a time operator not commuting with a suitable energy operator (the bounded or even discrete Hamiltonian of a system) has been formulated by Pauli (1933). It has recently been shown in detail (Busch et al. 1994) how Pauli's objection can be circumvented if the relevant time observable is defined by a positive operator valued (POV) measure rather than a projection valued (PV) spectral measure. POV measures are concepts for observables that are more general than PV measures. A POV measure is a PV measure, if it is multiplicative, if the corresponding operator is idempotent and not only Hermitian but also self-adjoint, and if the set of its eigenfunctions (if they exist) is orthogonal.

If H is the Hamiltonian of a system S, then $t \mapsto e^{-iHt}$, $t \in \mathbb{R}$, is a unitary representation V_t of the time translation group. If, furthermore, Θ is a time interval during which an event is (with some probability) expected to occur in a certain state of S, then a suitable dynamical variable $B(\Theta)$ satisfies

$$V_t^* B(\Theta) V_t = B(\Theta - t), \qquad V_t = e^{-iHt}. \tag{19}$$

Such a dynamical variable is a POV measure for a time observable (Busch et al. 1994). Its construction is possible in specific cases, but cannot be universally prescribed; it depends crucially on contexts given by the system considered. On the basis of B, a time operator (briefly: POV time) can be defined according to

$$T_B = \int t \, B \, (dt), \tag{20}$$

which is *not* self-adjoint and fulfills the commutation relation

$$i[H, T_B] = \mathbb{1} \tag{21}$$

on an appropriate domain, without contradicting Pauli's theorem (Busch et al. 1994).

From a different point of view, Misra and others (Misra 1978, Misra et al. 1979a,b; see also Suchanecki 1992) have introduced a time operator T_L on the basis of the Liouville representation of a dynamical system. A related definition of a time operator, based on the theory of stochastic processes, has been proposed by Tjøstheim (1976) and Gustafson and Misra (1976); see also Primas (1997a). For both approaches, the condition of a K-flow is crucial. In the Liouville representation, the time operator is a shift operator T_L

$$U_t^* T_L U_t = T_L + t\mathbb{1}, \qquad U_t = e^{-iLt}. \tag{22}$$

T_L does not commute with the Liouvillean L that can be formulated as the Poisson bracket $L\rho = \{H, \rho\}$,

$$i[L, T_L] = \mathbb{1}, \tag{23}$$

and hence does not contradict Pauli's theorem either (Misra 1978).

The two time operators are not the same, and their relationship with each other is not immediately clear. An important difference between them is given by the fact that POV time is in general not self-adjoint, whereas a time operator in the sense of Misra is a self-adjoint PV measure (for more details, see Atmanspacher and Amann 1998). Both T_B and T_L are defined as shift operators, but with respect to different generators of the dynamical evolution (H and L, respectively). If the eigenstates of a system are non-stationary, energy levels due to eigenvalues of H have a non-vanishing width ΔE corresponding to the fact that some dynamical variable, as required for the definition of POV time, is *not* a conserved quantity, i.e., its expectation value changes in time, $d\langle B\rangle/dt \neq 0$ (cf. Mandelstam and Tamm 1945).

This bandwidth can be expressed as the eigenvalue of a PV operator L (cf. Prigogine and Petrosky 1987, 1988). The interpretation of Misra's T_L as an observable is different from that of POV time. POV time refers to the "time of occurence" of an event in a non-stationary state as an observable canonically conjugate to a proper Hamiltonian. By contrast, Misra's T_L refers to the "eigentime" of a non-stationary state as an observable canonically conjugate to the Liouvillean considered as an operator whose eigenvalues are differences of energies, e.g., bandwidths ΔE. (Grelland (1993) has shown that those eigenvalues of L vanish if the eigenstates of H are stationary. See also Ban (1991)).

The non-commutative character of a time operator in the Liouville representation can be operationalized if one moves from T_L to an information operator M (Atmanspacher and Scheingraber 1987) according to

$$U_t^* M U_t = M - h_T t\mathbb{1}, \qquad U_t = e^{-iLt}, \tag{24}$$

where h_T is the empirically accessible (Grassberger and Procaccia 1983) Kolmogorov-Sinai (KS) entropy of the system (Kolmogorov 1958, Sinai 1959). It is important to realize that the definition of M is *more general* than that of T_L insofar as M can be defined even for non-mixing ergodic systems with $h_T = 0$, whereas a necessary condition for the existence of T_L is strong mixing, sometimes even $h_T > 0$. While T_L, if it exists, does not commute with the Liouvillean L, M commutes with L iff $h_T = 0$, and M does not commute with L iff $h_T > 0$ (cf. Atmanspacher and Scheingraber 1987, Atmanspacher 1997):

$$i[L, M] = h_T\mathbb{1}. \tag{25}$$

The KS entropy h_T as the key quantity in this framework can (somewhat roughly) be defined as the sum of positive Ljapunov exponents of the system (for more details, see Atmanspacher 1997). The Ljapunov exponents are basically derived from eigenvalues of a linear stability matrix which refers to a comoving (with the flow) coordinate system. Hence it is intuitively clear that the set of its eigenfunctions is orthogonal only locally in state space (at a given time). (Compare recent work of Grossmann and collaborators (Grossmann 1996) on the problem of non-orthogonal eigenfunctions of non-normal shear operators in turbulent flows.) In addition to other features (Davies and Lewis 1970, Davies 1970, Srinivas 1980, Ozawa 1984, Primas 1990b), this illuminates the use of POV measures, in particular POV time, to characterize K-flows in measurement theory (and otherwise).

An additional interesting feature in this context is the fact that h_T is an inverse relaxation time or, more precisely, an inverse predictability time (see Atmanspacher and Scheingraber 1987, Atmanspacher 1997). At a speculative level, one might think of a quantitative relationship between h_T and $1/\tau_\nu$ (in simple cases). This would allow us to characterize the duration of a measurement by the dynamical invariant (h_T) of a nonlinear process in a finite-dimensional state space (KS chaos) as well as a relaxation time of a linear process in an infinite-dimensional state space (Wiener chaos).

At the same time, it is interesting to think of the properties of KS chaos in terms of an information flow (Goldstein 1981, Shaw 1981, Farmer 1982, Caves 1994) which always accompanies the transition from ontic to epistemic states. The transition from an infinitely refined partition for ontic states to binary alternatives has been addressed by Atmanspacher (1989) for the act in which "virtual particles realize themselves" – another example for a generalized understanding of the process of measurement. It is of interest that K-flows entail an intrinsic and unique, but system-specific (contextual) definition of a so-called generating partition (Cornfeld et al. 1982; see also Crutchfield 1983) which is *not* imposed ad hoc or with the help of additional arguments. For such a generating partition, trajectories which are very close to each other initially need an average time of the order of the inverse of h_T to become sufficiently separated to be distinguishable with respect to the generating partition.

If measurement is generically related to K-flow properties (Lockhart and Misra 1986, Primas 1997a), then a time observable in the sense of Misra exists which does not commute with all other observables, i.e., which does not belong to the center of the algebra of contextual observables of the exo-system considered. At the level of the exo-object, the K-flow properties refer to the possible histories (trajectories) corresponding to the temporal evolution of the probability distribution of pure states. At this level of description it is appealing to apply the concept of consistent histories (Omnès

1992) in order to study temporal Bell inequalities (Leggett and Garg 1985, Paz and Mahler 1993, Mahler 1994). Analogous to the standard Bell inequalities, one can show that their temporal counterparts may violate the classical imagination of consistent histories satisfying probability sum rules. The interpretation of inconsistent histories in this sense is due to a temporal nonlocality which is bound to a system-specific time scale (relaxation time) and expresses itself as a lack of well defined temporal order (before-after relationships) within that time scale. Inconsistent histories are relevant at the level of endo-systems rather than exo-objects or exo-systems.

The precise connections between such a temporal nonlocality, the existence of a non-commuting time observable, and the relaxation properties of measurement processes have yet to be explored in detail. Some indications and perspectives have been sketched in Atmanspacher (1997). An important point is that there is more than one way to look at the process of measurement and corresponding topics such as quantum jumps etc. Possibility (1) is to look at the measurement process in terms of the evolution of expectation values of observables as a function of a parameter time, together with their temporal relaxation properties. A second way (2) is to consider measurement from the perspective of temporal nonlocality, associated with the absence of temporal order within the time interval which (1) provides for the relaxation. This time interval is related to the information flow rate represented by the KS entropy. Such a temporally nonlocal interval corresponds to the existence of a time operator (POV or Misra) outside the center of the relevant algebra of observables. There are other possibilities to define time operators (cf., e.g., Amann 1986, Isham 1993) which are outside the scope of this paper.

Let us finally return to the symmetry breaking in time due to the split into retarded and advanced solutions for the feedback between object and environment. Cramer (1980, 1986) has suggested considering this split in the framework of his transactional interpretation of quantum mechanics. After the completion of an emitter-absorber transaction, the resulting superposition of retarded and advanced components can be reinterpreted due to a purely retarded solution. (For a comprehensive presentation of this and other approaches toward questions about the arrow of time from the viewpoint of the philosophy of physics compare the recommendable book by Price (1996).) Before the transaction is completed, everything is spatially and temporally nonlocal, such that the two solutions are indistinguishable. It is interesting to look at this feature from the perspective of Misra's (1995) result that the Klein-Gordon evolution of massive particles (not of photons!) admits a time operator if conditions are satisfied that are set by the restriction of self-adjointness (cf. the discussion in Atmanspacher 1997). Since the characteristic time scale associated with the resulting time operator is mass-dependent, one might speculate that there are mass-dependent effects in EPR experiments with massive particles. So far, all uncontroversial conclusive evidence for nonlocal correlations in quantum systems has been gained in experiments with photons, although some experiments with massive particles have been carried out (cf. the collection of articles in Selleri (1988), see also new proposals of EPR type experiments with massive particles by Fry et al. (1995) and Freyberger et al. (1996)).

The symmetry between advanced and retarded solutions is unbroken at the level of ontic (pure) states of the holistic isolated endo-system. At present, we do not know in what precise sense their explicit distinction has to do with the Heisenberg cut. There certainly are connections, but their nature in detail is still unclear. A fascinating idea in this context, advocated by Costa de Beauregard (1987), suggests that symmetry breaking with respect to temporal direction goes hand in hand with the Cartesian cut between the material world of res extensa and its mental counterpart. Bringing this idea into contact with the argumentation in the present article leads to the speculation that the scientific concept of causality (the Aristotelian causa efficiens), referring to backward deterministic processes in the material world, could be complemented by the scientifically disregarded concept of finality (the Aristotelian causa finalis), consistent with forward deterministic, "goal-oriented" processes (Primas 1992) in the mental world. The maximum entropy principle that has been suggested (compare Sec. 3) to find most stable decompositions, i.e., most appropriate Heisenberg cuts, for a given endo-system, effectively represents an example of such a final concept. We may wonder whether this indicates any relation to the advanced counterpart of the retarded solution for proper observables in the environment.

References

Amann A. (1986): Observables in W*-algebraic quantum mechanics. *Fortschr. Phys.* **34**, 167–215.

Amann A. (1987): Jauch-Piron states in W*-algebraic quantum mechanics. *J. Math. Phys.* **28**, 2384–2389.

Amann A. (1995): Structure, dynamics and spectroscopy of single molecules: A challenge to quantum mechanics. *J. Math. Chem.* **18**, 247–308.

Amann A. (1996): Can quantum mechanics account for chemical structures? In *Fundamental Principles of Molecular Modeling*, ed. by W. Gans, J. Boeyens, and A. Amann (Plenum, New York), pp. 55–97.

Amann A. (1997): Fuzzy classical structures in genuine quantum systems. In *Fuzzy Logic in Chemistry*, ed. by D. Rouvray (Academic Press, San Diego), pp. 91–138.

Amann A. and Primas H. (1997): What is the referent of a non-pure quantum state? In *Potentiality, Entanglement, and Passion-at-a-Distance. Quantum Mechanical Studies in Honor of Abner Shimony*, ed. by R.S. Cohen, M. Horne, and J. Stachel (Kluwer, Dordrecht), pp. 9–29.

Atmanspacher H. (1989): The aspect of information production in the process of observation. *Found. Phys.* **19**, 553–577.

Atmanspacher H. (1994): Is the ontic/epistemic distinction sufficient to represent quantum systems exhaustively? In *Symposium on the Foundations of Modern Physics 1994*, ed. by K.V. Laurikainen, C. Montonen, and K. Sunnarborg (Editions Frontières, Gif-sur-Yvette), pp. 15–32.

Atmanspacher H. (1997): Dynamical entropy in dynamical systems. In *Time, Temporality, Now*, ed. by H. Atmanspacher and E. Ruhnau (Springer, Berlin), pp. 327–346.

Atmanspacher H. and Amann A. (1998): Positive operator valued measures and projection valued measures of non-commutative time operators. *Int. J. Theor. Phys.* **37**, 629–650.

Atmanspacher H. and Scheingraber H. (1987): A fundamental link between system theory and statistical mechanics. *Found. Phys.* **17**, 939–963.

Ban M. (1991): Relative number state representation and phase operator for physical systems. *J. Math. Phys.* **32**, 3077–3087.

Bratteli O. and Robinson D.W. (1981): *Operator Algebras and Quantum Statistical Mechanics Vol. 2* (Springer, New York).

Bratteli O. and Robinson D.W. (1987): *Operator Algebras and Quantum Statistical Mechanics Vol. 1.* (Springer, New York, 2nd revised edition).

Busch P., Grabowski M., and Lahti P.J. (1994): Time observables in quantum theory. *Phys. Lett.* A **191**, 357–361.

Caves C. (1994): Information, entropy, and chaos. In *Physical Origins of Time Asymmetry*, ed. by J.J. Halliwell, J. Pérez-Mercader, and W.H. Zurek (Cambridge University Press, Cambridge), pp. 47–89.

Cornfeld I.P., Fomin S.V., and Sinai Ya.G. (1982): *Ergodic Theory* (Springer, Berlin), pp. 250–252.

Costa de Beauregard O. (1987): *Time, the Physical Magnitude* (Reidel, Dordrecht).

Cramer J.G. (1980): Generalized absorber theory and the Einstein-Podolsky-Rosen paradox. *Phys. Rev. D* **22**, 362–376.

Cramer J.G. (1986): The transactional interpretation of quantum mechanics. *Rev. Mod. Phys.* **58**, 647–687.

Crutchfield J.P. (1983): *Noisy Chaos.* PhD Thesis, University of California at Santa Cruz, published by University Microfilms Intl., Minnesota, 1983, Sect. 5.2 (pp. 5.9–5.14) and Sect. 5.4 (pp. 5.19–5.26).

Davies E.B. (1970): On the repeated measurement of continuous observables in quantum mechanics. *J. Functional Analysis* **6**, 318–346.

Davies E.B. and Lewis E.T. (1970): An operational approach to quantum probability. *Commun. Math. Phys.* **17**, 239–260.

Dixmier J. (1977): *C*-Algebras* (North-Holland, Amsterdam).

Farmer D. (1982): Information dimension and the probabilistic structure of chaos. *Z. Naturforsch.* **37a**, 1304–1325.

Fick E. and Sauermann G. (1986): *Quantenstatistik dynamischer Prozesse. Band IIa: Antwort- und Relaxationstheorie* (Harri Deutsch, Thun).

Freyberger M., Aravind P.K., Horne M.A., and Shimony A. (1996): Proposed test of Bell's inequalities without a detection loophole by using entangled Rydberg atoms. *Phys. Rev.* A **53**, 1232–1244.

Fry E.S., Walther T., and Li S. (1995): Proposal for a loophole-free test of the Bell inequalities. *Phys. Rev.* A **52**, 4381–4395.

Ghirardi G.C., Rimini A., and Weber T. (1986): Unified dynamics for microscopic and macroscopic systems. *Phys. Rev. D* **34**, 470–491.

Gisin N. (1984): Quantum measurements and stochastic processes. *Phys. Rev. Lett.* **52**, 1657–1660.

Giulini D., Joos E., Kiefer C., Kupsch J., Stamatescu I.-O., and Zeh H.D. (1996): *Decoherence and the Appearance of a Classical World in Quantum Theory* (Springer, Berlin). See particularly pp. 114–134 in the contribution by Joos.

Goldstein S. (1981): Entropy increase in dynamical systems. *Israel J. Math.* **38**, 241–256.

Grassberger P. and Procaccia I. (1983): Estimation of the Kolmogorov entropy from a chaotic signal. *Phys. Rev. A* **28**, 2591–2593.

Grelland H. (1993): Tomita representations of quantum and classical mechanics in a bra/ket formulation. *Int. J. Theor. Phys.* **32**, 905–925.

Grossmann S. (1996): Instability without instability? In *Nonlinear Physics of Complex Systems*, ed. by J. Parisi, S.C. Müller, and W. Zimmermann (Springer, Berlin), pp. 10–22.

Guichardet A. (1970): *Symmetric Hilbert Spaces and Related Topics. Lecture Notes in Mathematics Vol. 261* (Springer, Berlin).

Gustafson K. and Misra B. (1976): Canonical commutation relations of quantum mechanics and stochastic regularity. *Lett. Math. Phys.* **1**, 275–280.

Heitler W. (1954): *The Quantum Theory of Radiation* (Clarendon Press, Oxford, 3rd revised edition).

Isham C.J. (1993): Canonical quantum gravity and the problem of time. In *Integrable Systems, Quantum Groups, and Quantum Field Theories*, ed. by L.A. Ibert and M.A. Rodriguez (Kluwer, Amsterdam), pp. 157–288.

Jost R. (1973): *Quantenmechanik II* (Verlag des Vereins der Mathematiker und Physiker an der ETH Zürich, Zürich).

Kolmogorov A.N. (1958): A new metric invariant of transitive dynamical systems and automorphisms in Lebesgue spaces. *Dokl. Akad. Nauk SSSR* **119**, 861–864.

Kubo R., Toda M., and Hashitsume N. (1985): *Statistical Physics II* (Springer, Berlin).

Leggett A.J. and Garg A. (1985): Quantum mechanics versus macroscopic realism: is the flux there when nobody looks? *Phys. Rev. Lett.* **54**, 857–860.

Lockhart C.M. and Misra B. (1986): Irreversibility and measurement in quantum mechanics. *Physica A* **136**, 47–76; cf. H. Primas, *Math. Rev.* **87k**, 81006 (1987).

Mahler G. (1994): Temporal Bell inequalities: a journey to the limits of "consistent histories". In *Inside Versus Outside*, ed. by H. Atmanspacher and G.J. Dalenoort (Springer, Berlin), pp. 196–205.

Mandelstam L.I. and Tamm I. (1945): The uncertainty relation between energy and time in non-relativistic quantum mechanics. *J. Phys. (USSR)* **9**, 249–254.

Misra B. (1978): Nonequilibrium entropy, Lyapounov variables, and ergodic properties of classical systems. *Proc. Natl. Acad. Sci. USA* **75**, 1627–1631.

Misra B. (1995): From time operator to chronons. *Found. Phys.* **25**, 1087–1104.

Misra B., Prigogine I., and Courbage M. (1979a): From deterministic dynamics to probabilistic descriptions. *Physica A* **98**, 1–26; cf. A.S. Wightman, *Math. Rev.* **82e**, 85066 (1982).

Misra B., Prigogine I., and Courbage M. (1979b): Lyapounov variable: entropy and measurement in quantum mechanics. *Proc. Natl. Acad. Sci. USA* **76**, 4768–4772.

Müller-Herold U. (1980): Disjointness of β-KMS states with different chemical potential. *Lett. Math. Phys.* **4**, 45–48.

Neumann J. von (1932): *Mathematische Grundlagen der Quantenmechanik* (Springer, Berlin). English: *Mathematical Foundations of Quantum Mechanics* (Princeton University Press, Princeton), 1955.

Omnès R. (1992): Consistent interpretations of quantum mechanics. *Rev. Mod. Phys.* **64**, 339–382.

Ozawa M. (1984): Quantum measuring processes of continuous observables. *J. Math. Phys.* **25**, 79–87.

Pauli W. (1933): Die allgemeinen Prinzipien der Wellenmechanik. *Handbuch der Physik*, *Vol. 24*, ed. by H. Geiger und K. Scheel, (Springer, Berlin), pp. 83–272, here: p. 140. Reprinted in S. Flügge, ed. (1958): *Encyclopedia of Physics, Vol. V, Part 1* (Springer, Berlin), pp. 1–168, here: p. 60.

Paz J.P. and Mahler G. (1993): Proposed test for temporal Bell inequalities. *Phys. Rev. Lett.* **71**, 3235–3239.

Pearle P. (1976): Reduction of the state vector by a nonlinear Schrödinger equation. *Phys. Rev. D* **13**, 857–868.

Pedersen G.K. (1979): *C*-Algebras and Their Automorphism Groups* (Academic Press, London).

Price H. (1996): *Time's Arrow and Archimedes' Point* (Oxford University Press, Oxford).

Prigogine I. and Petrosky T.Y. (1987): Intrinsic irreversibility in quantum theory. *Physica A* **147**, 33–47.

Prigogine I. and Petrosky T.Y. (1988): An alternative to quantum theory. *Physica A* **147**, 461–486.

Primas H. (1983): *Chemistry, Quantum Mechanics, and Reductionism. Perspectives in Theoretical Chemistry* (Springer, Berlin).

Primas H. (1990a): Mathematical and philosophical questions in the theory of open and macroscopic quantum systems. In *Sixty-Two Years of Uncertainty*, ed. by A.I. Miller (Plenum, New York), pp. 233–257.

Primas H. (1990b): The measurement process in the individual interpretation of quantum mechanics. In *Quantum Theory Without Reduction*, ed. by M. Cini and J.M. Lévy-Leblond (Adam Hilger, Bristol), pp. 49–68.

Primas H. (1990c): Induced nonlinear time evolution of open quantum objects. In *Sixty-Two Years of Uncertainty*, ed. by A.I. Miller (Plenum, New York), pp. 259–280.

Primas H. (1992): Time-asymmetric phenomena in biology. Complementary exophysical descriptions arising from deterministic quantum endophysics. *Open Systems & Information Dynamics* **1**, 3–34.

Primas H. (1994): Endo- and exo-theories of matter. In *Inside Versus Outside*, ed. by H. Atmanspacher and G.J. Dalenoort (Springer, Berlin), pp. 163–193.

Primas H. (1997a): The representation of facts in physical theories. In *Time, Temporality, Now*, ed. by H. Atmanspacher and E. Ruhnau (Springer, Berlin), pp. 241–263.

Primas H. (1997b): Individual description of dynamical state reductions in quantum mechanics. Unpublished report.

Selleri F., ed. (1988): *Quantum Mechanics Versus Local Realism* (Plenum, New York).

Sewell G.L. (1986): *Quantum Theory of Collective Phenomena* (Clarendon, Oxford).

Shaw R. (1981): Strange attractors, chaotic behavior, and information flow. *Z. Naturforsch.* **36a**, 80–112.

Sinai Ya.G. (1959): On the notion of entropy of a dynamical system. *Dokl. Akad. Nauk SSSR* **124**, 768–771.

Srinivas M.D. (1980): Collapse postulate for observables with continuous spectra. *Commun. Math. Phys.* **71**, 131–158.

Suchanecki Z. (1992): On lambda and internal time operators. *Physica A* **187**, 249–266.

Takesaki M. (1970): Disjointness of the KMS states of different temperatures. *Commun. Math. Phys.* **17**, 33–41.

Tjøstheim D. (1976): A commutation relation for wide sense stationary processes. *SIAM J. Appl. Math.* **30**, 115–122.

CONTEXTUAL BACKGROUND

Günter Mahler:
TEMPORAL NONLOCALITY

Quantum nonlocality or nonseparability arises whenever there are observables of a system which do not commute, i.e., whenever its algebra of observables is non-commutative. Although the nonlocal features are often illustrated by spatial nonlocality (correlations over large distances in position space), this is somewhat misleading; more precisely, nonlocality refers to those properties of the system whose associated observables do not commute.

Therefore, it is interesting to explore situations with a non-commutative time observable and ask whether they imply a kind of temporal nonlocality which is not much studied so far. After Pauli's (1933) statement on the impossibility of time operators in traditional quantum theory, it took some decades until non-commutative time operators were introduced for stochastic systems (Tjøstheim 1976, Gustafson and Misra 1976) and, shortly later, for low-dimensional deterministic systems (Misra 1978). However, these types of time operators are not elements of the irreducible algebra of observables of traditional quantum mechanics, so there is no straightforward way to use them for a demonstration of a basic quantum temporal nonlocality .

Although attempts exist in the literature to work out consequences of the time operators just mentioned (see, e.g., Atmanspacher and Amann 1998, and references given therein), much work remains to be done to embed them in a sound formal framework. However, there is an alternative line of research addressing temporal nonlocality from an entirely different starting point. It was initiated by Leggett and Garg's (1985) question "is the flux there when nobody looks?", and it makes essential use of the framework of the histories approach to quantum theory (Griffiths 1984).

Motivated by Leggett and Garg, Paz and Mahler (1993) studied temporal Bell inequalities constructed on the basis of temporal correlations rather than correlations as in standard Bell inequalities. They demonstrated, on the basis of computer simulations, that these temporal Bell inequalities are *violated in specific situations. The parameter ranges for which these viola-tions occur correspond to inconsistent histories (i.e., histories which do not satisfy the usual rules of probability theory). Whenever histories become inconsistent or, respectively, temporal Bell inequalities are violated, it does not make sense to talk about a well-defined dynamics of a system in terms of its trajectory. The history of a system is not an element of physical reality in such situations.*

In the following contribution, Mahler investigates state changes of closed quantum systems as confined to unitary transformations, sets of which de-fine a multi-step process. Two aspects of time can then be distinguished: parametrization and sequential order. The former relates transformation parameters to time, while the latter depends on the non-commutativity of the set. Parametrization of a commuting set can easily violate temporal Bell inequalities. Special sequences of non-commuting unitary transformations can be established so that parts of a system serve as reference for other parts. Then, quantum Zeno effects and quantum parallelism result. The constrained structure of Hilbert space responsible for these features should also underlie the expected computational efficiency of quantum networks.

References

Atmanspacher H. and Amann A. (1998): Positive-operator-valued measures and projec-tion-valued measures of non-commutative time operators. *Int. J. Theor. Phys.* **37**, 629–650.

Griffiths R.B. (1984): Consistent histories and the interpretation of quantum mechanics. *J. Stat. Phys.* **36**, 219–279.

Gustafson K. and Misra B. (1976): Canonical commutation relations of quantum me-chanics and stochastic regularity. *Lett. Math. Phys.* **1**, 275–280.

Leggett A.J. and Garg A. (1985): Quantum mechanics versus macroscopic realism: is the flux there when nobody looks? *Phys. Rev. Lett.* **54**, 857–860.

Misra B. (1978): Nonequilibrium entropy, Lyapounov variables, and ergodic properties of classical systems. *Proc. Natl. Acad. Sci. USA* **75**, 1627–1631.

Pauli W. (1933): Die allgemeinen Prinzipien der Wellenmechanik, *Handbuch der Physik*, *Vol. 24*, ed. by H. Geiger und K. Scheel (Springer, Berlin), pp. 83–272, here: p. 140. Reprinted in S. Flügge (ed.): *Encyclopedia of Physics, Vol. V, Part 1* (Springer, Berlin 1958), pp. 1–168, here: p. 60.

Paz J.P. and Mahler G. (1993): Proposed test for temporal Bell inequalities. *Phys. Rev. Lett.* **71**, 3235–3239.

Tjøstheim D. (1976): A commutation relation for wide sense stationary processes. *SIAM J. Appl. Math.* **30**, 115–122.

TEMPORAL NONLOCALITY

GÜNTER MAHLER

Oregon Center for Optics, Department of Physics,
University of Oregon, Eugene, OR 97401, USA;
permanent address:
Institut für Theoretische Physik, Universität Stuttgart,
Pfaffenwaldring 57, D–70550 Stuttgart, Germany

1. Introduction

How come time? By means of change. How come change? By comparison of properties (states). But how can properties be *and* change? These are, in a nutshell, the thoughts of the Greek philosopher Zeno of Elea, which led him to believe that change, and therefore time, was an illusion. It was Heraclitus, who held the opposing view: change is "real", states are an illusion. Both together support the common impression that the concepts of *state* and *change* contradict each other.

Let us not be deceived by the fact that we seem to control time with unprecedented accuracy, actually with the help of open quantum systems ("atomic clocks"). The fact that the experimentalist may have such an impressive time-measuring device in his laboratory can hardly be of any direct impact on the experiments he is about to perform. It is only convenient for him to use the time label for a description of his results.

The two most dramatic developments in the physics of this century are Einstein's theory of relativity and quantum mechanics. While the former has undoubtedly much to say about time with all its bewildering deliberations on eigentime, time dilatation, twin paradoxes and the like, the latter does not appear to refer to time at all except by means of a negative statement: *time is not an observable* (i.e., in this context, time is not an operator). And this statement is less surprising than it may seem at first sight. Change can operationally be recognized only by comparison with a previous or some standard reference state. That time "as such" is not an observable thus simply means that – in quantum mechanics as in classical physics – there is no absolute time flowing "out there". Time is change, without something to change there is no time.

But how can we define change? The state of any quantum system can conveniently be described by a wave vector (in Hilbert space) or, alternatively, by a vector composed of the expectation values of some appropriate set of basis operators, the so-called coherence vector (generalized Bloch vector; see, e.g., Hioe and Eberly 1981). For a closed quantum system the resulting unitary dynamics then amounts to a continuous rotation of this coherence vector in a high-dimensional abstract space. However, this space has no preferred direction, no marks. Just like there is no absolute motion in classical mechanics there is no absolute quantum motion. This constitutes a kind of "quantum relativism", supplementing the more familiar concept of classical relativism. Insofar as "change" cannot be pinned down, time has no operational meaning.

Two-time correlation functions describe how an observable of some given system at one parameter time t_1 is correlated with the observable at some other time t_2. These correlation functions can, in principle, be calculated from the Heisenberg equations of motion. One might expect that they could be interpreted by assuming a distribution of eigenvalue sequences associated with the respective parameter times (just like any simple expectation value of some observable could be explained in terms of some underlying distribution of eigenvalues). However, this is not the case. There is, in general, no "hidden history" (Paz and Mahler 1993). The evolution tends to be *non-local in time*, as can be tested by temporal Bell inequalities (Leggett and Garg 1985, Paz and Mahler 1993, Calarco and Onofrio 1997).

Quantum mechanics has many surprising features, but it is not the "dark" and entirely counter-intuitive concept as which it is often seen. Rather, it challenges our implicit assumptions and prejudices by enforcing a strict logic of "operational decidability". Zeno's 2000 years old question is with us again, but now in the quantum domain: how can a state be *and* change? We cannot use a previous or any other state as a reference, simply because such a state would merely be a hypothetical construct, not an "element of reality", unless the comparison of states is actually implemented, requiring memories and coupling devices etc. (see Sect. 3).

2. Composite Systems

Zeno's problem can be carried to the extreme when applied to the simplest systems which still allow for a study of states and change: systems with a binary state space. In the classical domain such systems exist at most in an approximate sense: two stable states may be found within a continuum of unstable ones, like in coin tossing. In the quantum domain we have fundamental $(n=2)$-level systems (spins) or more complex systems which can be made to behave like effective spins.

Here we study a fairly simple interacting array of such effective spins. We will assume that unitary transformations can be designed to act on any subspace related to individual spins or to certain pairs of spins. This "modularity" appears to be crucial for quantum control (cf. Section 8). It implicitly assumes distinguishability of the effective spins, which is likely to become impractical for larger networks.

2.1. STATES

The quantum network we are going to consider here will be composed of $M + 1$ physically different spins, $\mu = S, 1, 2, 3, ..., M$. S denotes the "system", the other M subsystems denote "memories" for reasons that will become clear in Section 7. The respective states are $|p(\mu)\rangle$, $p = 0, 1$. We presently restrict ourselves to $M = 2$. The corresponding product basis is $|p(S)q(1)r(2)\rangle \equiv |pqr\rangle$. Arranged in the order of increasing binary numbers we introduce the single-index notation $|s\rangle$, $s = 1, 2, ..., 2^3$, by identifying $|1\rangle = |000\rangle$, $|2\rangle = |001\rangle$, $|3\rangle = |010\rangle$, etc. This single-index representation will serve to simplify some algebra. However, we should be aware that, though any such representation is mathematically equivalent, the "interpretation" of the state $|s\rangle$ in terms of product states will be essential.

2.2. CLUSTER OPERATORS

There are $(2^3)^2 = 64$ orthogonal basis operators. One possible choice would be products of local transition operators, $\hat{P}_{pq}(\mu) = |p(\mu)\rangle\langle q(\mu)|$. For reasons that will become clear shortly it is more convenient to separate out the local unit operators $\hat{1}(\mu)$ so that the remaining operators become traceless. Such a scheme is provided by the Hermitian and unitary $SU(2)$-generators, $\hat{\lambda}_j(\mu)$,

$$\begin{aligned}
\hat{\lambda}_1(\mu) &= \hat{P}_{01}(\mu) + \hat{P}_{10}(\mu) , \\
\hat{\lambda}_2(\mu) &= i\hat{P}_{01}(\mu) - i\hat{P}_{10}(\mu) , \\
\hat{\lambda}_3(\mu) &= \hat{P}_{11}(\mu) - \hat{P}_{00}(\mu) , \\
\hat{\lambda}_0(\mu) &= \hat{P}_{11}(\mu) + \hat{P}_{00}(\mu) = \hat{1}(\mu) .
\end{aligned} \tag{1}$$

The $|p(\mu)\rangle$ are eigenstates of $\hat{\lambda}_3(\mu)$ with eigenvalue $\lambda_3 = -1$ $(p = 0)$ and $\lambda_3 = 1$ $(p = 1)$. The corresponding product operators $(j, k, l = 0,1,2,3)$ (Mahler and Weberruss 1995)

$$\hat{Q}_{jkl} = \hat{\lambda}_j(S)\hat{\lambda}_k(1)\hat{\lambda}_l(2) \tag{2}$$

with $\hat{Q}_{jkl}\hat{Q}_{jkl} = \hat{1}$ (for any j, k, l) and

$$\text{Tr}\{\hat{Q}_{jkl}\hat{Q}_{j'k'l'}\} = 2^3 \delta_{jj'}\delta_{kk'}\delta_{ll'} \tag{3}$$

then come in four classes, depending on the number m of subsystems they act on, i.e., the number of indices unequal zero. $\hat{Q}_{000} = \hat{1}$ is the only $(m{=}0)$-cluster operator. Examples for $(m{=}1)$-cluster operators (which we will need below) are, transcribed to the single-index space, $s = 1,2,...,8$,

$$
\begin{aligned}
\hat{Q}_{300} &= (\hat{P}_{55} + \hat{P}_{66} + \hat{P}_{77} + \hat{P}_{88}) - (\hat{P}_{11} + \hat{P}_{22} + \hat{P}_{33} + \hat{P}_{44}) , \\
\hat{Q}_{030} &= (\hat{P}_{33} + \hat{P}_{44} + \hat{P}_{77} + \hat{P}_{88}) - (\hat{P}_{11} + \hat{P}_{22} + \hat{P}_{55} + \hat{P}_{66}) , \\
\hat{Q}_{003} &= (\hat{P}_{22} + \hat{P}_{44} + \hat{P}_{66} + \hat{P}_{88}) - (\hat{P}_{11} + \hat{P}_{33} + \hat{P}_{55} + \hat{P}_{77}) , \\
\hat{Q}_{100} &= (\hat{P}_{15} + \hat{P}_{26} + \hat{P}_{37} + \hat{P}_{48}) + (\hat{P}_{51} + \hat{P}_{62} + \hat{P}_{73} + \hat{P}_{84}) , \\
\hat{Q}_{200} &= i(\hat{P}_{15} + \hat{P}_{26} + \hat{P}_{37} + \hat{P}_{48}) - i(\hat{P}_{51} + \hat{P}_{62} + \hat{P}_{73} + \hat{P}_{84}) .
\end{aligned}
\tag{4}
$$

Correspondingly, $(m{=}2)$- and $(m{=}3)$-cluster operators are

$$
\begin{aligned}
\hat{Q}_{033} &= (\hat{P}_{11} + \hat{P}_{44} + \hat{P}_{55} + \hat{P}_{88}) - (\hat{P}_{22} + \hat{P}_{33} + \hat{P}_{66} + \hat{P}_{77}) , \\
\hat{Q}_{303} &= (\hat{P}_{11} + \hat{P}_{33} + \hat{P}_{66} + \hat{P}_{88}) - (\hat{P}_{22} + \hat{P}_{44} + \hat{P}_{55} + \hat{P}_{77}) , \\
\hat{Q}_{330} &= (\hat{P}_{11} + \hat{P}_{22} + \hat{P}_{77} + \hat{P}_{88}) - (\hat{P}_{33} + \hat{P}_{44} + \hat{P}_{55} + \hat{P}_{66}) , \\
\hat{Q}_{333} &= (\hat{P}_{22} + \hat{P}_{33} + \hat{P}_{55} + \hat{P}_{88}) - (\hat{P}_{11} + \hat{P}_{44} + \hat{P}_{66} + \hat{P}_{77}) .
\end{aligned}
\tag{5}
$$

This subgroup of operators appears like a set of generalized $SU(2)$-operators of the form given in Eq. (1) with each single transition or projection operator replaced by a group of four. Such operators acting on multi-level subspaces would be hard if not impossible to implement in a simple one-particle system with eight states; they reflect the structure of the underlying product space.

Any operator $\hat{A}$ in the 2^3-dimensional Hilbert space of spin states can be represented as (summation over repeated indices),

$$\hat{A} = \frac{1}{2^3} A_{jkl}\hat{Q}_{jkl} , \tag{6}$$

with the $SU(2)$-parameters

$$A_{jkl} = \text{Tr}\{\hat{A}\hat{Q}_{jkl}\} . \tag{7}$$

(Tr means trace over the total Hilbert space.) In particular, the density operator $\hat{\rho}$ is here uniquely specified by the set of expectation values (note that $\hat{Q}_{jkl}$ is unitary),

$$-1 \leq K_{jkl} = \text{Tr}\{\hat{\rho}\hat{Q}_{jkl}\} \leq 1 , \tag{8}$$

with m-cluster operators defining m-particle correlations. For a pure state, $\hat{\rho} = |\psi\rangle\langle\psi|$, Eq. (8) reduces to

$$K_{jkl} = \langle\psi|\hat{Q}_{jkl}|\psi\rangle \ . \tag{9}$$

By definition, $K_{000} = 1$; the local Bloch vectors $K_{j00}, K_{0k0}, K_{00l}$ $(j,k,l > 0)$ are equivalent to the respective reduced density matrices. A pure *local* state has Bloch-vector length 1. This scheme is easily generalized to $M > 2$, as will be done in Sections 6 and 7. For so-called product states all these correlations factor into one-point functions, i.e., $K_{jkl} = K_{j00}K_{0k0}K_{00l}$, but in general they are independent. Local realism, however, postulates that an appropriate distribution of local variables (eigenvalues $\lambda_j = \pm 1$) can explain all the correlation functions rendering them dependent. The Bell inequalities are one way to test this far reaching assumption. We will return to this problem in Sections 4 and 7.

For later reference we also define symmetrized correlation functions *within* one and the same system μ:

$$C_{AB}^{(\mu)} = \frac{1}{2}(\mathrm{Tr}\{\hat{\rho}\hat{A}(\mu)\hat{B}(\mu)\} + \mathrm{Tr}\{\hat{\rho}\hat{B}(\mu)\hat{A}(\mu)\}) \ . \tag{10}$$

Restricting ourselves to traceless operators, this joint expectation value is independent of $\hat{\rho}$ (for two-dimensional Hilbert spaces) and can simply be written as the normalized scalar product between the two representing vectors (Mahler and Weberruss 1995), for $\mu = S$, e.g.,

$$C_{AB}^{(S)} = \frac{1}{26} A_{j00}B_{j00} \ . \tag{11}$$

C_{AB}^{S} is thus maximal (minimal), if A_{j00} and B_{j00} are parallel (anti-parallel).

2.3. UNITARY TRANSFORMATIONS

A unitary transformation of an operator $\hat{A}$,

$$\hat{A}' = \hat{U}\hat{A}\hat{U}^{+} \ , \tag{12}$$

with $\hat{U}^{+}\hat{U} = \hat{U}\hat{U}^{+} = \hat{1}$, reads in terms of the $SU(2)$-parameters,

$$A'_{jkl} = T_{j\,k\,l}^{j'k'l'} A_{j'k'l'} \ , \tag{13}$$

where

$$T_{j\,k\,l}^{j'k'l'} = \frac{1}{2^3} \mathrm{Tr}\{\hat{U}^{+}\hat{Q}_{jkl}\hat{U}\hat{Q}_{j'k'l'}\} \ . \tag{14}$$

There are different types: We may distinguish transformations which operate in certain subspaces only. (For $\hat{U} = \hat{1}$, T is just the unit matrix, see

Eq. (3).) The locally selective transformation $\hat{U}(S)$ in the ($n{=}2$)-dimensional local Hilbert space of S, e.g., is equivalent to a local rotation of the $SU(2)$-parameters with respect to the first index, generated by (Schlienz and Mahler 1995)

$$
\begin{aligned}
T_{j\,k\,l}^{j'k'l'} &= T_{jj'}^{(S)} \delta_{kk'} \delta_{ll'} \,, \\
T_{jj'}^{(S)} &= \tfrac{1}{2} \mathrm{Tr}_S \{ \hat{U}^+(S) \hat{\lambda}_j(S) \hat{U}(S) \hat{\lambda}_{j'}(S) \} \,.
\end{aligned}
\tag{15}
$$

(Here, Tr_S means trace over the subspace of S only.) As $T_{00}^{(S)} = 1$, $T_{jj'}^{(S)} = 0$ if either j or j' is zero, all parameters A_{0kl} are invariants. Correspondingly, a unitary transformation $\hat{U}(S,1)$ leaves the expectation values A_{00l} unchanged, etc. These invariants (conservation laws) are important characteristics of the respective transformations.

Unitary transformations are the only allowed type of changes (of states or observables) in closed quantum systems. Typically they are generated by the underlying Hamiltonian model. In the Schrödinger picture this unitary transformation is applied to $\hat{\rho}$, in the Heisenberg picture the inverse transformation (replacing $\hat{U}$ by $\hat{U}^+$ and vice versa) is applied to the observables.

3. The Model

Our network Hamiltonian $\hat{H}$ can be written in the same type of cluster representation as used before. The model parameters H_{jkl} are usually constrained to ($m{=}0,1,2$)-cluster terms (Mahler and Weberruss 1995). We assume to have explicit control over all these parameters for any parameter time t, which may even be modified in terms of pulses (Δt is some pulse length). Granted this access we can implement virtually any unitary transformation via

$$
\hat{U}(\Delta t) = \mathrm{e}^{-i\hat{H}\Delta t/\hbar}
\tag{16}
$$

though this may be seriously limited in practice. (Just to implement $\hat{U} = \hat{1}$ for an extended period of time, i.e. freezing any state, is often impossible.) Measurements will be postponed until the desired state has been prepared by a number of pulses.

In what follows we want to implement a web of correlations relating different steps in the preparation process to each other. Rather than solving respective equations of motion we focus on the following elementary unitary transformations.

3.1. LOCAL TRANSFORMATION ON (S)

Let us consider the one-parameter form

$$
\begin{aligned}
|0(S)\rangle &\longrightarrow \cos{(\alpha/2)}|0(S)\rangle - \mathrm{i}\sin{(\alpha/2)}|1(S)\rangle , \\
|1(S)\rangle &\longrightarrow -\mathrm{i}\sin{(\alpha/2)}|0(S)\rangle + \cos{(\alpha/2)}|1(S)\rangle ,
\end{aligned}
\tag{17}
$$

which can be generated by

$$
\hat{U}_\alpha(S) = \hat{Q}_{000}\cos{(\alpha/2)} - \hat{Q}_{100}\,\mathrm{i}\sin{(\alpha/2)} . \tag{18}
$$

According to Eq. (15) we find

$$
\begin{aligned}
T^{(S)}_{jj'} &= \cos^2{(\alpha/2)}\,\delta_{jj'} + \tfrac{1}{2}\sin^2{(\alpha/2)}\,\mathrm{Tr}_S\{\hat{\lambda}_1\hat{\lambda}_j\hat{\lambda}_1\hat{\lambda}_{j'}\} \\
&\quad + \tfrac{\mathrm{i}}{4}\sin\alpha\,\mathrm{Tr}_S\{\hat{\lambda}_1\hat{\lambda}_j\hat{\lambda}_{j'} - \hat{\lambda}_j\hat{\lambda}_1\hat{\lambda}_{j'}\} ,
\end{aligned}
\tag{19}
$$

so that $T^{(S)}_{11} = 1$, $T^{(S)}_{22} = T^{(S)}_{33} = \cos\alpha$, $T^{(S)}_{32} = -T^{(S)}_{23} = \sin\alpha$ (here and in the following all terms not explicitly given are zero). The matrix $T^{(S)}_{ij}$ defines a rotation around the 1-axis in the 2,3-plane. The inverse transformation simply means to replace α by $-\alpha$. For $\alpha = \pm\pi$ we have a so-called π-pulse, generated by $\hat{U}(S) = \mp\mathrm{i}\hat{Q}_{100}$. The operators $\hat{U}(S)_\alpha$ commute with each other for any α.

Consider as a first example the observable $\hat{A} = \hat{\lambda}_3(S)$, represented by $A_{300} = 2^3$ (cf. Eq. (6)). The inverse unitary transformation with phase $-\alpha_1$ leads to (Heisenberg picture)

$$
\begin{aligned}
A'_{200} &= -2^3\sin\alpha_1 , \\
A'_{300} &= 2^3\cos\alpha_1 .
\end{aligned}
\tag{20}
$$

The correlation function between this transformed $\hat{\lambda}_3(S)$ and the same operator transformed by a phase angle $-\alpha_2$ then is, according to Eq. (11),

$$
C^{(S)}_{33}(\alpha_1, \alpha_2) = \cos{(\alpha_2 - \alpha_1)}, \tag{21}
$$

thus depending on the relative phase only. Our second example refers to a density operator $\hat{\rho}$ with local Bloch vector $K_{300} = -r$, $|r| \le 1$ (see Eq. (8)). The local state parameters transformed by $\hat{U}$ with phase α_1 are (Schrödinger picture)

$$
\begin{aligned}
K'_{200} &= r\sin\alpha_1 , \\
K'_{300} &= -r\cos\alpha_1 .
\end{aligned}
\tag{22}
$$

This relation reminds us that the rotation is independent of whether the local state is pure, $|r| = 1$, or not, $|r| < 1$. The expectation value for this observable $\hat{A} = \hat{\lambda}_3(S)$ in both pictures is $A_{j00}K_{j00}/2^3 = -r\cos\alpha_1$.

3.2. DISCRETE PAIR TRANSFORMATION ON (S, μ)

This unitary transformation is a conditioned π-pulse and most intuitively illustrated by its effect on selected input states:

$$\text{Resonance:} \quad |0(S)0(\mu)\rangle \longleftrightarrow |0(S)1(\mu)\rangle \,,$$
$$\text{Off-resonance:} \quad |1(S)m(\mu)\rangle \longleftrightarrow |1(S)m(\mu)\rangle \,. \tag{23}$$

We may thus write

$$
\begin{aligned}
\hat{U}(S, 1) &= \hat{P}_{00}(S)\hat{\lambda}_1(1) + \hat{P}_{11}(S)\hat{1}(1) \\
&= \tfrac{1}{2}(\hat{Q}_{000} + \hat{Q}_{300} + \hat{Q}_{010} - \hat{Q}_{310}) \,.
\end{aligned}
\tag{24}
$$

The operators $\hat{U}(S, \mu)$ commute; their implementation requires pair inter-actions, which make the transition frequency in subsystem μ depend on the state of subsystem S (Mahler and Weberruss 1995, Obermayer et al. 1988). This transformation has become known as the (quantum-) controlled NOT (Barenco et al. 1995), as subsystem S acts as a control for a π-pulse on μ. The π-pulse, of course, will take some finite time Δt. Note that both transformation rules apply to a superposition state term by term.

In the following we will distinguish two aspects of time with respect to such a set of unitary transformations: parametrization proper and se-quential order. To some extent these two features are independent. While the former relates parameters of the transformation to time (duration), the latter depends on the non-commutativity of a composite transformation se-quence (ordered set). A third aspect of time would show up in the recording of actual measurement *information*.

4. Transformation Parameters: Temporal Bell Inequalities

The conventional 2-spin Bell inequalities are expressed in terms of (1-time) 2-particle correlation functions, each referring to a specific local measure-ment angle setting. Rather than rotating the respective measurement an-gles, we may as well apply the inverse rotation to the local spins (for fixed external measurement direction). In any case, the realization of different settings amounts to a physical process, so that the joint "existence" of those correlation functions may be questioned (Brody 1993).

For the temporal Bell inequalities we consider the multi-step process $(n = 1, 2, ..., 4)$

$$|\psi^{(n)}\rangle = \hat{U}_{(\alpha_n - \alpha_{n-1})}(S)|\psi^{(n-1)}\rangle \,. \tag{25}$$

All the transformations $\hat{U}_\alpha(S)$ commute and obey $\hat{U}_\alpha \hat{U}_\beta = \hat{U}_{\alpha+\beta}$. Applying Eq. (21) we find for the correlation (in the Heisenberg picture) between

$\hat{\lambda}_3(S)$ after step μ and after step ν (irrespective of the reference state $|\psi_H\rangle$)

$$C_{33}^{(S)}(\alpha_\mu, \alpha_\nu) = \cos(\alpha_\mu - \alpha_\nu) . \tag{26}$$

If we relate the phase angles to parameter time by

$$\alpha_\nu = g t_\nu \mod 2\pi , \tag{27}$$

Eq. (26) becomes a 2-time 1-particle correlation function. Here, g could be the coupling strength to an external optical driving field. This parametrization still allows any order of phase angles, as $t_{\nu+1} > t_\nu$ does not imply $\alpha_{\nu+1} > \alpha_\nu$. Then, if we assume a "hidden history", the following temporal Bell inequality would have to hold (Leggett and Garg 1985, Paz and Mahler 1993, Calarco and Onofrio 1997):

$$W = |C_{33}^{(S)}(t_1, t_2) + C_{33}^{(S)}(t_2, t_3) + C_{33}^{(S)}(t_3, t_4) - C_{33}^{(S)}(t_4, t_1)| \leq 2 . \tag{28}$$

Choosing equidistant time steps,

$$g(t_2 - t_1) = g(t_3 - t_2) = g(t_4 - t_3) = \pi/4 , \tag{29}$$

and observing that the sum of these three relative phase-shifts must equal $g(t_4 - t_1)$, one gets

$$W = 3\cos(\pi/4) - \cos(3\pi/4) = 2\sqrt{2} , \tag{30}$$

which clearly violates the inequality.

In the original EPR context (Aspect et al. 1982) the Bell inequalities can be seen as an entanglement test (Horodecki et al. 1996). Violations are a sufficient (but not necessary) condition for a two-particle state to be entangled (i.e., to deviate from a simple product state). For the two-time (single-subsystem) correlation function the concept of entanglement, which refers to a product space, has no meaning.

The violation of temporal Bell inequalities indicates that it is not possible to introduce a probability on the space of spin-up, spin-down sequences over parameter time t_ν (Paz and Mahler 1993). The well-defined two-time correlation functions are not based on specific and well-defined "histories". This is reminiscent of the two-spin correlation in an EPR state, which also cannot be explained in terms of a probability distribution on well-defined local states.

The determination of *individual* two-time correlation functions has been discussed by Calarco and Onofrio (1997) under the aspect of so-called quantum non-demolition (QND) experiments. Here we are interested in the whole set of correlation functions and wonder to what extent any non-locality may still be present within our composite system, where the memories $\mu = 1, 2, ...$ represent a reference (of correlations) for the central system S. Note that this referencing does not yet involve any measurement.

5. Transformation Sequence

We now follow a sequence of four unitary transformations and their effect on the network state in terms of pertinent correlation functions. At this stage we only need to specify the order of the actions, there is no need yet to refer quantitatively to parameter time t. Note that all these manipulations are strictly reversible.

Let the initial state be $|\psi^{(0)}\rangle = |1\rangle = |000\rangle$ so that the local Bloch vectors are given by

$$K_{300}^{(0)} = K_{030}^{(0)} = K_{003}^{(0)} = -1 . \qquad (31)$$

In the first step we apply the local transformation with a phase α_1 leading to

$$|\psi^{(1)}\rangle = \cos(\alpha_1/2) |1\rangle - i \sin(\alpha_1/2) |5\rangle . \qquad (32)$$

In the second step we execute the pair transformation on $(S, 1)$:

$$|\psi^{(2)}\rangle = \cos(\alpha_1/2) |3\rangle - i \sin(\alpha_1/2) |5\rangle . \qquad (33)$$

We note that the single-subsystem expectation values of subsystems S and 1 (see Eq. (4)) obey the relations

$$\begin{aligned}
K_{300}^{(2)} &= -K_{030}^{(2)} = -\cos\alpha_1 = K_{300}^{(0)} \cos\alpha_1 , \\
K_{100}^{(2)} &= K_{200}^{(2)} = 0 ,
\end{aligned} \qquad (34)$$

and as a consequence of the controlled-NOT logic (cf. Eq. (5)),

$$K_{330}^{(2)} = -1 . \qquad (35)$$

We thus see that the two systems S and 1 are strictly anticorrelated (the state $|\psi^{(2)}\rangle$ is actually an eigenstate of $\hat{Q}_{330}$!), while the lengths of the local Bloch vectors are $|\cos\alpha_1| < 1$, i.e., local properties are not dispersion-free ("fuzzy"). This is typical for non-classical correlations. There can be *strict* correlations between *fuzzy* subsystems.

In the third step we again apply the local transformation, now with phase $\alpha_2 - \alpha_1$, leading to

$$\begin{aligned}
|\psi^{(3)}\rangle = \; &\cos(\alpha_1/2) \cos(\alpha_2 - \alpha_1)/2 |3\rangle - i \cos(\alpha_1/2) \sin(\alpha_2 - \alpha_1)/2 |7\rangle \\
&- i \sin(\alpha_1/2) \cos(\alpha_2 - \alpha_1)/2 |5\rangle - \sin(\alpha_1/2) \sin(\alpha_2 - \alpha_1)/2 |1\rangle .
\end{aligned} \qquad (36)$$

In the fourth step the pair transformation on $(S, 2)$ implies

$$
\begin{aligned}
|\psi^{(4)}\rangle = & \cos(\alpha_1/2)\cos(\alpha_2 - \alpha_1)/2\,|4\rangle - i\cos(\alpha_1/2)\sin(\alpha_2 - \alpha_1)/2\,|7\rangle \\
& - i\sin(\alpha_1/2)\cos(\alpha_2 - \alpha_1)/2\,|5\rangle - \sin(\alpha_1/2)\sin(\alpha_2 - \alpha_1)/2\,|2\rangle .
\end{aligned}
$$
$$(37)$$

The state $\psi^{(4)}$ has been generated by

$$
|\psi^{(4)}\rangle = \hat{U}(S, 2)\hat{U}_{(\alpha_2 - \alpha_1)}(S)\hat{U}(S, 1)\hat{U}(S)_{\alpha_1}|\psi^{(0)}\rangle .
$$
$$(38)$$

As $\hat{U}(S, \mu)$ and $\hat{U}(S)$ do not commute for $\hat{U}(S) \neq \hat{Q}_{000}$ (i.e., $\alpha \neq 0$, see Eqs. (24),(18)), the ordering of the transformation is essential. One easily convinces oneself that now

$$
\begin{aligned}
K_{300}^{(4)} &= -K_{003}^{(4)} = -\cos\alpha_1\cos(\alpha_2 - \alpha_1) = K_{300}^{(2)}\cos(\alpha_2 - \alpha_1) , \\
K_{100}^{(4)} &= K_{200}^{(4)} = 0 .
\end{aligned}
$$
$$(39)$$

Again, the controlled-NOT logic leads to the strict anticorrelation

$$
K_{303}^{(4)} = -1 .
$$
$$(40)$$

Other pertinent two- and three-subsystems correlations are

$$
\begin{aligned}
K_{033}^{(4)} &= -K_{330}^{(4)} = \cos(\alpha_2 - \alpha_1) , \\
K_{333}^{(4)} &= -\cos\alpha_1 .
\end{aligned}
$$
$$(41)$$

The local state of subsystem 1 remains unchanged, as to be expected:

$$
K_{030}^{(4)} = \cos\alpha_1 = K_{030}^{(2)} .
$$
$$(42)$$

All these expectation values refer to the cluster operators and thus to the way in which the total system is partitioned. Note that this does not imply that the state (and thus the correlations) can be decomposed ("factored") in the same way. The advantage of the present cluster representation is that any reduced "perspective" is easily singled out by restricting attention to the appropriate type of expectation values.

6. Reduced Description: Object System S

The description reduced to the object subsystem S is based on the local Bloch vector K_{j00} only. Starting from the ground state, $K_{300}^{(0)} = -1$, this vector is subject to the continuous rotation as given by Eq. (17). We see that each controlled NOT operation implies a projection on the 3-axis ($K_{100} = K_{200} = 0$). After the first projection, $K_{300}^{(2)} = -\cos\alpha_1$, after the second projection, $K_{300}^{(4)} = -\cos\alpha_1\cos(\alpha_2 - \alpha_1)$. This can easily be generalized to $M > 2$ projections with $\nu = 1, 2, ..., M$ and $\alpha_{\nu+1} - \alpha_\nu = \pi/M$. The result is

$$K_{3000...}^{(2M)} = -\cos^M(\pi/M) \tag{43}$$

with

$$
\begin{aligned}
K_{3000...}^{(2M)} &= 1 && \text{for } M = 1, \\
K_{3000...}^{(2M)} \to K_{3000...}^{(0)} &= -1 && \text{for } M \to \infty.
\end{aligned}
\tag{44}
$$

A π-pulse, decomposed into a series of M identical transformations, each followed by a controlled NOT operation, tends to freeze-in the initial state. When the phase α_ν is taken to be a monotonic function of parameter time t (see Eq. (27)), the quantum Zeno effect (Misra and Sudarshan 1977) results. (For a more recent investigation see Power and Knight (1996), for an experimental test see Itano et al. (1990).) Contrary to common belief, this result has nothing to do with recurrent measurements (information retrieval). (Similar conclusions have also been reached by Home and Whittaker (1993) and by Pascazjo and Namiki (1994).) It is just a consequence of the internal correlation structure generated by a sequence of unitary transformations in an $(M+1)$-spin network (each projection requires a "fresh" 2-level system to do the controlled NOT).

However, the finite time Δt, needed to perform an individual controlled NOT operation, may be neglected only if $M\Delta t \ll T$. M therefore cannot go to infinity and the freezing can only be approximate. This restriction holds also if the controlled NOT operations are replaced by real measurements.

It is interesting to note that the reduced density matrix (or Bloch vector) of subsystem S is, at any time t, identical with the density matrix of an *ensemble* of non interacting spins (all with the same initial state and subject to the same local unitary transformation) but *actually measured* at each time t_ν, $\nu = 1, 2, ..., M$. For each ensemble member the series of measurements constitutes a "decision tree", with each measurement result given by one of the eigenvalues of $\hat{\lambda}_3(S)$, $K_{3000...}' = \pm 1$ (see Fig. 1). The ensemble average over these trajectories leads back to the behavior realized here by just one single object! This is what one may call *quantum parallelism*. It will only hold, though, as long as neither the single system S nor the surrounding memories are actually measured.

7. Reduced Description: Memories 1, 2, 3, 4

By contrast to the object system S, the other subsystems are each addressed by unitary transformations only once. Due to the built-in logic the state of subsystem 1 is strictly anticorrelated with S after preparation step

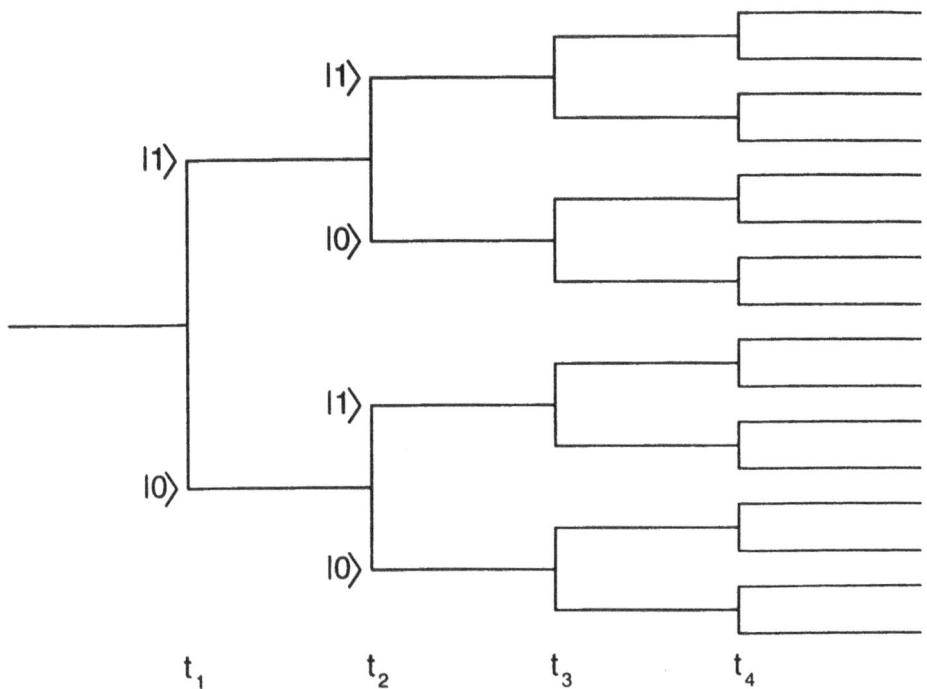

Figure 1. Alternative histories (decision tree) as realized in an *ensemble* of driven spins S under *actual* measurement at times t_ν, $\nu = 1, 2, 3, 4$. Between the measurements the respective states are superpositions.

2, subsystem 2 is anticorrelated with S after step 4. This means that an actual projective measurement performed on these subsystems would also reveal the respective states of S. When the continuous local transformation of S is interpreted to happen in parameter time t (see Eq. (27)), the subsystems $\mu \neq S$ indeed act as a kind of "memory". They allow delayed measurements on S. One may argue that this fact is the origin of the quantum Zeno effect discussed in the last section: in order to get the freezing tendency of measurements, it is sufficient to *be able to measure* ("virtual watchdog" effect). In discrete systems as discussed here, even *potential* information about the state appears to suppress change. But as fundamental limitations do not allow strictly continuous (in time) system-memory correlations (cf. Sect. 4), state and change do coexist, each a little bit "fuzzy" in the operational sense.

But not only this: The correlation between memory 1 and 2 must, by construction, reflect the correlation between the states of S taken at $\alpha_1 = gt_1$ and $\alpha_2 = gt_2$, respectively. Comparing our result for $K_{033}^{(4)}$ (Eq. (41)) with $C_{33}^{(S)}$ given by Eq. (21), we see that $K_{033}^{(4)}$ and $C_{33}^{(S)}$ are identical, i.e., a measurement of $K_{033}^{(4)}$ can be used to infer the unperturbed $C_{33}^{(S)}$. In this way we relate the 2-time 1-particle correlation back to a 1-time 2-particle correlation function.

The measurement of $K_{033}^{(4)}$ would have to be an ensemble measurement. The fact that the associated non-unitary transformation will destroy the state we are measuring is of no concern here. What matters is that – after the unitary preparation steps – we can find $K_{033}^{(4)}$ and thus $C_{33}^{(S)}$ experimentally.

Suppose we continue the procedure with respect to an additional memory subsystem 3. We get, as before,

$$
\begin{aligned}
K_{3000}^{(6)} &= -K_{0003}^{(6)} = K_{3000}^{(4)} \cos{(\alpha_3 - \alpha_2)} \,, \\
K_{3003}^{(6)} &= -1 \,,
\end{aligned}
\tag{45}
$$

while

$$
\begin{aligned}
K_{0033}^{(6)} &= \cos{(\alpha_3 - \alpha_2)} \,, \\
K_{0303}^{(6)} &= \cos{(\alpha_3 - \alpha_2)} \cos{(\alpha_2 - \alpha_1)} \,.
\end{aligned}
\tag{46}
$$

As step 5 involves only a further local unitary transformation on S and step 6 the pair transformation on $(S, 3)$, all expectation values K_{0kl0} will remain unaffected (see Sect. 2), in particular

$$
K_{0330}^{(6)} = K_{0330}^{(4)} = \cos{(\alpha_2 - \alpha_1)} \,.
\tag{47}
$$

Thus the two 1-time 2-particle correlation functions K_{0330}, K_{0033} "coexist", and – due to the built-in logic – correspond to the two sequential 2-time 1-particle correlations (using the identification Eq. (27)). As their joint measurability cannot be questioned, the measurement status of the inequality Eq. (28) may seem to be on a much better footing than that of the original Bell inequalities. This is not the case.

The scheme can be extended to any finite number of memory spins M. In Fig. 2 we have sketched the situation for $M = 4$. The pair correlations (shown by broken lines) are implemented step by step by means of unitary transformations on pair subspace (S, ν), $\nu = 1, 2, 3, 4$, at time t_ν (solid lines):

$$
\begin{aligned}
K_{03300}^{(4)} &= \cos{(\alpha_2 - \alpha_1)} \,, \\
K_{00330}^{(6)} &= \cos{(\alpha_3 - \alpha_2)} \,, \\
K_{03030}^{(6)} &= \cos{(\alpha_3 - \alpha_2)} \cos{(\alpha_2 - \alpha_1)} \,, \\
K_{00033}^{(8)} &= \cos{(\alpha_4 - \alpha_3)} \,, \\
K_{00303}^{(8)} &= \cos{(\alpha_4 - \alpha_3)} \cos{(\alpha_3 - \alpha_2)} \,, \\
K_{03003}^{(8)} &= \cos{(\alpha_4 - \alpha_3)} \cos{(\alpha_3 - \alpha_2)} \cos{(\alpha_2 - \alpha_1)} \,.
\end{aligned}
\tag{48}
$$

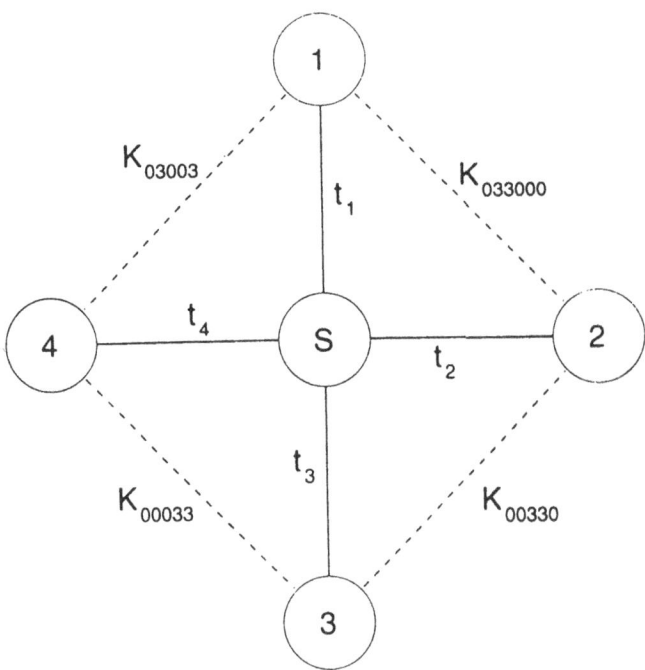

Figure 2.   Build-up of pair correlations (broken lines). The memory subsystems $\nu = 1, 2, 3, 4$ never interact directly, only with the center S at times t_ν (solid lines). The pair correlations depend on the order of the preparation steps.

The correlations decay with "step distance", i.e., the number of *intermediate* rotation and coupling steps to other memories (cf. also Fig. 1). This is why K_{03003} has the product form of Eq. (48) rather than the "unperturbed form" $\cos(\alpha_4 - \alpha_1)$. In order to get this unperturbed form one would have to change the order in the preparation process. However, the Bell inequality Eq. (28) refers to a circle in phase space, so that correlations in product form cannot be avoided, and the set of transformations is non-commuting.

Identifying any of these sets of inter-memory correlation functions with the appropriate two-time correlation function of subsystem S within the network, we see that the temporal Bell inequalities would no longer be violated. Even for the individual network the various possible histories can be considered as "elements of reality" in the following sense: subsequent measurements of the memories (in any time order) would project the subsystem S into a specific history with a probability as realized in the directly measured ensemble of driven S-systems. The yet unselected histories suffice to generate the quantum Zeno effect (for which intermediate measurement information is not needed); they should be distinguished from the so-called "consistent histories" (Omnès 1992) defined on given sets of projections on some closed system under consideration.

8. Quantum Computation and Quantum Complexity

In the preparation of the desired final state (Sect. 5) we have argued along lines of thought which have become quite popular in the context of quantum computation. In fact, the series of unitary transformations is equivalent to formal quantum-gate operations (Barenco et al. 1995). Our purpose, though, has been quite different. We tried to explore the role of time in quantum mechanics, while quantum computation is concerned with specific algorithms. In this context it has been shown that certain problems scale more favorably when carried out on a quantum system than on *any* classical computer (Ekert and Jozsa 1996). In some way, quantum complexity appears to reduce computational complexity. One may wonder whether from the present considerations we may also gain some new insight into the origin of the impressive efficiency of quantum computation.

Up to now there is no clear and generally accepted answer to this question. Hilbert space is large indeed, but mere size does not present new qualities. The superposition principle is clearly at the heart of quantum mechanics, but as such it also applies to classical linear waves. Entanglement (i.e., a specific non-classical feature of the superposition principle for composite systems) disappears if the system is described as a whole (single-index space, see Sect. 2.1), what we can always choose to do. It thus seems more adequate to expect clues from the strange web of correlations, as can already be realized in our simple example.

Our $SU(2)$ product description might have supported the impression that any motion of the total system is just a rotation, and *in turn*, any rotation should correspond to a possible motion. While the former is true, the·· latter is wrong: admissible rotations are severely constrained, which becomes obvious already for a 3-level system, much more so for our present 8-level system. Additional constraints are built in by the selection of transformations which can actually be implemented (cf. Eqs. (4,5)). It is such constraints which make up a useful machinery; systems with large, unrestricted state spaces (like a free gas) are "useless".

The quantum parallelism discussed in Sections 6 and 7 can be seen as a consequence of these internal constraints. Rather than being based on entanglement as such (though entanglements are present), this parallelism emerges as the inter-subsystem correlations establish a "quantum reference frame". The resulting set of undecided alternatives ("no collapse") drastically differs from our everyday experience and has even been taken as a "proof" for the many-worlds hypothesis (see Deutsch 1997). In some final step, though, measurements cannot be avoided.

The implementation of "useful" quantum algorithms would require orders of magnitude more preparation steps und much larger networks. The resulting graphs would no longer have the simple topology as sketched in Figure 1. Nevertheless, the same basic rules are at work.

9. Summary

In this paper we have neither touched upon the problem of time-asymmetry (Primas 1992) nor on attempts to define quantum-mechanical clock variables (Pegg 1991), nor on alternative ways to introduce time operators (Misra 1995). We have started from the operational point of view that time is not a (quantum-mechanical) observable, but rather given to us as a classical external parameter, i.e., time coordinates, which we may conveniently use like spatial coordinates to mark different locations. (Ironically, one cannot even talk about non-locality without a local reference frame.) Incidentally, this is also the starting point of any quantum field theory.

Zeno's paradox is related to the old question of how change can be stated if only the present (the state) is real. This question becomes even more pronounced in quantum mechanics insofar as this theory is much less tolerant toward unjustified implicit assumptions than classical mechanics.

It seems that change could easily be introduced, at least in a formal way, by means of two-time correlation functions (explicitly referring to parameter time). However, their non-classical nature does not allow to conceive these functions as fingerprints of an underlying history (i.e., sequence of eigenvalues of some observable associated with the parameter time). Without external reference change and thus time have no meaning.

How can one establish an external reference? Do we need to resort to an open system with actual measurement devices? We have demonstrated that much weaker conditions are sufficient. Parts of a (closed) composite system can be made to become references for other parts: correlations and the built-in logic will then generate an internal time structure. We have studied such inter-subsystem correlations generated by a set of unitary transformations (the only type of state change allowed for a closed system). Time does enter here as a means to define the *order* in which those transformations have been applied and as their quantitative *parametrization*. In this way a "logic" has been implemented by which "delayed measurements" could be carried out on an individual network generating a respective measurement record ("history"). After complete measurement of the memories all local properties would be well-defined (i.e. "facts", cf. Primas 1997), and the subsystem S would be projected onto a specific history.

Alternatively, all correlation functions (in particular those between pairs of subsystems) could be measured in an ensemble, which, by means of the same logic, could then be identified with the original two-time correlations within subsystem S. For these correlations, the temporal Bell inequality is no longer violated. This should hold even *before* any measurement on the memories has been performed, i.e., when the measurement decisions are still suspended. Then there are just probabilities for various possible histories as shown in Fig. 1 for an ensemble. Not only concrete measurement outcomes, but already the mere *order* of possible measurements exhibits a trend towards locality in time. As Primas (1993) noted: "... objects are carriers of patterns, they arise in interaction with the rest of the world and are always contextual and inherently fuzzy." Here this contextuality has been realized entirely via correlations *within a closed* quantum network. "Change" emerges as the correlations refer to subsequent parameter time steps in the preparation sequence.

Acknowledgments: I thank Howard Carmichael for many fruitful discussions and for his hospitality during my visit at the Oregon Center for Optics, and Ilki Kim for useful comments. Financial support by the University of Oregon and the Deutsche Forschungsgemeinschaft is gratefully acknowledged.

References

Aspect A., Grangier P., and Roger G. (1982): Experimental realization of Einstein–Podolsky–Rosen–Bohm–Gedankenexperiment: a new violation of Bell's inequalities. *Phys. Rev. Lett.* **49**, 91–94.

Barenco A., Bennett C.H., Cleve R., DiVincenzo D.P., Margolus N., Shor P., Sleator T., Smolin J.A., and Weinfurter H. (1995): Elementary gates for quantum computation. *Phys. Rev. A* **52**, 3457–3467.

Brody T. (1993): The Bell inequality I: Joint measurability. In *The Philosophy Behind Physics*, ed. by L. de la Peña and P. Hodgson (Springer, Berlin), pp. 205–222. Originally published in: *Revista Mexicana de Fisica* **35** *Suplemento* (1989) 52–70.

Calarco T. and Onofrio R. (1997): Quantum nondemolition measurements on two-level atomic systems and temporal Bell inequalities. *Appl. Phys. B* **64**, 141–144.

Deutsch D. (1997): *The Fabric of Reality* (Viking Penguin, New York).

Ekert A. and Jozsa R. (1996): Quantum computation and Shor's factoring algorithm. *Rev. Mod. Phys.* **68**, 733–753.

Hioe F.T. and Eberly J.H. (1981): N-level coherence vector and higher conservation laws in quantum optics and quantum mechanics. *Phys. Rev. Lett.* **47**, 838–841.

Home D. and Whittaker M.A.B. (1993): A unified framework for quantum Zeno processes. *Phys. Lett. A* **173**, 327–331.

Horodecki R., Horodecki M., and Horodecki P. (1996): Teleportation, Bell's inequalities and inseparability. *Phys. Lett. A* **222**, 21–25.

Itano W.M., Heinzen D.J., Bollinger J.J., and Wineland D.J. (1990): Quantum Zeno effect. *Phys. Rev. A* **41**, 2295–2300.

Leggett A.J. and Garg A. (1985): Quantum mechanics versus macroscopic realism: is the flux there when nobody looks? *Phys. Rev. Lett.* **54**, 857–860.

Mahler G. and Weberruss V.A. (1995): *Quantum Networks: Dynamics of Open Nanostructures* (Springer, Berlin).

Misra B. (1995): From time operator to chronons. *Found. Phys.* **25**, 1087–1104.

Misra B. and Sudarshan E.C.G. (1977): The Zeno's paradox in quantum theory. *J. Math. Phys.* **18**, 756–763.

Obermayer K., Teich W.G., and Mahler G. (1988): Structural basis of multistationary quantum systems. II. Effective few-particle dynamics. *Phys. Rev. B* **37**, 8111–8121.

Omnès R. (1992): Consistent interpretations of quantum mechanics. *Rev. Mod. Phys.* **64**, 339–382.

Pascazjo S. and Namiki M. (1994): Dynamical quantum Zeno effect. *Phys. Rev. A* **50**, 4582-4592.

Paz J.P. and Mahler G. (1993): Proposed test for temporal Bell inequalities. *Phys. Rev. Lett.* **71**, 3235–3239.

Pegg D.T. (1991): Time in a quantum mechanical world. *J. Phys. A* **24**, 3031–3040.

Power W.L. and Knight P.L. (1996): Stochastic simulations of the quantum Zeno effect. *Phys. Rev. A* **53**, 1052–1059.

Primas H. (1992): Time-asymmetric phenomena in biology: complementary exophysical descriptions arising from deterministic quantum endophysics. *Open Systems and Information Dynamics* **1**, 3–34.

Primas H. (1993): The Cartesian cut, the Heisenberg cut, and disentangled observers. In *Symposia on the Foundations of Modern Physics, Wolfgang Pauli as a Philosopher*, ed. by K.V. Laurikainen and C. Montonen (World Scientific, Singapore), pp. 245-269.

Primas H. (1997): The representation of facts in physical theories. In *Time, Temporality, Now*, ed. by H. Atmanspacher and E. Ruhnau (Springer, Berlin), pp. 243–265.

Schlienz J. and Mahler G. (1995): Description of entanglement. *Phys. Rev. A* **52**, 4396–4404.

CONTEXTUAL BACKGROUND

H. Narnhofer and W. Thirring:
MACROSCOPIC PURIFICATION OF STATES

About a decade and a half ago, Griffiths (1984) introduced an approach to interpret quantum mechanics using the notion of histories. This approach has been taken up by others, e.g., Omnès (1992) and Gell-Mann and Hartle (1993), sometimes in combination with decoherence processes, see the recent book by Giulini et al. (1996). (The idea of decoherence goes back to an early paper by Pascual Jordan (1949), who used the term "dephasing" instead of decoherence.) Up to now, these approaches have been refined and worked out to some detail, but some central questions remain problematic. For instance, a detailed understanding of key features of measurement is still lacking, and it is not yet clear how classical observables emerge in terms of decohering histories.

Primas has often emphasized some crucial points in this regard in unpublished manuscripts. In particular, superselection rules for classical observables have to provide disjoint states (in different superselection sectors), not only orthogonal states as they result from decaying off-diagonal terms in the density matrix of the (open) system under consideration. Zurek (1994) has sketched some ideas on the role that predictability might play for identifying those subspaces of Hilbert space that correspond to superselection sectors and contain the orthogonal states of what he calls the "pointer basis". Zurek argues that histories of sequences of events corresponding to projections onto those states are consistent.

The idea of using predictability arguments suggests that K-flows might play an important role in achieving a more rigorous treatment of decohering histories. Narnhofer and Thirring describe some first steps in such a direction from the viewpoint of algebraic quantum theory. They do not address the measurement problem as such insofar as their work makes essential use of the "reduction postulate" within the traditional Copenhagen interpretation. In this framework, they consider large quantum systems (with infinitely many degrees of freedom) satisfying the K-property, thus *providing decaying time correlations, and interacting with an environment. If the time interval between successive measured events (put in in terms of state reductions) is long enough, these events form a consistent history. Narnhofer and Thirring show that for an infinite spin chain under unitary time evolution, such a sequence of events makes a mixed state pure on the classical quantities.*

References

Gell-Mann M. and Hartle J. (1993): Classical equations for quantum systems. *Phys. Rev. D* **47**, 3345–3382.

Giulini D., Joos E., Kiefer C., Kupsch J., Stamatescu I.-O., and Zeh H.D. (1996): *Decoherence and the Appearance of a Classical World in Quantum Theory* (Springer, Berlin).

Griffiths R.B. (1984): Consistent histories and the interpretation of quantum mechanics. *J. Stat. Phys.* **36**, 219–279.

Jordan P. (1949): On the process of measurement in quantum mechanics. *Philosophy of Science* **16**, 269–278.

Omnès R. (1992): Consistent interpretations of quantum mechanics. *Rev. Mod. Phys.* **64**, 339–382.

Zurek W.H. (1994): Preferred sets of states, predictability, classsicality, and environment-induced decoherence. In *Physical Origins of Time Asymmetry*, ed. by J.J. Halliwell, J. Pérez-Mercader, and W.H. Zurek (Cambridge University Press, Cambridge), pp. 175–212.

MACROSCOPIC PURIFICATION OF STATES BY INTERACTIONS

H. NARNHOFER AND W. THIRRING
Institut für Theoretische Physik, Universität Wien,
Boltzmanngasse 5, A-1090 Wien, Austria

1. Introduction

Ever since Schrödinger's seminal 1935 paper (Schrödinger 1935) quantum physicists are wondering why we always find states where classical observables have definite values but never mixtures of them. Mixed states like the canonical state are frequently used, we live in a 3K universe, but we observe always pure states on macroscopic observables. Of course, an animal such as Schrödinger's cat is too complicated to be described mathematically, so as classical observables we rather think of the mean magnetization

$$\vec{m} = \lim_{N \to \infty} \frac{1}{N} \sum_{i=1}^{N} \vec{\sigma}_i$$

of a magnet or the center of mass velocity

$$\vec{v} = \lim_{N \to \infty} \frac{\sum_{i=1}^{N} m_i \vec{v}_i}{\sum_{i=1}^{N} m_i}$$

of a large object. We take the limit $N \to \infty$ because the behavior of a system with $N \sim 10^{24}$ is closer to that with $N = \infty$ than to a finite system with, e.g., $N \sim 10$. The limiting observables can be characterized mathematically as the center of the algebra of observables, and states which are pure over the center are called factor states. This means they assign a definite c-number to each element of the center, they become dispersion-free.

The key feature we need is the so-called K-property which occurs for quantum systems if they are infinite and implies that for factor states the time correlation functions factorize for long times. This does not hold for

states which are mixed over the classical observables and will be the origin
of the purification of states by successive interactions with the environment.
The literature with proposals how to resolve the cat puzzle is too extensive
to be discussed in detail here (for a survey see, e.g., d'Espagnat 1995). We
only want to point out such a mechanism typical for large quantum systems
which to our knowledge has not been investigated in detail so far.

The first hint about what distinguishes factor states physically is found
in Narnhofer and Robinson (1975) where it is shown that the canonical (i.e.,
KMS) states which are also factor states enjoy the following property: If
one disturbs the Hamiltonian H by any sequence of perturbations h_n such
that $H + h_n$ generates the same time evolution as H in the limit $n \to \infty$,
then the associated canonical state converges to the one associated to H
exactly if the latter is a factor. In one direction this theorem is obvious since
if we take $h_n \neq c\mathbf{1}$ from the center this will not change the time evolution
but it does change the canonical state. The converse of this statement is
not so trivial.

However, this kind of stability is not exactly what one wants: it does not
explain a dynamical purification of the state. Ideally one should show that
the unavoidable interactions with the environment drive the state into a
factor state. That this is so was shown by Narnhofer (1993), if the interactions are considered as measurements and one accepts the usual reduction
of the state postulate.

In this contribution we first derive this result in detail using the language of the many histories interpretation of quantum mechanics (Griffiths
1984, Gell-Mann and Hartle 1991, Omnès 1992) which gives a precise formulation of the reduction postulate. We consider histories that result from
measurements with long time intervals. Here we assume that the time evolution of the system (without outside interactions) is a K-system and this
property guarantees that our histories satisfy the so-called consistency condition. Therefore we are free to interpret them with classical probability
theory, otherwise we are neutral with respect to the ideology which goes
along with this interpretation.

Secondly, we use a certain model for the measuring device to demonstrate that the same result can be derived by a unitary time evolution with
a time dependent Hamiltonian. We consider the simplest case of a mixture
of two factor states and show that for almost each history the ratio of the
two contributions (or its inverse) decreases exponentially with the number
of outside interactions (again for a K-time evolution in between). We speak
of interactions rather than measurements since we do not want to enter the
discussion when an interaction qualifies as measurement. To determine the
history one has finally to make a reading of some pointers. Since pointer
variables are not quantum mechanical observables there is no difference to

the same situation in a classical system where the reading of a pointer just means an update of information.

The observables being measured are a complete set of orthogonal projectors and replace the classical partitioning of phase space. Thus we are dealing with the quantum generalization of symbolic dynamics where the trajectory is replaced by a list of the gates through which the trajectory went at a discrete sequence of times. For some dynamical systems this list, if it becomes infinitely long, coincides with that of a single trajectory only, corresponding to a state with maximal purification (i.e., sharp initial conditions). Our results cannot be derived for general quantum systems but for infinite systems where inequivalent representations and a center of the algebra can occur. This center is interpreted as consisting of the classical observables. Triviality of the center, i.e., if the state is a factor state, is thus equivalent to the fact that the classical, i.e., macroscopic, observables are dispersion-free. Furthermore these observables must be chaotic in as much as we need decaying time correlations. This can be proved for so-called K-systems, both classical and quantum mechanical. For infinite systems it is known that (reversible) time evolutions exist that have the K-property.

2. Purification by Interaction

In the many histories interpretation of quantum mechanics one assigns probabilities W to histories consisting of a sequence of events. The latter arc represented by projection operators $P_k(t) \in \mathcal{A}$, $k = 1, \ldots, r$, where the Heisenberg representation is used for the time dependence and $\mathcal{A}$ is the algebra of observables. For finite quantum systems a state is given by a density matrix ρ and the key formula for the probabilities is

$$
\begin{aligned}
W(\underline{\alpha}) &= \mathrm{Tr}\, P_{\alpha_n}(t_n)\ldots P_{\alpha_1}(t_1)\rho P_{\alpha_1}(t_1)\ldots P_{\alpha_n}(t_n) \\
&= \mathrm{Tr}\, \sqrt{\rho}\, P_{\alpha_1}(t_1)\ldots P_{\alpha_n}(t_n)P_{\alpha_{n-1}}(t_{n-1})\ldots P_{\alpha_1}(t_1)\sqrt{\rho} \quad (1)
\end{aligned}
$$

where $\underline{\alpha} = (\alpha_1 \ldots \alpha_n)$, $\alpha_i \in \{1, \ldots, r\}$. If the P_k are a complete set of orthogonal projections, $P_k P_{k'} = \delta_{kk'} P_k$, $\sum_{k=1}^{r} P_k = 1$, and the probabilities are correctly normalized:

$$
W(\underline{\alpha}) \geq 0, \qquad \sum_{\underline{\alpha}} W(\underline{\alpha}) = 1. \quad (2)
$$

For the infinite systems we are concerned with, the density matrix ρ will not exist but the notion of a state as a positive linear functional ω as a generalization of $\mathrm{Tr}\, \sqrt{\rho}\, A \sqrt{\rho} = \omega(A)$ carries over. For these systems in physical situations the time correlations are expected to decay for an extremal time invariant (factor) state ω such that (in Thirring (1996) and in Narnhofer and Thirring (1994) this is proved for K-systems)

$$W(\underline{\alpha}) \to \prod_i \omega(P_{\alpha_i}) \qquad \text{if } t_{i+1} - t_i \to \infty \ \forall \ i. \tag{3}$$

For two t_i this follows from mixing but for more than two t_i we need the K-property which is not satisfied for finite systems where the time evolution is quasiperiodic. Actually one can expect the limit to be reached exponentially fast with a typical relaxation time. Extremal invariance means that ω is not a combination

$$\mu\omega_1 + (1 - \mu)\omega_2, \qquad 0 < \mu < 1, \tag{4}$$

of two other invariant states and the factorization Eq. (3) is lost by convex combinations. Nevertheless consistency of histories is preserved since it can be expressed in short by $\omega(P_{\underline{\alpha}'}P_{\underline{\alpha}}) = \delta_{\underline{\alpha}',\underline{\alpha}}$. It is necessary for an interpretation in terms of classical probabilities and is satisfied as soon as the time correlation functions factorize. If it holds for every extremal invariant state it is preserved by convex combinations. Extremal invariance requires that $\mathcal{A}$ does not contain an element constant in time $z \neq c\mathbf{1}$ ($\mathbf{1}$ is the unit operator) since with it one could decompose ω into

$$\omega_1(A) = \frac{\omega(z^* A z)}{\omega(z^* z)} \qquad \text{and} \qquad \omega_2(A) = \frac{\omega(\sqrt{1 - z^* z}\, A \,\sqrt{1 - z^* z})}{\omega(1 - z^* z)}$$

(we may assume $z^* z < 1$ by replacing z by $z/\|z\|$). For infinite quantum systems with suitable interactions, constant elements belong to the center which is the classical part of the system. On the other hand, for equilibrium (KMS) states all elements of the center are constant. The extremal invariant states are the ones where z is represented by a c-number, thus one assigns a definite value to the classical quantities and one deals with a factor state. In the example of the spin chain to be studied in Sect. 3 the classical quantity is the mean magnetization

$$\vec{m} = \lim_{N \to \infty} \frac{1}{2N} \sum_{i=-N}^{N} \vec{\sigma}_i \, ,$$

and classically pure states are the ones where all spins except a finite number of them point in the same direction $\vec{m}$. For the mixed states of the form Eq. (4) the histories W are also convex combinations

$$W_\omega(\underline{\alpha}) = \mu W_{\omega_1}(\underline{\alpha}) + (1 - \mu)W_{\omega_2}(\underline{\alpha}) \, . \tag{5}$$

We will show that for long histories $n \to \infty$ and $t_{i+1} - t_i \to \infty$ they purify in the sense that in Eq. (5) either W_{ω_1} or W_{ω_2} dominates such that

$W_{\omega_i}/W_{\omega_j} < \varepsilon$ for arbitrarily small ε. Which one dominates, depends on the history. Of course, one cannot dominate over the other for all histories since

$$\sum_{\underline{\alpha}} W_{\omega_1}(\underline{\alpha}) = \sum_{\underline{\alpha}} W_{\omega_2}(\underline{\alpha}) = 1.$$

We shall elaborate only on the simplest nontrivial case with two states $\omega_{1,2}$ and two projectors P, $1 - P$. The generalization to several states or projectors is easy and shows the same features. Denote $\omega_{1,2}(P) = p_{1,2}$, then for a history with ℓ projectors P and $n - \ell$ projectors $(1 - P)$, $0 \le \ell \le n$, we get from Eq. (3)

$$W(\underline{\alpha}) = W(n, \ell) = \mu p_1^\ell (1 - p_1)^{n-\ell} + (1 - \mu)p_2^\ell (1 - p_2)^{n-\ell}. \tag{6}$$

The amount of mixing is given by the ratio

$$R = \frac{\mu}{1 - \mu} \left(\frac{p_1}{p_2}\right)^\ell \left(\frac{1 - p_1}{1 - p_2}\right)^{n-\ell}$$

of the two contributions. The mixing is noticeable if for some small number $\varepsilon > 0$ we have $\varepsilon < R < 1/\varepsilon$ or

$$\frac{\varepsilon(1 - \mu)}{\mu} < a^\ell b^{n-\ell} < \frac{1 - \mu}{\varepsilon \mu} \tag{7}$$

where $a := p_1/p_2$ and $b := (1-p_1)/(1-p_2)$. For definiteness we may assume $a > 1$, thus $b < 1$. If we scale ℓ with n, $\ell = n\lambda$, $0 \le \lambda \le 1$, the condition Eq. (7) becomes

$$\frac{1}{\ln \frac{a}{b}} \left(\ln \frac{1}{b} + \frac{1}{n} \ln \frac{1 - \mu}{\mu} + \frac{\ln \varepsilon}{n}\right) < \lambda$$

$$\lambda < \frac{1}{\ln \frac{a}{b}} \left(\ln \frac{1}{b} + \frac{1}{n} \ln \frac{1 - \mu}{\mu} - \frac{\ln \varepsilon}{n}\right) \tag{8}$$

Thus, for the mixed histories, λ is in an interval of length $2 \ln \varepsilon/(n \ln \frac{a}{b})$. We are interested in long histories, $n \to \infty$, for which this is a small number, and want now to calculate which fraction of the histories is mixed. Since at each event the histories can take two courses there are 2^n histories altogether. Their probabilities (Eq. (6)) depend only on n and ℓ. There are $n!/(\ell!(n - \ell)!)$ histories with this probability. For $n \to \infty$ we calculate with Stirlings's formula the density d of histories with $\lambda = (1 + \delta)/2$ to be

$$d(\delta) = \sqrt{\frac{n}{\pi}} \, e^{-n\delta^2}, \qquad \int_{-1}^{1} d\delta \, d(\delta) = 1 + O(e^{-n}).$$

Then Eq. (8) tells us that for $n \to \infty$ the fraction of histories more mixed than ε is less than

$$\frac{4}{\sqrt{\pi n}} \frac{\ln 1/\varepsilon}{\ln a/b} e^{-n\delta_m^2} , \qquad (9)$$

where

$$\delta_m = \left| \frac{\ln 1/b - \ln a}{\ln 1/b + \ln a} \right|$$

is the minimum of $|\delta|$ in the interval defined by Eq. (8). We have $\delta_m > 0$ and therefore exponential decrease unless $a = 1/b$, which happens if $p_1 = p_2$ or $p_1 = 1 - p_2$. In the first case, $\omega_1 = \omega_2$ for the algebra generated by P and $1 - P$, and there is no mixing. In the second case, $\omega_1 = \omega_2$ combined with a reflection, and in this special case the fraction decreases only as $1/\sqrt{n}$. In any case, for $n \to \infty$ most histories are pure in the sense that within ε only those states contribute for which the classical quantities have a definite value.

Since the ℓ-dependence of W goes with $(p/(1-p))^\ell$ for the most probable history we have $\ell = 0$ or $\ell = n$ depending on whether $p < 1/2$ or $p > 1/2$. Thus, the most likely answer to each experiment is always the most probable one.

To show that we are not just doing abstract mathematics, let us estimate the orders of magnitude we are talking about. Consider the center of mass velocity of a tennis ball. A mixed history over this observable would mean a zig-zag motion of the ball. (It would certainly upset the players and the referee.) A dispersion-free state over the classical observables remains so by unitary time evolution and Eq. (9) shows that any state is driven into a factor state by successive measurements. For large systems one does not have to keep repreparing the original state because it is reestablished by relaxation. So provided the time between measurements is longer than the relaxation time by a sequence of measurements, the state will eventually look like a factor state. Typical relaxation times in normal matter are $10^{-8} - 10^{-12}$ sec and the time between the observations should be much longer. This would be the case if we observe with a high speed camera with 10^3 pictures/sec. If we observe for 10 sec, n becomes 10^4 and if δ_m is, say, $1/10$, then the purification factor in Eq. (9) is e^{-10^3} and thus very effective.

3. The Spin Chain

In this section we illustrate the abstract ideas of Sect. 2 by a standard model of a measuring apparatus. We consider the simplest nontrivial physical example which shows the necessary features; an infinite spin chain $\mathcal{A} = \{\vec{\sigma}_i\}$, where the index i ranges over the integers. The shift $\sigma_i \to \sigma_{i+1}$ is taken as discrete evolution. For the benefit of readers who prefer to think in terms of wave functions we shall now use the Schrödinger representation. For each direction $\vec{n}_i$, $\vec{n}_i^2 = 1$, there is a vector $|\vec{n}_i\rangle_i$ in Hilbert space such that $\vec{\sigma}_i$ points in this direction, $\vec{\sigma}_i \cdot \vec{n}_i|\vec{n}_i\rangle_i = |\vec{n}_i\rangle_i$. For all spins together the corresponding vector is the tensor product $\bigotimes_i |\vec{n}_i\rangle_i$. If there were N spins these vectors spanned a 2^N-dimensional space but for $N = \infty$ the space is huge (nonseparable). We shall work in separable subspaces of this monster which are obtained, such as the Fock space, by letting $\mathcal{A}$ act on a reference vector $|\Omega_1\rangle : \mathcal{H}_1 = \mathcal{A}|\Omega_1\rangle$. For $|\Omega_1\rangle$ we choose a polarized state where all spins point in the same direction

$$|\vec{n}\rangle : |\Omega_1\rangle = \bigotimes_{i=-\infty}^{\infty} |\vec{n}\rangle_i \ . \tag{10}$$

In $\mathcal{H}_1$ we get an irreducible representation π_1 of $\mathcal{A}$. Of course, for individual spins $\pi_1(\sigma_i)$ acts like σ_i but weak limits such as the mean magnetization

$$\vec{M} = \lim_{N \to \infty} \frac{1}{2N+1} \sum_{i=-N}^{N} \pi_1(\vec{\sigma}_i) = \vec{n}\mathbf{1} \tag{11}$$

depend on the representation. Considering another reference state $|\Omega_2\rangle = \bigotimes_{i=-\infty}^{\infty} |m\rangle_i$, the magnetization would turn out to be $\vec{m}$. Thus the two representations are not unitarily equivalent, $U^{-1}\pi_1(\sigma_i)U = \pi_2(\sigma_i)$ would imply

$$U^{-1}\vec{n} \cdot \mathbf{1}U = \vec{m} \cdot \mathbf{1}$$

which is impossible since U cannot change the unity $\mathbf{1}$. The $|\Omega_{1,2}\rangle$ define states $\omega_{1,2}(\cdot) = \langle\Omega_{1,2}| \cdot |\Omega_{1,2}\rangle$ and the mixed state $\mu\omega_1 + (1-\mu)\omega_2$ is obtained by a vector in the orthogonal sum of π_1 and π_2

$$|\Omega_S\rangle = \sqrt{\mu}\,|\Omega_1\rangle \oplus \sqrt{1-\mu}\,|\Omega_2\rangle \in \mathcal{H}_1 \oplus \mathcal{H}_2 =: \mathcal{H}_S, \qquad \pi = \pi_1 \oplus \pi_2, \tag{12}$$

$\langle\Omega|\pi(\vec{\sigma}_i)|\Omega\rangle = \mu\vec{n} + (1-\mu)\vec{m}$. This representation is reducible, there are two "superselection sectors" (Landsman 1991), the magnetization $\vec{M}$ is in the center and is not a multiple of unity

$$\lim_{N \to \infty} \frac{1}{2N+1} \sum_{i=-N}^{N} \pi(\vec{\sigma}_i) = \mu\vec{n}\mathbf{1}_1 \oplus (1-\mu)\vec{m} \cdot \mathbf{1}_2 \ . \tag{13}$$

If we identify, with a poetic license, $\vec{M}$ with Schrödinger's cat, if $\vec{n}$ means alive and $\vec{m}$ means dead, then the vectors of $\mathcal{H}_1$ ($\mathcal{H}_2$) represent the cat alive (dead) whereas in $|\Omega\rangle$ (with $\mu = 1/2$) the cat is half dead and half alive. We shall now show how by a succession of measurements $|\Omega\rangle$ purifies in the sense that the dominant components turn into either $\mathcal{H}_1$ or $\mathcal{H}_2$.

First we have to construct for the measuring device a classical system which can store the information contained in $\mathcal{A}$. Since a measurement of σ_i can only have two outcomes, pointers with two positions suffice. Since the Hilbert space description is useful also for classical systems we represent the state of the device measuring σ_i by a two-dimensional vector $\begin{pmatrix} u_i \\ d_i \end{pmatrix}$, u and d meaning pointer up or down. The measuring array for all spins is again an infinite tensor product $\otimes_i \begin{pmatrix} u_i \\ d_i \end{pmatrix}_i \in \mathcal{H}_A$ and we start with a state $|\Omega_A\rangle$ where all u_i are zero.

For the time evolution we take a shift U, and then consider an instantaneous measurement of the direction $\vec{s}$ of a spin. The corresponding proposition is the projector

$$P_k = \frac{1}{2}(1 + \vec{\sigma}_k \cdot \vec{s}) = |s\rangle_k \ {}_k\langle s| \ .$$

If the answer is one we have the pointer unchanged, if the answer is zero we turn the pointer up. This turning of the pointer is effected by an operator τ,

$$\tau_k \begin{pmatrix} u \\ d \end{pmatrix}_k = \begin{pmatrix} d \\ u \end{pmatrix}_k .$$

Thus the effect of measuring $\vec{\sigma}_1$ is $V_1 = P_1 + (1 - P_1)\tau_1$, or written in full detail with operators in $\mathcal{H} = \mathcal{H}_S \otimes \mathcal{H}_A$

$$V_1 = \pi_1(P_1) \oplus \pi_2(P_1) \otimes 1 + (1 - \pi_1(P_1)) \oplus (1 - \pi_2(P_1)) \otimes \tau_1 \ . \qquad (14)$$

Note that V_1 is unitary, and in $\mathcal{H}_S \otimes \mathcal{H}_A$ there is no reduction of the full state vector. The time evolution U between the measurements shifts by one unit $U\pi_{1,2}(P_k) = \pi_{1,2}(P_{k+1})U$, $U\tau_k = \tau_{k+1}U$, so that the full time evolution of $|\Omega\rangle = |\Omega_S\rangle \otimes |\Omega_A\rangle$ after n time units is

$$|\Omega(n)\rangle = V_1 U V_1 U \ldots V_1 U |\Omega\rangle = V_1 V_2 \ldots V_n |\Omega\rangle \qquad (15)$$

since $U^k|\Omega\rangle = |\Omega\rangle$. The results of the measurements are encoded in the $\mathcal{H}_A$-part of $|\Omega(n)\rangle$, so we decompose them in an orthogonal basis of $\mathcal{H}_A$,

$$|\Omega(n)\rangle = \sum_{\alpha_i=0}^{1} v(\underline{\alpha})\tau_1^{\alpha_1}\tau_2^{\alpha_2} \ldots \tau_n^{\alpha_n}|\Omega_A\rangle, \qquad v(\underline{\alpha}) \in \mathcal{H}_S. \qquad (16)$$

Wherever $\alpha_i = 0$ the corresponding spin is in direction $\vec{s}$, for $\alpha_i = 1$ we have $-\vec{s}$. If we have such a situation, the system is left with a wave function $v(\underline{\alpha})$ which has a component in $\mathcal{H}_1$ and one in $\mathcal{H}_2$:

$$v(\underline{\alpha}) = v_1(\underline{\alpha}) \oplus v_2(\underline{\alpha}). \qquad (17)$$

To calculate the length of the $v_{1,2}(\underline{\alpha})$ we have to use $P_1|n\rangle_1 = |s\rangle_1\langle s|n\rangle$, $(1 - P_1)|n\rangle = |-s\rangle\langle -s|n\rangle$. If we introduce $|\langle s|n\rangle|^2 = p_1$, thus $|\langle -s|n\rangle|^2 = 1 - p_1$, and similarly $|\langle s|m\rangle|^2 = p_2$, $|\langle -s|m\rangle|^2 = 1 - p_2$, and if $\underline{\alpha}$ contains ℓ zeros and $n - \ell$ ones, we have

$$\|v_1(\underline{\alpha})\|^2 = \mu p_1^\ell (1 - p_1)^{n-\ell}, \qquad \|v_2(\underline{\alpha})\|^2 = (1 - \mu)p_2^\ell(1 - p_2)^{n-\ell}. \quad (18)$$

Thus with $W(\underline{\alpha}) = \|v(\underline{\alpha})\|^2$ we arrive exactly at Eq. (6), and the decomposition Eq. (16) displays the 2^n histories. Remember that we had a unitary time evolution and there was no collapse of the wave function after each measurement. Only at the end we read the configuration of classical pointers.

Remarks

1. In terms of convergence of states, our result can be expressed as follows. The vector $|\Omega\rangle$ gives (for $\mu \neq 0, 1$) a mixed state over the algebras of system $\otimes$ apparatus since the system is represented reducibly. It evolves by a unitary evolution into the vector $|\Omega(n)\rangle$ such that the state $\omega_n(\cdot) = \langle \Omega(n)| \cdot |\Omega(n)\rangle$ of the system $\otimes$ apparatus remains pure for $\mu = 0$ or 1, otherwise it remains mixed. Reading the pointer in a position $\underline{\alpha}$ changes the state to $\omega_{n,\underline{\alpha}}(\cdot) = \omega_n(P_{\underline{\alpha}}\cdot)W_\omega(\underline{\alpha})^{-1}$ where P_α projects onto the vector $\tau_1^{\alpha_1} \ldots \tau_n^{\alpha_n}|\Omega_A\rangle$. For $n \to \infty$ (and making the history $\underline{\alpha}$ infinite) this converges weakly to a pure state. The limit $\lim_{n\to\infty} \omega_n = \sum_{\underline{\alpha}} \omega_{n,\underline{\alpha}}W(\underline{\alpha})$ is a mixed state even for $\mu = 0$ or 1. This is in accordance with the result of Hepp (1972) who observed that in a similar situation of an infinite quantum system weak limits of pure states may be mixed.

2. It is irrelevant that the states $\omega_{1,2}$ in the example were pure. For asymptotic Abelian systems they only have to be extremally invariant to possess the required cluster properties. If we restrict ourselves to canonical temperature states, then only at phase transition points one has to be aware that they themselves are mixtures of states with the same temperature but different values of a central element. This corresponds to the fact that at transition points space clustering becomes critical (no rigorous results are available about time clustering).

3. Talking about macroscopic quantum systems we mean many degrees of freedom and not just large size. Hence our considerations are not applicable to the proposal of Leggett and Garg (1985). What we really need is that the relaxation time of the system is shorter than the time between measurements.

4. Summary

It has to be emphasized that we actually stay within the Copenhagen interpretation of quantum mechanics. Though we use the language of the history interpretation, we do not have to worry about inconsistent histories since we are dealing only with consistent histories. The long time limit gives us the consistency which is expected for a semiclassical limit, and we study how the purification of the state comes about from the histories. Our derivation is based on the reduction of the wave function postulate which we do not consider as a mystic departure from the unitary time evolution but as the effect of the measurement. It may seem strange that all different ways of measuring some quantity should lead to the same state but the reduced state corresponds in some sense to the softest measurement (Narnhofer 1993). It is the state for which the measured quantity has the observed value and the smallest distance (measured by the relative entropy) from the original state. For one simple model of a measuring apparatus we have shown that the unitary evolution of the joint system leads to the same result. Our goal is to see what the special properties of large quantum systems imply for the histories.

To get the optimal knowledge of the state of a finite quantum system one has to repeat the experiment for other members of an ensemble of equally prepared systems to gather enough statistics. For our infinite system one can also redo the experiment with the same system after an appropriate relaxation time and get the same result provided the initial state was pure on the macroscopic part. In all quantum systems some quantities will remain fluctuating even in an optimally refined state. This remains true for infinite quantum systems but we have seen that by repeated measurements the classical observables will assume definite values. Thus, e.g., below the phase transition, domains of a magnet will be magnetized in a definite direction. Even if nobody looks at them there will be enough "events" (i.e., interactions with the environment) to purify the state over the classical part. However, a quantum mechanically pure state over all the microscopic observables will not be obtainable here because for these systems all observable projections are infinite dimensional. One dimensional projections in Hilbert space do not belong to the algebra of observables.

References

d'Espagnat B. (1995): *Veiled Reality* (Addison-Wesley, Reading, Mass.).

Gell-Mann M. and Hartle J. (1991): Alternative decohering histories in quantum mechanics. In *Proc. of the 25th Int. Conf. on High Energy Physics, Vol. 2*, ed. by K.K. Phua and Y. Yamaguchi (World Scientific, Singapore), pp. 1303–1310.

Griffiths R.B. (1984): Consistent histories and the interpretation of quantum mechanics. *J. Stat. Phys.* **36**, 219–279.

Hepp K. (1972): Quantum theory of measurement and macroscopic observables. *Helv. Phys. Acta* **45**, 237–248.

Landsman N.P. (1991): Algebraic theory of superselection sectors and the measurement problem in quantum mechanics. *Int. J. Mod. Phys. A* **30**, 5349–5371.

Leggett A.J. and Garg A. (1985): Quantum mechanics versus macroscopic realism: is the flux there when nobody looks? *Phys. Rev. Lett.* **54**, 857–860.

Narnhofer H. (1993): Stability of pure thermodynamic phases in quantum statistics. In *Phase Transitions*, ed. by R. Kotecky (World Scientific, Singapore), pp. 150–158.

Narnhofer H. and Robinson D.W. (1975): Dynamical stability and pure thermodynamic phases. *Commun. Math. Phys.* **41**, 89–97.

Narnhofer H. and Thirring W. (1994): Clustering for algebraic K-Systems. *Lett. Math. Phys.* **30**, 307–316.

Omnès R. (1992): Consistent interpretations of quantum mechanics. *Rev. Mod. Phys.* **64**, 339–382.

Schrödinger E. (1935): Die gegenwärtige Situation in der Quantenmechanik. *Die Naturwissenschaften* **23**, 807–812, 823–828, 844–849.

Thirring W. (1996): The histories of chaotic quantum systems. *Helv. Phys. Acta* **69**, 706–716.

CONTEXTUAL BACKGROUND

Alfred Rieckers:

MACROSCOPIC QUANTUM PHENOMENA AT THE SQUID

In his attempts toward an algebraic quantum mechanics incorporating the classical features of the various types of material systems as well as their quantal properties, Primas considered problems on varying levels of difficulty. One first category of problems was the emergence of classical observables themselves: how can they be extracted from the formalism in a mathematically rigorous way? It transpired that even this first step gives rise to formidable technical intricacies. In one way or another, infinite quantum systems always come into play when classical observables are constructed. Infinite quantum systems, however, are difficult to handle, and even the most advanced techniques only gave a fairly restricted number of fully understood classical quantities.

A second category of problems refers to the dynamics of classical observables. In classical physics, the motion of observables is usually described by some kind of differential equation, e.g., of the Hamiltonian type. If one tries to develop a generalized quantum mechanics of the aforementioned kind, one has to establish classical equations of motion for the classical part of the quantum system in question. This seems to entail almost insuperable difficulties. In most cases the very question of the existence of a global dynamics in infinite quantum systems is highly delicate. The next problem then is how these dynamics, if they exist, can be transferred to the classical part of the system. A third question relates to the explicit form of the equations of motions in the classical part.

There are only few complete results in this regard. One of the rare, pioneering examples is the motion of the phase difference in superconducting quantum interference devices (SQUID), as worked out by Alfred Rieckers (1984) together with Ullrich (Rieckers and Ullrich 1985, 1986). Rieckers was able to demonstrate how a classical phase difference can be constructed and how to derive its equation of motion.

As the change of the phase is related to a gauge transformation, a sec-

ond type of "motion" comes into play. Just like time evolution, gauge trans-
formations figure as one-parameter automorphism groups of the observable
algebra. In appropriate Hilbert space representations, these automorphisms
can be represented by unitary one-parameter groups. The respective gen-
erator of such a one-parameter unitary group is sometimes used as an "ob-
servable": an energy observable in the case of the time evolution, and a
particle number operator in the case of gauge transformations.

Primas was skeptical about using such observables which are not con-
tained in the original observable algebra. Rieckers, on the other hand, is
almost a protagonist of such observables outside the original observable al-
gebra. Can rigorous mathematical arguments provide some clue concerning
the incorporation of "external" observables into a given observable algebra?

In the following contribution, Rieckers deals with an extension of his
earlier model. He starts with an outline of his theory of the Josephson
junction. He then uses the above-mentioned generators of gauge transfor-
mations as "particle number operators" for the two junction electrodes,
in order to supplement a well established capacitance term for the ring
SQUID. This gives, as he puts it, a "non-commutative dressing" for the
macroscopic phase difference dynamics. Including an additional inductive
term and expressing the macroscopic phase difference by the magnetic flux,
he comes to a microscopic foundation of the SQUID Hamiltonian which is
commonly used for the so-called "macroscopic wave function". This lends to
his contribution its particular "reductionistic" flavor: macroscopic quantum
theory, which has been partially confirmed by experiments, is evaluated as
a higher integration level over the microscopic degrees of freedom.

Rieckers thinks that the inclusion of "exterior" observables is not only
justified by its connection to practical (experimental) superconductor phys-
ics but also by mathematical-structural arguments. First, he argues, the
spectrum of the exterior observables coincides with the Arveson spectrum
of the corresponding symmetry group in the original observable algebra
and, second, the "spectral properties" of these exterior observables can be
defined as new faces in the original set of states, where one has spectral
subspaces in any case. He concludes, very much in Primas' vein, with some
philosophical considerations on the ontological status of macroscopic quan-
tum phenomena within his extended framework.

References

Rieckers A. (1984): On the classical part of the mean field dynamics for quantum lattice
    systems in grand canonical representations. *J. Math. Phys.* **25**, 2593–2601.
Rieckers A. and Ullrich M. (1985): Extended gauge transformations and the physical
    dynamics in a finite temperature (BCS) model. *Acta Phys. Austr.* **56**, 131–152.
Rieckers A. and Ullrich M. (1986): On the microscopic derivation of the finite-temperature
    (Josephson) relation in operator form. *J. Math. Phys.* **27**, 1082–1092.

MACROSCOPIC QUANTUM PHENOMENA AT THE SQUID

ALFRED RIECKERS
Institut für Theoretische Physik, Universität Tübingen,
Auf der Morgenstelle 14, D-72076 Tübingen, Germany

1. Introduction

The possibility of a coherent superposition of two macroscopically different states is discussed, e.g., in connection with the most popular Schrödinger "cat paradox" (Schrödinger 1935, Audretsch and Mainzer 1990). The usual point of view takes it for granted that the attributes "living" and "dead" of the system "cat" are so different that their combination into a state, in which none of them is actualized, seems paradoxical. Thus from this point of view the fundamental problem arises how to deal with quantum theory in a manner avoiding this apparent paradox.

As has been emphasized by Primas (1990a) the "cat paradox" inherits different sub-problems, which should be clearly separated from each other. One of the most involved aspects concerns the fact that a (living) cat is not a closed system. In our present investigation we shall deal with closed systems only and concentrate our discussion on the consequences which originate from the macroscopic size of the system. For the theoretical description we shall employ algebraic quantum theory, which is strongly advocated by Primas (1990a) as the most comprehensive and best investigated codification of quantum mechanics.

In an analogous process of reduction and simplification, which leads Primas from the "cat" to a molecule with a soft photon cloud, we take up some questions of the "cat paradox" in terms of certain macroscopically different states of a Josephson junction, which are to picture basic features of the SQUID (superconducting quantum interference device). In this way we join actual developments in the field of superconductivity which are both of theoretical and practical interest. For basic theoretical discussions our strictly microscopic approach seems to be better suited than semi-phenomenological theories, which often use concepts like the macroscopic wave function without any reference to the constituting atomic particles, here the electrons.

The decisive difference between our model of the "cat" and most other idealizations of this theoretically intractable system lies in the fact that one expects from experimental and theoretical reasons a positive answer to the "cat paradox" : Coherent superpositions of states of the Josephson junction, which are macroscopically different in a very precise sense, seem to be possible. This makes the situation, however, not easier. By contrast, one has now to refine the entire discussion as to explain, why *usually* macroscopic coherence is broken, whereas in very *special* systems (exhibiting a quantum condensate) it may nevertheless be realized. And then one has to consolidate these extraordinary quantum phenomena in the large with our realistic interpretation of the macroscopic world.

In the following we try to expound a consistent model treatment of the SQUID which incorporates basically all required levels of description, employing somewhat advanced techniques of algebraic quantum theory. Especially, we have to perform net limits over increasing sets of volumina for observables and states in different kinds of topologies. We hope to convince the reader that these versions of the thermodynamic limit constitute reasonable and even unavoidable forms of concept formation, which lead from the microscopic stage of the theory to higher hierarchical levels. We see the adequate way of introducing a concise notion of a "macroscopic quantum phenomenon" just in the careful elaboration of the various hierarchical levels of the Josephson junction.

The author is deeply indebted to Professor Hans Primas for drawing, many years ago, his attention to the problem of concept formation in quantum mechanics. In quantum mechanical calculations certain forms of "approximations" have in fact a much deeper meaning than being close to the "exact" numerical value of an observable quantity. As is emphasized and formulated in various ways in Primas (1981, 1987, 1990b), every pattern of a holistic system, like a quantum system, is conceptually and observationally obtained by neglecting features, which are declared irrelevant. Thus the pattern depends on a particular point of view. In cases of central importance in mathematized theories, one can formalize the neglect of information in the empirical data processing by a limit over a directed parameter set. The selection of the parameter set as, e.g., an increasing length or a decreasing coupling constant, reflects the special interest of the observing subject. Irrelevant features are then discarded in a controlled manner by replacing a system with a fixed parameter value by the limiting system.

Especially the thermodynamic limit of a quantum mechanical many-body system is a powerful and rather universal tool for shaping those concepts which are relevant for certain forms of macroscopic observation. The use of *nets* of local quantum observables allows for the introduction of temperature, chemical potential, entropy, and order parameters as quantities, which are robust against local perturbations and compatible with all local and (nonlocal) limiting observables, that is "classical" in the framework of a quantum theory. Spatial structures may be incorporated by choosing directed subfamilies of the set of all finite regions or combining such directed families of local regions. In this way a Josephson junction, which consists of two macroscopic superconducting electrodes, is formulated in terms of a thermodynamic limit over a directed set of pairs of local regions. It introduces the new classical concepts of "right" and "left" for the originally indistinguishable electrons. Thus, only by neglecting local quantum correlations one gains the components of a superconducting network experimental physicists are concerned with.

For a theoretical physicist who is strictly employing traditional quantum theory, the wholeness of a quantum system is never broken. But this narrow version of reductionism prevents not only the notion of "objects" in a quantum world, characterized by robust classical observables, but also the notion of "macroscopic quantum effects" in superconductor electronics.

The richness, on the other hand, which is displayed by a quantum theory supplemented with additional ways of concept formation – and which the present models are to illustrate – should not be considered due to artificial tricks. It belongs to the special merits of Professor Primas' scientific work to have pointed out in terms of many practical examples that without discarding irrelevant aspects quantum mechanics would be void of objects, of molecules, of any directly observable phenomenon. While Niels Bohr's "phenomenon" (Bohr 1935) includes the principal incompatibility of certain measurement devices and by this interprets the overall entanglement of quantum properties, Hans Primas' "directly observable phenomenon" includes additional forms of context dependences, which abstract from – in this context – unobservable or irrelevant correlations (Primas 1981, 1987).

These latter forms of abstraction are not only needed for the incorporation of the measurement process into quantum mechanics but are quite generally indispensable for connecting the theory of quantum systems with any experimental language. Since an abstraction from features which originally have been covered by the theoretical frame is – like a mathematical limit – intimately connected with this frame, its addition to the traditional quantum formalism should be viewed still as a reductionistic, microscopic theory. In particular, a uniform theoretical language allows us to describe the dynamical interrelations between the higher and lower hierarchical levels. The comparison with the mathematical procedure of completing a topological space and extending, e.g., the rational to the real numbers, makes evident, that the new kind of reductionistic quantum theory is an immensely more flexible and comprehensive theory. It covers both the partially discrete microscopic quantum values and the continuum of macro-properties.

2. Superconducting Electrons and Their Equilibrium States

2.1. SINGLE ELECTRONS

The atomic constituents of our model are quasi-free electrons in the effective, periodic potential of a crystal. Their one-particle Hilbert space – a separable subspace of the nonseparable Hilbert space of all almost periodic functions – is spanned by the Bloch wave functions $u_{k,\sigma}$ with $(k,\sigma) \in \mathcal{K} \times \{\uparrow, \downarrow\}$. These electrons are dressed Galileons (Primas 1981) displaying features of the surrounding crystal, especially the symmetry breakdown from the Euclidean group to the space group of the lattice.

We consider only a single band, and take the momenta from a denumerable but dense set $\mathcal{K}$ in the vicinity of the Fermi surface. In $u_{k,\sigma}$ the momentum, spin, and energy ϵ_k have sharp values, whereas the position is completely undetermined.

The classical observables, *which characterize the object* "(bare) electron" are the absolute value $1/2$ of the spin, the rest mass, the negative elementary charge, and the magnetic moment. (Other classical observables would arise from more comprising classification schemes of the elementary particles.) By contrast to momentum, position, energy, and the spin in 3-direction, the classical observables of the object "bare electron" exhibit always a distinct value. The conservation laws for these kinds of classical observables, which never have been observed violated, are individually valid: the bare electron may exchange energy with another system but cannot exchange its charge or rest mass (in the nonrelativistic regime).

The "superselection rules" of these microscopic classical observables, which interdict coherent superpositions of states from different "sectors", are considered strict laws of nature.

As explained by Primas (1981), elementary particles become always dressed by the influence of the environment (crystal, electromagnetic field). Nevertheless, the classical observables of the bare elementary particles constitute the invariant features of the dressed Galileons in varying surroundings.

2.2. ELECTRONS IN A FIXED VOLUME

We characterize a system of finitely many electrons in a fixed volume by a finite subset $\Lambda \subset \mathcal{K}$ of momenta. The unique CAR-algebra $\mathcal{A}_\Lambda$, which is generated by the annihilation (creation) operators $c_{k,\sigma}^{(*)}$, $k \in \Lambda$, satisfying the C(anonical) A(nticommutation) R(elations)

$$\{c_{k\sigma}, c_{k'\sigma'}\} = 0, \ \{c_{k\sigma}, c^*_{k'\sigma'}\} = \delta_{kk'}\delta_{\sigma\sigma'} \, \mathbb{1} \tag{2.1}$$

– with $\{\cdot,\cdot\}$ the anti-commutator – is isomorphic to the finite-dimensional, full matrix algebra

$$\mathcal{A}_\Lambda = \bigotimes_{k\in\Lambda} (\mathbb{M}^{(2)} \otimes \mathbb{M}^{(2)})_k \, . \tag{2.2}$$

The generalized BCS-Hamiltonian (Bardeen et al. 1957) with k-dependent pairing interaction ($\epsilon_k \in \mathbb{R}$, $g_{kk'} = \overline{g_{k'k}} \in \mathbb{C}$)

$$H_\Lambda = \sum_{\sigma,k\in\Lambda} \epsilon_k c^*_{k\sigma} c_{k\sigma} - \sum_{k,k'\in\Lambda} \frac{g_{k,k'}}{|\Lambda|} c^*_{k\uparrow} c^*_{-k\downarrow} c_{-k'\downarrow} c_{k'\uparrow} \tag{2.3}$$

describes the effective interaction between the electrons including lattice effects and is attractive on the average.

The sum of the occupation number operators $n_{k,\sigma} := c^*_{k\sigma} c_{k\sigma}$ is the total number operator

$$N_\Lambda := \sum_{\sigma,k\in\Lambda} n_k \, , \tag{2.4}$$

and the reduced Hamiltonian is defined by

$$H^r_\Lambda = H_\Lambda - \mu N_\Lambda \, , \tag{2.5}$$

with $\mu \in \mathbb{R}$ the chemical potential.

For a macroscopic lattice the cardinality $|\Lambda|$ of Λ is huge, but sometimes one declares its finiteness as important in order to describe a "realistic" system.

Usually, non-Hermitian elements are included in the observable algebra. In the same spirit we take the full field algebra $\mathcal{A}_\Lambda$ into account, and not only elements $A \in \mathcal{A}_\Lambda$ which are gauge invariant, i.e., satisfy

$$\alpha_\theta(A) := e^{i\theta N_\Lambda} A e^{-i\theta N_\Lambda} = A, \ \forall\, \theta \in [0, 2\pi) \, . \tag{2.6}$$

Since Eq. (2.6) is equivalent with the compatibility of A with N_Λ, it should be satisfied by all $A \in \mathcal{A}_\Lambda$ which describe genuine physical observables as, e.g., H_Λ (which is the case for our models). For calculational purposes and for the description of transitions between states, however, non-gauge invariant elements – as the $c^{(*)}_{k,\sigma}$ – have to be included. This makes the center of our type I C^*-algebra $\mathcal{A}_\Lambda$ (Pedersen 1979) trivial, no matter how large Λ is chosen.

Nevertheless, most many-body physicists pretend to do thermodynamics with such an irreducible type I observable algebra and introduce – motivated by maximal disorder arguments – the grand canonical density operator

$$\sigma_\Lambda^\beta = \exp\left[-\zeta_\Lambda - \beta H_\Lambda^\tau\right], \quad \zeta_\Lambda \in \mathbb{R}. \tag{2.7}$$

The classical physical observables "chemical potential" and "natural temperature" are incorporated into the theoretical description by means of the real parameters μ and $\beta = 1/k_B T$ in Eq. (2.7). That means, however, that these notions are bound to the special form (2.7) of the equilibrium density operator. Every small perturbation of our system would change the state and consequently would destroy the concepts of "chemical potential" and "temperature". This formalism is thus not even capable of describing experiments at fixed chemical potential and temperature. All the more it is not capable of describing different thermodynamical phases, phase transitions, or spontaneous symmetry breakdown. (Note that σ_Λ^β is gauge invariant for all β!)

2.3. ELECTRONS IN A VARIABLE VOLUME

In macroscopic many-body physics, the volume is usually treated finite but unspecified. Thus, there is the implicit assumption that certain properties one is interested in do not depend on whether the system occupies a volume of 1 or 2 cm^3. A class of (not only microscopic) properties of the system, which are different in the two scales of volume, is neglected. This kind of abstraction is formalized in terms of the index set $\mathcal{L} := \{\Lambda \subset \mathcal{K}; |\Lambda| < +\infty\}$, which is directed by the inclusion relation. (There is an order isomorphism between $\mathcal{L}$ and a special subset of finite volumina.) By the uniqueness of the CAR-algebra (Bratteli and Robinson 1981) one knows that, for $\Lambda \subset \Lambda'$, $\mathcal{A}_\Lambda$ is $\star$-isomorphic to a subalgebra of $\mathcal{A}_{\Lambda'}$. Therefore, the elements of $\mathcal{A}_\Lambda$ may be, e.g., multiplied by those in $\mathcal{A}_{\Lambda'}$. Concerning the observables, the system with Λ is part of the system with $\Lambda' \supset \Lambda$, i.e., $\mathcal{A}_\Lambda \subset \mathcal{A}_{\Lambda'}$ (if the embedding isomorphism is dropped).

On the side of the states there is no canonical extension of a state (given by the density operator) ρ_Λ of $\mathcal{A}_\Lambda$ to a state $\rho_{\Lambda'}$ on $\mathcal{A}_{\Lambda'}$, for $\Lambda \subset \Lambda'$. The quantum entanglement expresses itself in the state language. (Cf. also the state reduction by a measurement in the EPR-paradox.) One has, however, the well determined restriction map $\rho_{\Lambda'} \to \rho_{\Lambda'|\Lambda}$. In $\rho_{\Lambda'|\Lambda}$ the expectations with $\rho_{\Lambda'}$ are formed only for observables in $\mathcal{A}_\Lambda$, where again $\Lambda \subset \Lambda'$.

The variability of the volume, expressed here by the variable momentum set Λ, leads directly to *algebraic quantum theory*:

$$\mathcal{A}_0 := \bigcup_{\Lambda \in \mathcal{L}} \mathcal{A}_\Lambda \tag{2.8}$$

is a $\star$-algebra (by means of the embedding mechanism), but has no distinguished Hilbert space to act upon. (The inductive limit of the local Hilbert spaces is too large.) Thus the states of $\mathcal{A}_0$ are the abstract algebraic states, given by the set $S(\mathcal{A}_0) \subset \mathcal{A}_0^*$ of all positive, normalized, linear functionals on $\mathcal{A}_0$, and are a convex subset of the dual Banach space $\mathcal{A}_0^*$. Since $\mathcal{A}_0$ has a unit, $S(\mathcal{A}_0)$ is w^*-compact. As part of $\mathcal{A}^*$ one has also a norm topology in $S(\mathcal{A}_0)$. Every $\varphi \in S(\mathcal{A}_0)$ is uniquely determined by its net of restrictions $\{\varphi_{\Lambda'} \in S(\mathcal{A}_{\Lambda'}); \Lambda' \in \mathcal{L}\}$, where the compatibility condition $\varphi_{\Lambda'|\Lambda} = \varphi_\Lambda$ holds for all pairs $(\Lambda, \Lambda') \in \mathcal{L} \times \mathcal{L}$ with $\Lambda \subset \Lambda'$. Reversely, every compatible net of local states determines a unique state on $\mathcal{A}_0$.

The harmless looking transition

$$S(\mathcal{A}_\Lambda), \ \Lambda \text{ fixed}, \ \longrightarrow \ S(\mathcal{A}_0) \tag{2.9}$$

is nothing else than the transition from microscopic, traditional quantum theory to the new hierarchic level of macroscopic many-body quantum physics with superimposed classical structures.

Only for the sake of a simplified mathematical language one takes the norm closure of the local algebra $\mathcal{A}_0$,

$$\mathcal{A} := \overline{\mathcal{A}_0}^{\|\cdot\|}, \tag{2.10}$$

called "quasi-local" algebra, and easily finds $S(\mathcal{A}_0) = S(\mathcal{A})$.

For variable Λ the parameters in H_Λ of Eq. (2.3) have to be specified in their dependence on Λ. Gerisch and Rieckers (1997) have formulated assumptions which express a "not too strong deviation" from a homogeneous model with $\epsilon_k = \epsilon$ and $g_{k,k'} = g > 0$ and which define "our model class". Interchanging the thermodynamic and the perturbation limit (starting from the homogeneous model), the limiting dynamics and its KMS-states could be established. Some additional reasoning (unpublished so far) provides the limiting relation

$$\lim_{\Lambda' \in \mathcal{L}} \sigma_{\Lambda'|\Lambda}^\beta =: \rho_\Lambda^\beta = \int_0^{2\pi} \rho_\Lambda^{\beta,\vartheta} \frac{d\vartheta}{2\pi}, \ \forall \Lambda \in \mathcal{L}. \tag{2.11}$$

Here the inhomogeneous mean field Hamiltonian

$$H_\Lambda^{r,\beta,\vartheta} = \sum_{k \in \Lambda} \left\{ (\epsilon_k - \mu) n_k - \Delta_k \left[e^{-i(\vartheta + \delta\vartheta_k)} c_{k\uparrow}^* c_{-k\downarrow}^* + e^{i(\vartheta + \delta\vartheta_k)} c_{-k\downarrow} c_{k\uparrow} \right] \right\} \tag{2.12}$$

determines the local, pure phase states

$$\rho_\Lambda^{\beta,\vartheta} = \exp\left[-\xi_\Lambda - \beta H_\Lambda^{r,\beta,\vartheta} \right], \tag{2.13}$$

where $\xi_\Lambda \in \mathbb{R}$ adjusts normalization. By the compatibility of the nets $(\rho_\Lambda^\beta)_{\mathcal{L}}$ and $(\rho_\Lambda^{\beta,\vartheta})_{\mathcal{L}}$ the decomposition formula in Eq. (2.11) lifts to the global states on $\mathcal{A}$:

$$
\omega^\beta = \int_0^{2\pi} \omega^{\beta,\vartheta} \, \frac{d\vartheta}{2\pi} \, . \tag{2.14}
$$

The $\Delta_k = \Delta_k(\beta)$ vanish for $0 \le \beta \le \beta_c$, but are positive for $\beta_c > \beta$, where β_c is the critical temperature of the system. The $\delta\vartheta_k$ are temperature independent functions of the complex $g_{k,k'}$ which average to zero.

In order to reveal some structural aspects of Eq. (2.11), a state limit based on a rather involved argumentation, let us introduce the needed state space concepts.

3. Concepts in State Space

Although the iterated composition of observable algebras was the guiding idea to obtain the state space $S(\mathcal{A})$ in the preceding section, we consider now S as the basic structure (and drop the symbol $\mathcal{A}$).

A "property" E of the system is now identifiable with the set of states which display it. In an operational language a "property" is a filter which the states displaying the property pass unchanged. We shall not enter into the ontological discussion of these formulations (see, however, Sect. 6), nor do we claim that all conceivable "properties" fit into the mathematical structure we are dealing with. For our examples of "properties" the filtering picture seems to be adequate. Especially one concludes from the filtering picture that E is closed under mixtures and decompositions, i.e., is a face of S. A "property" should be topologically closed, and many arguments speak for closedness in the norm topology. Then the properties are norm-detectable (Rüttimann 1979), projective (Alfsen and Shultz 1976), and spectral (Alfsen and Shultz 1976).

Being projective, E is the image of a filtering operation (P-projection) and has a well defined orthogonal complement $E^\perp$. The infimum $E_1 \wedge E_2$ of two "properties" is obtained by the set intersection. Together with $\perp$, it determines the supremum relation. With these relations the set $\mathcal{E}(S)$ of all "properties" is a complete orthomodular lattice, a quantum logic.

Special "properties" $F \in \mathcal{E}(S)$ satisfy the split-face relations:

(1) $F^\perp = \bigcup\{E \in \mathcal{E}(S); \ E \wedge F = \emptyset\}$;
(2) for each $\varphi \in S$ there are unique states $\varphi_1 \in F$ and $\varphi_2 \in F^\perp$ with

$$
\varphi = \lambda\varphi_1 + (1 - \lambda)\varphi_2, \ \lambda \in [0,1] \, .
$$

The set of all split-faces constitutes a complete Boolean sublattice $\mathcal{F}(S) \subset \mathcal{E}(S)$, the set of all "classical properties" of the system.

Theorem (3.1). *Denoting $S_\Lambda := S(\mathcal{A}_\Lambda)$, $\Lambda \in \mathcal{L}$, we have: $\mathcal{F}(S_\Lambda)$ is trivial (equal to $\{\emptyset, S_\Lambda\}$) and $\mathcal{F}(S)$ is of overcountable cardinality (for the latter statement, cf. Pedersen 1979).*

Since $\mathcal{A}$ is isomorphic to an observable algebra of an infinite, discrete spin system, we recognize that the thermodynamic limit (in the weak form of Sect. 2.3) endows even a discrete quantum system with a continuous structure of "classical properties".

Definition (3.2).

(i) For $\varphi \in S$ set $E_\varphi := \wedge \{E \in \mathcal{E}(S); E \ni \varphi\}$ and $F_\varphi := \wedge \{F \in \mathcal{F}(S); F \ni \varphi\}$. A state φ is pure, iff E_φ is an atom in $\mathcal{E}(S)$, and φ is called a factor state, if F_φ is an atom in $\mathcal{F}(S)$.

(ii) For $\varphi, \psi \in S$ write $\varphi \perp \psi$, if $E_\varphi \perp E_\psi$ (i.e., $E_\varphi \leq E_\psi^\perp$) and call then φ and ψ "orthogonal". Similarly write $\varphi \Diamond \psi$, if $F_\varphi \wedge F_\psi = \emptyset$, and call then φ and ψ "disjoint".

(iii) For two properties $E_{1,2}$ write $C(E_1, E_2)$, if $E_1 = (E_1 \wedge E_2) \vee (E_1 \wedge E_2^\perp)$, and call these properties "compatible".

(iv) For $\varphi_i \in S$, $i = 1, 2, 3$, denote $\mathcal{K}(\varphi_1, \varphi_2, \varphi_3)$, if $E_{\varphi_i} \wedge E_{\varphi_j} = \emptyset$, $E_{\varphi_i} \vee E_{\varphi_j} = E \in \mathcal{E}(S)$, $\forall 1 \leq i \neq j \leq 3$, and call these states in "coherence relation".

Since $\mathcal{E}$ is orthomodular, the compatibility relation according to Definition (3.2.iii) is symmetric. If $\mathcal{K}(\varphi_1, \varphi_2, \varphi_3)$ holds, then at least two pairs of incompatible state properties are involved. This demonstrates the non-classical nature of the coherence relation. Since $\mathcal{K}(\varphi_1, \varphi_2, \varphi_3)$ implies $F_{\varphi_1} = F_{\varphi_2} = F_{\varphi_3}$, coherence is broken for classically different states (Raggio and Rieckers 1983, Zanzinger 1993).

In the state language, a metric observable is expressed in terms of a "scaled family of properties".

Definition (3.3).

(α) A "scaled family of properties" is a set $\{E_\lambda ; \lambda \in \mathbb{R}\} \subset \mathcal{E}(S)$ with

(i) $\lambda \leq \lambda'$ implies $E_\lambda \leq E_{\lambda'}$;

(ii) $E_\lambda = \bigwedge_{\lambda' > \lambda} E_{\lambda'}$;

(iii) $\bigwedge_\lambda E_\lambda = \emptyset$, $\bigvee_\lambda E_\lambda = S$.

(β) If in (α) $\{E_\lambda ; \lambda \in \mathbb{R}\} \subset \mathcal{F}(S)$, one has a "scaled family of classical properties".

The observables in a convex state space approach are bounded affine functions $\hat{A} : S \to \mathbb{C}$ and are here identified with their unique linear extensions $\hat{A} : \mathcal{A}^* \to \mathbb{C}$, that are precisely the elements of $\mathcal{A}^{**}$. For every

$E \in \mathcal{E}(S)$ there is a unique $\hat{E} \in \mathcal{A}^{**}$ with $\hat{E}(\varphi) = 1$, iff $\varphi \in E$. (E is "detected" by $\hat{E}$.) A Mackey observable O consists of the detecting observables $\{\hat{E}_\lambda ; \lambda \in \mathbb{R}\}$ of a scaled family of properties. Only Mackey observables with bounded support ($\equiv$ closure of the points of increment) correspond to self-adjoint elements in the universal enveloping von Neumann algebra $\mathcal{M}^u$ which is Banach space isomorphic to $\mathcal{A}^{**}$.

The empirical classical observables of many-body systems, especially the thermodynamic state variables, are typically introduced in terms of a scaled family of classical properties, since it is often empirically possible to prepare and to identify those states rather directly with one and the same classical property (temperature, entropy). This does basically apply to microscopic quantum properties as well, but is mostly not expressed in this way. (In the Stern-Gerlach-experiment there would be one nontrivial face for "spin smaller than -1/2" and the whole state space for "spin smaller than 1/2".)

For $\varphi \in S$ and a Mackey observable O denote by φ_O the Borel measure on $\mathbb{R}$ which is determined by the distribution $\lambda \to \hat{E}_\lambda(\varphi)$.

Definition (3.4). *For $\varphi, \psi \in S$ the transition probability is introduced by*

$$T_S(\varphi, \psi) := \inf_O \left[\int_{\mathbb{R}} \left(\frac{d\varphi_O}{d\mu} \frac{d\psi_O}{d\mu} \right)^{1/2} d\mu \right]^2 ,$$

where O runs over all Mackey observables, and the positive Borel measure μ dominates both φ_O and ψ_O (Cantoni 1975).

Basic properties of T_S combined with new results have been elaborated by Gerisch et al. (1997a). From there we take the following illustration of disjointness.

Proposition (3.5). *Let be $\varphi, \psi \in S$.*

(i) For each $B \in \mathcal{A}$ define the perturbation $\varphi_B \in S$ of φ by

$$\langle \varphi_B ; A \rangle = \begin{cases} \langle \varphi ; B^* A B \rangle / \langle \varphi ; B^* B \rangle & \text{if } \langle \varphi ; B^* B \rangle > 0 \\ \langle \varphi ; A \rangle & \text{otherwise} \end{cases}$$

for all $A \in \mathcal{A}$. Then

$$\varphi \lozenge \psi, \text{ iff } T_S(\varphi_B, \psi) = 0, \ \forall B \in \mathcal{A}.$$

(ii) For each $\Lambda \in \mathcal{L}$ define $\mathcal{A}_\Lambda^\perp := \cup \{\mathcal{A}_{\Lambda'}^e ; \Lambda' \cap \Lambda = \emptyset\}$, where $\mathcal{A}_{\Lambda'}^e$ is the even subalgebra of $\mathcal{A}_{\Lambda'}$, consisting of elements which are invariant under gauge transformations with gauge angle π. Define $\varphi_{\Lambda^\perp}$ as the restriction of φ to $\mathcal{A}_\Lambda^\perp$. Then $\varphi \lozenge \psi$, if there exists an absorbing directed subset $\mathcal{L}' \subset \mathcal{L}$ such that $\lim_{\Lambda \in \mathcal{L}'} T(\varphi_{\Lambda^\perp}, \psi_{\Lambda^\perp}) = 0$, where the transition probabilities are defined on $S(\mathcal{A}_\Lambda^\perp)$.

While $\varphi \perp \psi$, iff $T_S(\varphi, \psi) = 0$, $\varphi \Diamond \psi$ signifies vanishing transition probabilities also for all quasi-local perturbations of the involved states. Proposition (3.5.ii) also stresses the asymptotic character of disjointness; the behavior of the states outside of each bounded Λ is relevant for this notion.

Let us recall that by means of the relations $\perp$ and $\Diamond$, respectively, the notion of "orthogonal" and "subcentral" measures on S (equipped with the w^*-topology) can be completely defined in the state language: These two types of decomposition measures of a given state are characterized by the orthogonality and disjointness, respectively, of sub-states which are obtained by integrating over nonintersecting Borel subsets with positive measures (Bratteli and Robinson 1979). For $\omega \in S$ we denote by μ_ω the finest subcentral measure with bary center ω, the so-called "central measure". Its support consists of pair-wise disjoint factor states.

The classical properties of S also provide natural *conceptual frames* to express restricted conditions for the dynamics or symmetry transformations in the state picture. Especially, one is mostly interested in state transformations only for a given temperature or energy range. For $F \in \mathcal{F}(S)$ we denote the set of (logical) symmetries by

$$\Gamma_F := \{ \nu : F \to F ; \text{ affine, bijective } \} . \tag{3.1}$$

We identify each $\nu \in \Gamma_F$ with its isometric extension to the Banach space $X_F := LH(F)$ This extension is not so popular in the state picture, but in the observable picture it corresponds completely to the extension of a transformation from the Jordan algebra of all self-adjoint elements (the expressions for genuine observables) to the corresponding C^*-algebra. A symmetry group in the state picture is then a representation of a group G by elements of Γ_F. If G is locally compact and Abelian we may employ the Arveson spectral theory to $\nu_G := \{\nu_g \in \Gamma_F ; g \in G\}$ (Arveson 1974). One knows (Pedersen 1979) that for each γ in the Arveson spectrum $Sp(\nu_G) \subset \hat{G}$ there exists a net $\{\psi_i ; i \in I\} \subset X_F$ which constitutes a generalized eigen-element by the value γ in the sense that $\lim_I \|(\nu_g - \langle \gamma|g \rangle)\psi\| = 0$, where $\langle \gamma|g \rangle$ denotes the application of the character γ to g.

Especially $\nu_g(\psi) = \langle \gamma|g \rangle \psi$, $\forall g \in G$, characterizes an eigen-element $\psi \in X_F$ by the value γ. Since $\langle \gamma|g \rangle$ is complex, for γ different from the identity, eigen-elements from X_F may be states only if they are invariants.

Since the dual Banach space $(X_F)^* =: \mathcal{M}_F$ is a von Neumann algebra (and not only a Jordan algebra), one can use its algebraic structure to construct eigen-elements of symmetry groups, both in the Heisenberg picture and in the present case of an algebraic Schrödinger picture. In the typical situation of (relativistic and non-relativistic) quantum field theory one starts with a time invariant state $\omega = \nu_t(\omega)$, $\forall t \in \mathbb{R}$. The dynamical evolutions for weak perturbations of ω are described by the restriction ν_t^ω of ν_t to F_ω, cf., e.g., Dubin and Sewell (1970). The transformations ν_t^ω are here linearly extended to $LH(F_\omega) := X_\omega$. Defining the right multiplication by $\langle R(C)\omega\,;\,A\rangle := \langle\omega\,;\,AC\rangle, \forall C, A \in (X_\omega)^* =: \mathcal{M}_\omega$, let us call $R(C)\omega$ a "direct excitation" of ω by C and the conjugate functional $\overline{R(C)\omega}$ (defined by $\langle\overline{R(C)\omega}\,;\,A\rangle := \overline{\langle R(C)\omega;\,A^*\rangle}$) a "dual excitation" of ω by $C \in \mathcal{M}_\omega$. These excitations are in X_ω and are candidates for eigen-elements of the Schrödinger dynamics.

The spectrum $Sp\,(\nu_{\mathbb{R}}^\omega)$ is always symmetric around zero, but the direct excitations may, in special cases, lead to positive spectral values only. This would identify ω as a ground state.

We first employ the Arveson spectral theory in terms of eigen-excitations in the state picture. In Sect. 5.2 we will use the more general notion of spectral subspaces.

4. Junction States and Dynamics

4.1. THE COMPOSITE MACRO-OBJECT "JUNCTION"

The electrodes Σ_a and Σ_b of a Josephson junction are separated by an insulator, but they are so close together that single electrons and pairs have a considerable tunneling probability. This leads to a normal and a superconducting tunneling current with very peculiar features (Barone and Paterno 1982). In the formalism of finite quantum·systems the electrons of Σ_a cannot be distinguished from those of Σ_b, both systems are quantum mechanically entangled.

Formally we start with the sets of momenta $\mathcal{K}_a$ and $\mathcal{K}_b$ which may be different for different superconducting materials. The set of tuples

$$\mathcal{L} := \{\, \Lambda = (\Lambda_a\,,\,\Lambda_b)\,;\, \Lambda_{a,b} \subset \mathcal{K}_{a,b}\,,\, |\Lambda_{a,b}| < \infty\,\} \qquad (4.1)$$

is directed under the partial ordering relation $\Lambda \leq \Lambda'$ which holds if $\Lambda_a \subset \Lambda_a'$ and $\Lambda_b \subset \Lambda_b'$. The local algebras $\mathcal{A}_\Lambda := \mathcal{A}_{\Lambda a} \otimes \mathcal{A}_{\Lambda b}$ describe the observables of a local composite system. A certain quantum entanglement arises, as already mentioned, by the indistinguishability of the electrons of the composite system. The latter is expressed, e.g., by the fact that a one-particle observable cannot be connected with a specific particle index.

The family of local algebras is combined into the quasilocal algebra

$$\mathcal{A} = \overline{\bigcup_{\Lambda \in \mathcal{L}} \mathcal{A}_\Lambda}^{\|\cdot\|} = \overline{\mathcal{A}_0}^{\|\cdot\|} = \mathcal{A}_a \otimes \mathcal{A}_b\,, \qquad (4.2)$$

which is again a CAR-algebra, comprising all electrons of the composite system. At this level we have classical observables, which are connected with subsystem indices a and b, as we shall see immediately.

The state space $S = S(\mathcal{A})$ of the junction considered here is – as in the case of one superconductor – affine isomorphic to the set of all compatible families of local density operators $(\rho_\Lambda)_{\mathcal{L}}$.

The uncoupled pure phase states are

$$\omega^{\beta,\vartheta} = \omega^{\beta,\vartheta_a} \otimes \omega^{\beta,\vartheta_b}, \quad \vartheta := (\vartheta_a, \vartheta_b), \tag{4.3}$$

and the limiting Gibbs state is

$$\omega^\beta = \int \omega^{\beta\vartheta} d\vartheta, \quad d\vartheta := d\vartheta_a \, d\vartheta_b / (2\pi)^2. \tag{4.4}$$

Equation (4.4) is the central decomposition (described according to Proposition 3.5) of ω^β in terms of the parameters ϑ. It is a non-trivial double integral if $\beta > \beta_{ca}, \beta_{cb}$ (what we assume in the following). If one has first calculated ω^β by means of the thermodynamic limit, the central decomposition provides the pure phase states as its unique, classically pure (i.e., factorial) components.

For fixed β we select the folium $F_\beta = F_{\omega^\beta}$ as the frame for the subsequent model investigations. For every pair of Borel sets $(B_a, B_b) \subset \mathbb{R} \times \mathbb{R}$ we form the sub-states (dominated by ω^β)

$$\omega^\beta(B_a, B_b) := \int_{B_a \times B_b} \omega^{\beta,\vartheta} \, d\vartheta \, / \, |B_a \times B_b| \tag{4.5}$$

and folia

$$F_\beta(B_a, B_b) := F_{\omega^\beta(B_a, B_b)}. \tag{4.6}$$

The family of folia in Eq. (4.6) defines the classical pair phase variables (θ_a, θ_b). Here we take the phase values in $\mathbb{R}$ (implicitly assuming the modulo-2π identification). The marginal variable θ_a is given by the scaled family of classical properties $F_\beta((-\infty, \vartheta_a], \mathbb{R})$, the phase difference $\theta_b - \theta_a$ by the family $F_\beta(\vartheta) := F_\beta((\vartheta_a, \vartheta_b) | \vartheta_b - \vartheta_a \leq \vartheta)$. This prescription leads to a unique central operator in the von Neumann algebra $[LH(F_\beta)]^* = X_\beta^* = \mathcal{M}_\beta$ which will be denoted by $\hat{\theta}_b - \hat{\theta}_a$.

We see that in the thermodynamic limit one obtains subsystem indices for the classical observables without any anti-symmetrization. Our theoretical description predicts such kinds of classical observables. This implies the empirical possibility to identify the subsystems by different values of such classical observables. In the thermodynamic limit we have a well-defined notion of (macroscopic) composite objects.

Rather than two of them, one can, in an analogous manner, form even a continuous array of macro-objects, imaging our ideas of continuous, classical, geometric shapes (related to the macroscopic position variable) for the macroscopic world. This is not in contradiction to the quantum entanglement of the microscopic particles constituting the macro-structure, as is illustrated by the above introduction of the classical phase variables.

We conclude that the thermodynamic limit over a multicomponent net of local regions (as described above) is in accordance with our permanent identification of macro-objects by localizing their classical properties.

For two types of electrons we introduce the two-component gauge angle $\delta = (\delta_a, \delta_b) \in \mathbb{R}^2$. In $\mathcal{A}_\Lambda$ one sets

$$\alpha_\delta(A) := \exp\left[i\,\delta_a\,N_\Lambda^a + i\,\delta_b\,N_\Lambda^b\right] A \exp\left[-i\,\delta_a\,N_\Lambda^a - i\,\delta_b\,N_\Lambda^b\right] . \quad (4.7)$$

By duality, $\alpha_\delta^{\beta*}$ is a unique operator in X_β, and we define $\kappa_\delta^\beta := \alpha_{-\delta}^{\beta*}$ as the gauge transformations in the Schrödinger picture. We easily obtain $(n, m \in \{0, 1\})$

$$\kappa_\vartheta^\beta R\left(c_{k\sigma}^{a*n}\,c_{k'\sigma'}^{bm}\right)\omega^\beta = \exp\left(in\,\delta_a - im\,\delta_b\right)R(\cdot)\,\omega^\beta \quad (4.8)$$

as typical eigen-elements. Introducing

$$s_{a/b}^* := \exp\left[i\,2\hat{\theta}_{a/b}\right] \in \mathcal{M}_\beta , \quad (4.9)$$

one finds (directly by duality, starting from the local Heisenberg picture)

$$\kappa_\vartheta^\beta R\left(s_a^{*n}\,s_b^m\right)\omega^\beta = \exp\left(i\,2n\,\delta_a - i\,2m\,\delta_b\right)R(\cdot)\,\omega^\beta \quad (4.10)$$

as interesting classical direct excitations of ω^β with a paired particle structure, where $n, m \in \mathbb{Z}$.

4.2. UNCOUPLED JUNCTION DYNAMICS

For our model class there exists the (context dependent) limiting dynamics

$$\langle \nu_t^\beta(\varphi) ; A\rangle := \lim_{\Lambda \in \mathcal{L}} \langle \varphi ; \tau_t^\Lambda(A)\rangle,\ \varphi \in X_\beta,\, A \in \mathcal{A}, \quad (4.11)$$

where τ_t^Λ is the local Heisenberg dynamics defined by $H_{\Lambda_a}^a + H_{\Lambda_b}^b$ (compare Eq. (2.3)). For the simultaneous diagonalization of the "mean field Hamiltonians" (compare Eq. (2.12)) with undetermined ϑ we introduce the *gauge covariant* quasi-particle operators in $\mathcal{M}_\beta$,

$$\gamma_{k0}^a = u_k^a\,c_{k\uparrow}^a - v_k^a\,e^{-i\delta\vartheta_k}\,s_a\,c_{-k\downarrow}^{ax},\ \gamma_{k1}^a = u_k^a\,c_{-k\downarrow}^a + v_k^a\,e^{-i\delta\vartheta_k}\,s_a\,c_{k\uparrow}^{a*}, \quad (4.12)$$

and form the analogous operators for Σ_b. We have used the usual abbreviations

$$E_k^a = \left((\epsilon_k^a - \mu_a)^2 + \Delta_k^{a2} \right)^{1/2} , \tag{4.13}$$

$$u_k^a = \left[\left(1 + \frac{\epsilon_k^a - \mu_a}{E_k^a} \right) / 2 \right]^{1/2} ,$$

$$v_k^a = \left[\left(1 - \frac{\epsilon_k^a - \mu_a}{E_k^a} \right) / 2 \right]^{1/2} . \tag{4.14}$$

With $\delta g_k^a := \lim_{k' \to \infty} (g_{kk'}^a - g^a)$ and Δ^a as the homogeneous gap (for the model with $\epsilon_k^a = \epsilon^a$, $g_{kk'}^a = g^a$) we obtain the (absolute values of the) inhomogeneous gaps

$$\Delta_k^a = \left| 1 + \frac{\delta g_k^a}{g^a} \right| \Delta^a . \tag{4.15}$$

One finds the correct gauge behavior for particle-like excitations ($n, m \in \{0, 1\}$)

$$\kappa_\delta^\beta R \left(\gamma_{k\sigma}^{a*n} \gamma_{k'\sigma'}^{bm} \right) \omega^\beta = \exp \left(in \, \delta_a - im \, \delta_b \right) R(\cdot) \omega^\beta . \tag{4.16}$$

The straightforward definition of translations and spin-rotations reveal that γ_{k0}^{a*} excites a momentum k and spin $\uparrow$, whereas γ_{k1}^{a*} excites a momentum $-k$ and spin $\downarrow$. The Schrödinger dynamics has the energy excitation values ($n, m \in \{0, 1\}$)

$$\nu_t^\beta R \left(\gamma_{k\sigma}^{a*n} \gamma_{k'\sigma'}^{bm} \right) \omega^\beta = \exp \left[it \left(n \, E_k^a - m \, E_{k'}^b \right) \right] R(\cdot) \omega^\beta \tag{4.17}$$

and

$$\nu_t^\beta R \left(s_a^{*n} \, s_b^m \right) \omega^\beta = \exp \left[it \left(n2\mu_a - m2\mu_b \right) \right] R(\cdot) \omega^\beta , \tag{4.18}$$

for $n, m \in \mathbb{Z}$.

In the low temperature limit $\beta \to +\infty$, Eq. (4.17) provides non-vanishing functionals for $m = 0$ only, which makes the "direct excitation spectrum" of quasi-particles positive.

The classical pairs are energetically on the Fermi surface and represent condensed Cooper pairs with a rotating, macroscopic phase angle. An interesting aspect of the energy eigenvalues E_k^a is their accumulation at the homogeneous energy $E^a = [(\epsilon^a - \mu_a)^2 + \Delta^{a2}]^{1/2}$. Gerisch and Rieckers (1997) have claimed that E^a rather than $\inf_k E_k^a$ is the measurable gap.

It is an interesting feature of this situation that, due to the quantum condensate, the limiting grand canonical equilibrium state is not equivalent to the limiting canonical equilibrium state, only the latter being a KMS-state for the non-reduced dynamics ν_t^β. (For the KMS-condition in the Schrödinger picture, cf. Roos (1980).) In the grand canonical equilibrium the fixing of the particle number average causes the macroscopic phase rotation due to Eq. (4.18).

We have used an interpretational language for the excitation functionals in X_β which fits to common usage for Hilbert space quantum field theory but requires justification. Since perturbed *states* φ_B (introduced in Proposition 3.5), where B is a non-Hermitian excitation operator as in Eqs. (4.17) and (4.18), have canceling energy values, they do not exhibit the energy spectrum. In order to find state expressions which show the energy spectrum, let us try transition probabilities and calculate, e.g.,

$$T_S \left(\omega^\beta, \omega^\beta_{\gamma^{a*}_{k\sigma}} \right) = \left(1 + e^{\beta E^a_k} \right)^{-1} . \qquad (4.19)$$

In this formulation the thermal occupation probabilities of the Fermi distribution arise from the transition probabilities relative to the equilibrium state and have no longer an absolute character. Observe that E^a_k itself depends on the temperature T. One finds that Eq. (4.19) is not differentiable in T at the critical point.

Concerning the condensed Cooper pair spectrum the formula analogous to Eq. (4.19) gives a trivial result since $\omega^\beta_{s^*_a} = \omega^\beta$. Thus for *states* the contents of condensed pairs seem irrelevant in usual, nonextended quantum field theory. One should compare this with the wave functions incorporating an arbitrary amount of condensed particles in Landau and Lifshitz (1980).

4.3. WEAK COUPLING DYNAMICS

According to basic principles of quantum mechanical perturbation theory, weak coupling is to be expressed in terms of the eigen-excitations of the uncoupled system. Thus for the weak coupling of Σ_a and Σ_b one typically looks for polynomials in the excitation operators used in Eqs. (4.17) and (4.18). The more interesting part deals with the condensed Cooper pairs and has the leading expression

$$h_s = \lambda_s \left(s^*_a s_b + s_a s^*_b \right), \quad \lambda_s = g_s \frac{\Delta^a}{g^a} \frac{\Delta^b}{g^b} , \qquad (4.20)$$

where g_s depends on the one-pair tunneling frequency. Defining

$$Z = \exp\left[-\beta h_s/2\right] , \qquad (4.21)$$

the equilibrium state with condensed pair tunneling is ω^β_Z.

The transition probability between the uncoupled and the weakly perturbed equilibrium state is (Gerisch et al. 1997b)

$$T_S \left(\omega^\beta, \omega_Z^\beta\right) = \frac{I_0 \left(\beta \lambda_s\right)^2}{I_0 \left(2\beta \lambda_s\right)} , \tag{4.22}$$

with I_0 as the modified Bessel function of first kind and zero order. The purely classical transition probability in Eq. (4.22) vanishes for $\beta \to +\infty$ and indicates in this way disjointness of ω^β and ω_Z^β. This means that only for $\beta < +\infty$ the tunneling of condensed pairs is a weak perturbation in the precise sense that ω_Z^β generates the same folium F_β as ω^β: the classical properties are the same for both states. Also a possibly unbounded expression of γ-fields as a perturbation would not change the equilibrium folium at finite β.

We shall stick henceforth to the simple classical perturbation due to Eq. (4.20). The perturbed Schrödinger dynamics in F_β is then identical with ν_t^β. For

$$C := \frac{Z^2}{\langle \omega^\beta; Z^2 \rangle} = \exp\left[-\xi_s - \beta h_s\right]$$

(as right-perturbation), we find the equilibrium time evolution

$$\nu_t^\beta(\omega_Z^\beta) = \nu_t^\beta R(C) \omega^\beta = R(C_t) \omega^\beta , \tag{4.23}$$

where

$$\begin{aligned} C_t &= \exp\left[-\xi_s - 2\beta \lambda_s \cos\left(\hat{\theta}_a(t) - \hat{\theta}_b(t)\right)\right] \\ \hat{\theta}_a(t) &= \hat{\theta}_a + 2\mu_a t . \end{aligned} \tag{4.24}$$

Thus, the perturbed equilibrium state depends on t, iff $\mu_a - \mu_b \neq 0$. Then the exchange of condensed pairs is submitted to the rotation of the macroscopic phase difference.

By the physical meaning of the pair-field operators, the tunneling supercurrent from Σ_a to Σ_b should have the shape

$$\begin{aligned} I_s(t) &= \rho \lambda_s \left[s_a(t) s_b^*(t) - s_a^*(t) s_b(t)\right] \\ &= 2\rho \lambda_s \sin\left(\hat{\theta}_b(t) - \hat{\theta}_a(t)\right) . \end{aligned} \tag{4.25}$$

The factor ρ depends on the way $I_s(t)$ is theoretically derived from the time dependent cosine-potential in the exponent of C_t. In most cases, a time-differentiation is applied, and then ρ is proportional to the voltage $\mu_b - \mu_a$. In the method of Rieckers and Ullrich (1986), a differentiation with respect to the gauge angle is involved, and ρ does not contain $\mu_b - \mu_a$. Only in the latter case does $I_s(t)$ describe both the dc- and ac-Josephson supercurrents.

5. SQUID-States and Dynamics

5.1. DC-SQUID

In the dc-SQUID two Josephson junctions are connected in parallel by superconducting wires. The magnetic flux threading through the loop of the wires and through the two junctions influences the two macroscopic phase differences. The additive combination of the two (numerical) supercurrents gives an interference pattern, which depends very sensitively on the flux. The suggestion of Mercereau (1969) that this kind of interference constitutes quantum interference on the macroscopic scale has been refuted by Leggett (1980).

The direct explication of Mercereau's idea would be to view the dc-SQUID as a constructive realization of a superposition for two states of one Josephson junction (Rieckers 1990). If these states have a rather sharp phase difference, the expectation of $I_s(t)$ in this superposition imitates in fact the usual calculation for the interference current. The superposition is, however, not quantum coherent, since the two states differ in their classical macro-observable "phase-difference" (cf. the discussion after Definition 3.2).

This is true as long as we cling to the state space $S = S(\mathcal{A})$. For a more complete theoretical treatment we shall refine the state space later on.

5.2. SPECTRAL SUBSPACES AND PROPERTIES

The Arveson (1974) spectral theory is based on the notion of a spectral subspace and is partially influenced by the theory of invariant subspaces (cf., e.g., Helson 1964). It is just this kind of a spectral theory, which is suited to find the connection with the lattice of "properties".

Let us indicate for shortness by ν the Schrödinger dynamics of Sect. 4.3 in the Banach space $X_\beta \equiv X$ with dual $\mathcal{M}_\beta \equiv \mathcal{M}$. The general statements of the subsequent considerations remain true if we replace ν by κ, the symbol for the gauge transformations in X.

Let ν_f be the integration of the dynamics over $f \in L^1(\mathbb{R})$. Then for $\varphi \in X$ the local spectrum is defined by

$$Sp(\varphi) := \left\{ E \in \mathbb{R} ; \ \nu_f(\varphi) = 0 \text{ implies } \hat{f}(E) = 0, \ \forall f \in L^1(\mathbb{R}) \right\} . \quad (5.1)$$

The Arveson spectrum $Sp(\nu)$ is then introduced as

$$Sp(\nu) := \text{closure} \bigcup \left\{ Sp(\varphi) ; \varphi \in X \right\} . \quad (5.2)$$

For a Borel set $B \subset \mathbb{R}$, we have

$$X^\nu(B) := \overline{\left\{ \varphi \in X ; Sp(\varphi) \subset B \right\}}^w . \quad (5.3)$$

The spectral subspace $X^\nu(B)$ is ν-invariant and norm closed (since $X^\nu(B)$ is weakly closed by definition).

According to Prosser (1963) the linear subspace $X^\nu(B)$ is associated with a "property" $E^\nu(B)$ in F_β, iff it is left-invariant (i.e., $L(\mathcal{M})X^\nu(B) \subset X^\nu(B)$). In this case,

$$E^\nu(B) = X^\nu(B) \cap F_\beta , \qquad (5.4)$$

and reversely one has the two expressions:

$$X^\nu(B) = L(\mathcal{M}) E^\nu(B) = \left\{ R(\hat{E}^\nu(B))\varphi ; \varphi \in X \right\} . \qquad (5.5)$$

We observe that $E^\nu(B)$ is nonempty only if B contains zero, the only possible spectral value of states. We take this as an indication that the Schrödinger spectral values are only interpretable relative to given equilibrium states.

Take, e.g., for $E^\nu(B)$ in Eq. (5.5) a property E_ω induced by $\omega \in F_\beta$. If ω is pure, then $E_\omega = \{\omega\}$. $Sp(\omega) = \{0\}$, iff ω is ν-invariant. The corresponding left-invariant subspace contains all left-excitations of ω, including the eigen-excitations, some of them being zero. If, on the other extreme, ω is faithful, then $E_\omega = F_\omega = F_\beta$ and the left-invariant subspace is equal to X, representing the whole Arveson spectrum $Sp(\nu)$. In the latter case, local right-excitations of ω may also have interesting spectral properties.

Typical pure equilibrium states are ground states, and only in this case there is, by the Arveson-Borchers theorem, a general scheme to obtain the energy as a scaled family of properties starting from the spectral subspaces which one has in any case (Bratteli and Robinson 1979).

In our junction model of Sect. 4.3 we do not have a positive energy or particle number, even not for $\beta \to +\infty$. This is prevented already by the spectral values of the condensed pair excitations.

We focus our interest just on these collective excitations and study their spectral subspaces, especially for the gauge transformations. For $M \subset \mathbb{Z}^2$ and $n = (n_a, n_b) \in \mathbb{Z}^2$ we introduce

$$X_s^\kappa(M) = LH\left\{ R(s_a^{*n_a} s_b^{*n_b})\omega^\beta ; n \in M \right\}^{-w} \qquad (5.6)$$

according to the Arveson spectral theory. We consider the filtering of the n-values as empirically possible. In the hitherto developed formalism there exists no corresponding property, since

$$X_s^\kappa(M) \cap F_\beta = \{\omega^\beta\} . \qquad (5.7)$$

Because ω^β is mixed, Eq. (5.7) does not describe a face. Besides this, the information on condensed pairs has disappeared completely in Eq. (5.7).

5.3. REFINEMENT OF PROPERTIES

Denoting $X_s := X_s^\kappa(\mathbb{Z}^2)$, one can show that $X = X_\gamma \otimes X_s$ is a tensor product of Banach spaces. Here X_γ is generated by the left-right-γ-excitations of ω^β. (This structure can also be derived from X as the predual of a factorizing von Neumann algebra $\mathcal{M} = \mathcal{M}_\gamma \otimes \mathcal{M}_s$.)

Since X_s with its spectral subspaces is already contained in our theoretical frame, we get a refined description when we consider the filtering of the spectral values n – by adding (or subtracting, respectively) n condensed pairs to (from) ω^β – as preparing a new state ω_n^β. This new state preparation splits

$$\omega^\beta \text{ into the set } \left\{\omega_n^\beta ; n \in \mathbb{Z}^2\right\} \qquad (5.8)$$

and thus extends X_s to X_s^e, the structure of which has to be determined. Within X_s^e the singleton $\{\omega_n^\beta\}$ is viewed as a new "property". Belonging to pairwise different spectral values, the $\{\omega_n^\beta\}$ should be pairwise orthogonal "properties". In the extension X_s^e of X_s the left-right-excitations of all ω_n^β should be contained . Furtheron the positive, normalized part of X_s^e should be a state space S_s^e of a C^*-algebra and thus should exhibit the three-ball property and orientability of Alfsen and Shultz (1976). This leads then to the conclusion

$$X_s^e \text{ is isomorphic to } \mathcal{T}(\mathcal{H}_s) , \qquad (5.9)$$

the trace class operators of the Hilbert space $\mathcal{H}_s = L^2(T^2)$, with $T = \mathbb{R}/2\pi$. Since the trace class operators belong to traditional quantum theory, the n-filtering has converted the excitation space of a classical, macroscopic theory into that of an irreducible type I von Neumann algebraic quantum theory. In the extended X_s^e all quantum features deserve the attribute "macroscopic".

In the same way as the $\omega^{\beta,\vartheta}$ of Eq. (4.3) generate the phase variables (θ_a, θ_b), the ω_n^β generate the number observables (N_a^s, N_b^s) of the condensate as a pair of scaled families of "properties" in X_s^e. In the $L^2(T^2)$-representation space the number observables lead to the unbounded operators $(\hat{N}_a^s, \hat{N}_b^s) = (\frac{\partial}{i\partial\vartheta_a}, \frac{\partial}{i\partial\vartheta_b})$ and are canonical conjugates of $(\hat{\theta}_a, \hat{\theta}_b)$.

The differential operators for the particle numbers appear frequently in the literature about the Josephson contact, often without much argumentation. It is of interpretational importance that they can be derived by an extension procedure of a microscopic particle number operator (Rieckers and Ullrich 1985). Our indicated argumentation refers to macroscopic manipulations only and stresses the macroscopic character of the condensate particle numbers (N_a^s, N_b^s).

The filtering of the γ-particle number extends X_γ to X_γ^e. Empirically the quasi-particles constitute the normal component of the tunneling current which identifies in principle the γ-character of counted electrons.

Energy values provide again a new kind of state preparations for the two types of particles. The superconducting energies $(2\mu_a N_a, 2\mu_b N_b)$ do not lead to a refinement of X_s^e, but the quasi-particle energies require a new X_γ^e, the nature of which will not be investigated here.

The refined state space S^e has a new facial structure $\mathcal{E}(S^e)$ with trivial classical part $\mathcal{F}(S^e) = \{\emptyset, S^e\}$. It decomposes according to

$$S^e = S_\gamma^e \otimes S_s^e . \tag{5.10}$$

The states in S_s^e are the states of a macroscopic quantum theory with unrestricted superposition principle (belonging to an irreducible type I von Neumann algebra). Here one has coherent superpositions of states with different macroscopic phases, supporting Mercereau's point of view. Leggett's point is, however, also justified since the quantum features of such a state superposition should be more manifest.

5.4. MACROSCOPIC TRANSITION PROBABILITIES

The extended state space has more non-trivial transition probabilities than the original state space. Let us calculate transition rates for the tunneling of condensed pairs. We start from the uncoupled equilibrium state ω^β which still constitutes the frame of reference for our excitations and perturbations. Defining $s(l) := s_a^{*l} s_b^l, l \in \mathbb{Z}$, we describe the result of l transitions of condensed pairs from Σ_b to Σ_a by $\omega_{s(l)}^\beta$. Denoting by $\hat{\theta} := \hat{\theta}_b - \hat{\theta}_a$ the macroscopic phase difference, a phase filtering or perturbation, respectively, leads to the state ω_f^β where $f = f(\hat{\theta})$. Because of the macroscopic phase rotation, $\nu_t(\omega_f^\beta)$ depends non-trivially on time. These states are in S_s^e (for all t), and they are perturbations of an extreme state in S_s^e. Thus, they are extreme, too, constituting a decisive effect of the extension procedure. Their transition probabilities may be calculated in terms of vector representatives according to traditional quantum theory (Gerisch et al. 1997b), providing, e.g., with $\mu = \mu_b - \mu_a \neq 0$,

$$T_{S^e}\left(\omega_{s(l)}^\beta, \nu_t \omega_f^\beta\right) = \left| \sum_{k \in \mathbb{Z}} (-\mathrm{i})^k f_k \, \mathrm{e}^{\mathrm{i} k \mu t} \, J_{l-k}\left(2\lambda_s \sin(\mu t)/\mu\right) \right|^2 , \tag{5.11}$$

where f_k are the Fourier coefficients of f and J_{l-k} are the Bessel functions of first kind and order $l - k$. Expansion to first order in λ_s and specializing to $f = 1$ produces

$$T_{S^e}\left(\omega_{s(l)}^\beta, \nu_t\,\omega^\beta\right) \simeq \frac{1}{(l!)^2}\,(\lambda_s\,\sin(\mu t)/\mu)^{2l}\,. \tag{5.12}$$

We calculate therefrom the transition rate for $l = 1$:

$$\frac{d}{dt}\,T_{S^e}\left(\omega_{s(l)}^\beta, \nu_t\,\omega^\beta\right) \approx \lambda_s^2\,\sin(2(\mu_b - \mu_a)\,t)\,. \tag{5.13}$$

We obtain thus an ac-Josephson effect in terms of the algebraic transition probabilities of the extended theory, but there is no non-trivial stationary transition rate for $\mu_b - \mu_a = 0$.

5.5. MACROSCOPIC TIME AND ENERGY

The rotation

$$\hat{\theta}(t) = \hat{\theta}(0) + 2(\mu_b - \mu_a)\,t \tag{5.14}$$

of the macroscopic phase difference may be interpreted as a clock which is synchronized with our original time parameter. In the non-extended formalism this clock is purely macroscopic and classical. Its frequency can be measured by the Josephson microwave radiation and is used to define a voltage standard. Used as a clock the voltage $\mu_b - \mu_a$ must be given and kept fixed. In the extended formalism of Sect. 5.3, the dynamics according to Eq. (5.14) is unchanged, but the corresponding electrostatic energy generating the phase dynamics

$$H_{el} = (\mu_b - \mu_a)\left(\hat{N}_b^s - \hat{N}_a^s\right) \tag{5.15}$$

is now an observable which is incompatible with the phase measurements in order to read the macroscopic clock.

If one takes into account the capacitance C of the junction, one has the additional energy

$$H_C = \frac{1}{2C}\left(2e\left(\hat{N}_b^s - \hat{N}_a^s\right)\right)^2, \tag{5.16}$$

which produces a non-c-number term in the commutator with $\hat{\theta}$. Thus we have now

$$\hat{\theta}(t) = \hat{\theta}(0) + 2(\mu_b - \mu_a)\,t + B\left(\hat{N}_b^s - \hat{N}_a^s\right)t\,, \tag{5.17}$$

where B is a positive c-number. Then

$$\left[\hat{\theta}(t_1), \hat{\theta}(t_2)\right] = iB(t_2 - t_1);\quad t_1, t_2 \in \mathbb{R}\,. \tag{5.18}$$

Here the readings of the clock at different times are incompatible with each other. There is a discussion in the literature, how multitime correlation functions of the phase observable may be experimentally determined (Leggett and Garg 1985, Tesche 1990). In any case we have here a simple illustration (in not so simple a context) of various aspects of time operators (Atmanspacher and Amann 1998).

5.6. THE RING-SQUID

The ring-SQUID consists of a superconducting loop with one Josephson junction inserted. The new variable is the magnetic flux Φ through the loop which is quantized into integer multiples of the flux quantum $\Phi_0 = h/2_e$. This flux quantization as well as the relation

$$\theta = 2\pi n - 2\pi \frac{\Phi}{\Phi_0} \tag{5.19}$$

originate from a macroscopic, spatially inhomogeneous theory for the condensate which is outside the scope of a microscopic, quantum mechanical treatment at present. By Eq. (5.19) the total flux is incorporated into our spatially homogeneous, macroscopic quantum theory. If one applies an external flux Φ_x, which is a c-number expressing no back-action of the system, this induces a supercurrent I_s such that

$$\Phi = \Phi_x + LI_s \tag{5.20}$$

is screened to a quantized value (L is the self-inductance of the SQUID). The corresponding energy is then

$$H_L = \frac{I^2 L}{2} = \frac{1}{2L} \left(\Phi - \Phi_x\right)^2 . \tag{5.21}$$

Substituting $\hat\theta$ by Φ in h_s of Eq. (4.20) and combining this energy with Eqs. (5.21) and (5.16), we arrive at the Hamiltonian

$$H_{SQ} = \frac{1}{2M} \left(\frac{\hbar \partial}{i \partial \Phi}\right)^2 - \frac{\hbar I_1}{2e} \cos\left(2\pi \frac{\Phi}{\Phi_0}\right) + \frac{1}{2L} \left(\Phi - \Phi_x\right)^2 , \tag{5.22}$$

where appropriate constants have been introduced. The value $\Phi_x = \Phi_0/2$ shifts the minimum of the parabola to a maximum of the negative cosine and produces an approximate double-well potential. The wells can be made shallow with two equal minima at $\Phi = \pm\Phi_0$. Thus we have arrived at the exact analog of a one-particle Schrödinger equation for the condensate dynamics.

6. Conclusions and Interpretational Remarks

The preceding model discussions exemplify once more that standard methods of algebraic many-body physics allow for an appropriate introduction of macroscopic classical observables. Emphasizing the state picture with its complete orthomodular lattice of "properties" (norm closed faces), we have stressed in Sect. 3 that the basic concepts of "classical properties", "disjointness" (non-overlapping classical properties), and "central decomposition" (filtering a state into its components with pure classical properties) may be formulated entirely in terms of the convex state space approach.

The central decomposition of the superconducting equilibrium states of one BCS-model introduces the macroscopic phase variable, the different values of which indicate disjoint pure-phase states. In our inhomogeneous generalized BCS-models this decomposition parameter is the average of the microscopic relative phases between the destruction and creation parts of the quasi-particle annihilation operators in Eq. (4.12). Whereas for the present models only indirect arguments lead to the existence of and analytical expression for the limiting Gibbs states, a direct local approximation of their central decompositions has been carried out by Fleig (1983) for the homogeneous pair approximation.

This tedious analysis anticipated the large deviation method for non-commutative models (Werner 1992) not only for the free energy but even for its tangent functionals (equilibrium states). It expresses "decoherence" in a most detailed manner using only internal features of the system. The limiting distribution singles out the leading term in the spectral decompositions of the local grand canonical density matrices, which is degenerate below the critical temperature. The integration over the macroscopic phase arises from the conservation of pairs at infinity which are the condensed Cooper pairs with no local degrees of freedom. In spite of the proved convergence for the net of local equilibrium states, the decisive features of the macroscopic phase ordering effect are seen only after the thermodynamic limit has been performed. The conceptual emergence of the collective hierarchical level is in this sense discontinuous.

For the composite macro-object "Josephson junction" the macroscopic phase difference is introduced as a scaled family of classical properties in the state picture. The operators in the representation von Neumann algebra over the uncoupled equilibrium state are mainly used for a concise formulation of the state perturbations which are obtained as elements in the linear hull of the state space. In this frame we were able to elaborate the Arveson spectral theory in the Schrödinger picture. For finite temperatures the spectral subspaces for the energy and particle number values are associated with "properties" only in a refined state space.

The central purpose of our present discussion has been the analysis of the new observables in the refined theory which are given by the now existing energy and particle number "properties". As is indicated by the two types of their eigenvalues in Eqs. (4.10) and (4.16–4.18), these observables consist of local and collective parts. We have concentrated our discussion on the collective parts. These constitute another class of macroscopic quantities, robust against local perturbations and compatible with all quasilocal observables, but incompatible with some of the macroscopic classical observables of the original theory. Especially, the energy and the particle number differences of the condensate do not commute with the macroscopic phase difference (and not with the macroscopic time measured by the macroscopic phase difference).

The incorporation of these quantities as genuine observables leads to a context dependent quantum theory for which the usual representation (especially temperature) dependent many-body theory and the local traditional quantum theories are subtheories in the sense of an isomorphic embedding of their observable algebras. On the other hand, the extended quantum theory factorizes into the extended quasiparticle quantum theory and the extended condensate quantum theory as new kinds of subtheories, as is indicated by Eq. (5.10).

In the latter subtheory the ring-SQUID dynamics finds its systematic location (after the macroscopic phase difference has been substituted by the flux variable). Observe that in the ring-SQUID Hamiltonian (Eq. (5.22)) the unbounded capacitance term is a "weak perturbation" insofar as it does not alter the context given by the uncoupled equilibrium state of the junction, but a "strong perturbation" insofar as it drives the macroscopic flux without reference to the BCS-dynamics. We have then two groups of macroscopic quantities referring to the condensate: number, charge, and voltage on one side, and phase, flux, and current on the other. What does it mean that the two groups of theoretical expressions for observables are non-commutative?

Leggett and Garg (1985) discussed the noninvasive measurability of such macroscopic observables in connection with the problem of "macrorealism". Despite some unprecise formulations (Hofmann and Rieckers 1996), and in spite of dealing with a two-dimensional caricature of the SQUID only, their article has stimulated many interesting theoretical and experimental investigations (Ballentine 1987, Peres 1988, Tesche 1990).

Let us emphasize that within our more complete formalism, which comprises both microscopic and macroscopic degrees of freedom of the SQUID-models, the analysis of measurement processes attains a new quality. The theory predicts two basically distinct types of measurement procedures for each of the mentioned observables. For example, there is a phase dependent quasi-particle dynamics (in the pure phase representation) without back-action on the macroscopic phase. This is conceptually at the same level as neutron scattering by the SQUID-flux without disturbing the flux (Leggett and Garg 1985). Such a possibility of a noninvasive measurement appears in our comprehensive frame even as a built-in feature of a macroscopic phase, since the latter is obtained as an average over fluctuating phases: a microscopic perturbation may influence the fluctuating phases, but not their average. An analogous effect is caused by the macroscopic charge or voltage, respectively, at the capacitance of the SQUID, which deforms the trajectory of a microscopic charged particle without being disturbed.

Using microscopic test particles, we thus have noninvasive, and hence compatible, measurement procedures for the macroscopic charge and flux. (Note that Tesche (1990) gives the flux of our system explicitly the attribute "macroscopic".) In the operator expression for the charge of the condensate the differentiation to the flux indicates, however, a second type of charge measurement process, which is connected with a change of the flux. In our more comprehensive formalism we have strong arguments that this differential operator possesses – in sharp contrast to the Liouville operators in classical statistical mechanics – the status of an observable, since it is on the same footing as the charge for γ-particles .

If one describes the ring-SQUID in terms of a macroscopic Schrödinger equation governed by the Hamiltonian in Eq. (5.22), one is dealing with one macroscopic degree of freedom only. For the discussion of the measurement process one should, however, not forget about the microscopic degrees of freedom.

Our first main conclusion is that the mathematically (in terms of non-distributivity or non-commutativity, respectively) expressed incompatibility of two "properties" does in principle not exclude their noninvasive, simultaneous measurability (a well-defined connection with the empirical level presupposed). This is demonstrated by our examples of macroscopic quantum "properties".

Since a noninvasive measurement does not alter the physical status of the measured "property" (given by the physical influence on other systems), we consider this alone as a sufficient justification not to change the inter-pretation of the ontological status of the "property" depending on whether one interprets the situation before or after the measurement. Thus, for flux as our macroscopic quantum observable it seems not consistent to ap-ply Margenau's formulation (Margenau 1950) that the flux is "actualized" only by the performance of the measurement (cf. the title of Leggett and Garg 1985). It is also not adequate to use Bohr's "uncontrollable influence of the measurement process" on the pre-existing values of the observables (Bohr 1927, Murdoch 1987).

We would suggest that the concept of "macro-realism" should by based on the possibility of noninvasive measurements alone, in contradistinction to Leggett and Garg (1985), who also require "definite values" (our interpretation of the unclear formulation (A1) in their paper). That a certain range of observable values – mathematized as a spectral subspace – is realized may be a realistic macroscopic "property". Within this frame of "macro-realism" one may conceive of (at least) two empirical interpretations of the mathematical incompatibility of macroscopic "properties", which should not be confused with each other.

First, one could imagine that the conjugate observables have an intrinsic fuzziness in all realizable states and that their ranges of fuzziness are inversely interrelated like position and frequency for a macroscopic wave-packet. Let us appeal to the macroscopic classical electromagnetic wave-packet which has no sharp position nor frequency but nevertheless *has* objective localization and frequency properties with well-determined effects on test particles (to be used for noninvasive measurement procedures).

Second, one could form the idea that the mathematical incompatibility refers to certain incompatible measurement procedures only and does not exclude the simultaneous realization of distinct values. This would mean here that those measurements which are directly related to the incompatible operator expressions, i.e., the macroscopic parts, are mutually exclusive. This means that one operates at the level of the SQUIDs and neglects the microscopic test particles. These kinds of measurement processes are discussed in the literature in great detail (Tesche 1990) and whole arrays of several SQUIDs are put forward, which are interconnected by mutual inductive influences. As has been emphasized by Peres (1988), the incompatibility of measurement procedures is to be discussed quite generally in classical terms, e.g., by means of classical electrodynamics. That would explain quantum features by the (here artificial) restriction of the measurement devices to the (endo-) world of SQUIDs.

Both interpretations comply with the conclusion that in the universe of SQUIDs the quantum laws are valid. For their experimental confirmation this world must be completely decoupled from other influences. Though experimentalists try hard, they cannot entirely avoid any dissipative noise in SQUIDs. It seems, however, feasible to reduce the thermal noise so much – and to meet other experimental and fabricational requirements – as to observe the coherent oscillations of the flux states between localized configurations in the mentioned potential minima of Section 5.6. In between the localized configurations, quantum dynamics predicts delocalized wave functions. This would prove macroscopic quantum coherence (which is not to be confused with quantum optical coherence) as a typical nonclassical feature of the SQUID-world.

Macroscopic quantum coherence would give a bias for the fuzzy variable picture, which in terms of supercurrents and charges – by definition parts of a macroscopic collective phenomenon – is not at all easy to grasp. Let us repeat that in any case one can maintain macro-realism in the version indicated above, solely based on the possibility of noninvasive measurement procedures.

Having developed an – in our sense – realistic interpretation of the *macroscopic* quantum world of SQUIDs we wonder whether this does not also provide arguments for an ontic interpretation of the *microscopic* quantum domain Professor Primas has been advocating for so long.

References

Alfsen E.M. and Shultz F.W. (1976): Noncommutative spectral theory for affine function spaces on convex sets. *Mem. Am. Math. Soc.* **172**, 1–120.

Arveson W. (1974): On groups of automorphisms of operator algebras. *J. Funct. Analysis* **15**, 217–243.

Atmanspacher H. and Amann A. (1998): Positive-operator-valued measures and projection-valued measures of non-commutative time operators. *Int. J. Theor. Phys.* **37**, 629–650.

Audretsch J. and Mainzer K., eds. (1990): *Wieviele Leben hat Schrödingers Katze?* (BI Wissenschaftsverlag, Mannheim).

Ballentine L.E. (1987): Realism and quantum flux tunneling. *Phys. Rev. Lett.* **59**, 1493–1495.

Bardeen J., Cooper L.N., and Schrieffer J.R. (1957): Theory of superconductivity. *Phys. Rev.* **108**, 1175–1204.

Barone A. and Paterno G. (1982): *Physics and Application of the Josephson Effect* (Wiley, New York).

Bohr N. (1927): The quantum postulate and the recent development of atomic theory. In *Atti del Congresso dei Fisici, Como, 1927* (Zanichelli, Bologna), Vol. 2, pp. 565–588. See also N. Bohr: *Atomic Theory and the Description of Nature* (Cambridge University Press, Cambridge, 1934), pp. 52–91.

Bohr N. (1935): Can quantum-mechanical description of physical reality be considered complete? *Phys. Rev.* **48**, 696–702.

Bratteli O. and Robinson D.W. (1979): *Operator Algebras and Quantum Statistical Mechanics I* (Springer, Berlin).

Bratteli O. and Robinson D.W. (1981): *Operator Algebras and Quantum Statistical Mechanics II* (Springer, Berlin).

Cantoni V. (1975): Generalized transition probability. *Commun. Math. Phys.* **44**, 125–128.

Dubin D.A. and Sewell G. (1970): Time-translations in the algebraic formulation of statistical mechanics. *J. Math. Phys.* **11**, 2990–2998.

Fleig W. (1983): On the symmetry breaking mechanism of the strong coupling BCS-model. *Acta Phys. Austriaca* **55**, 135–153.

Gerisch Th. and Rieckers A. (1997): Limiting dynamics, KMS-states, and macroscopic phase angle for weakly inhomogeneous BCS-models. *Helv. Phys. Acta* **70**, 727–750.

Gerisch Th., Rieckers A., and Zanzinger S. (1997a): Operator algebraic transition probabilities and disjointness. Preprint, Tübingen.

Gerisch Th., Rieckers A., and Zanzinger S. (1997b): Algebraic transition probabilities for superconductors and the Josephson junction. Preprint, Tübingen.

Helson H. (1964): *Lectures on Invariant Subspaces* (Academic, London).

Hofmann F. and Rieckers A. (1997): Phase dynamics at the SQUID and macro-realism. *Int. J. Theor. Phys.* **37**, 537–543.

Landau L.D. and Lifshitz E.M. (1980): *Lehrbuch der Theoretischen Physik IX.* (Akademie, Berlin), Sect. 26.

Leggett A.J. (1980): Macroscopic quantum systems and the quantum theory of measurement. *Prog. Theor. Phys. Suppl.* **69**, 80–100.

Leggett A.J. and Garg A. (1985): Quantum mechanics versus macroscopic realism: is the flux there when nobody looks? *Phys. Rev. Lett.* **54**, 857–860.

Margenau H. (1950): *The Nature of Physical Reality* (McGraw-Hill, New York).

Mercereau J.E. (1969): Macroscopic quantum phenomena. In *Superconductivity I,II*, ed. by R.D. Parks (Dekker, New York), pp. 393–421.

Murdoch D. (1987): *Niels Bohr's Philosophy of Physics* (Cambridge University Press, Cambridge).

Pedersen G.K. (1979): *C*-Algebras and Their Automorphism Groups* (Academic, London).

Peres A. (1988): Quantum limitations on measurement of the magnetic flux. *Phys. Rev. Lett.* **61**, 2019–2021.

Primas H. (1981): *Chemistry, Quantum Mechanics, and Reductionism* (Springer, Berlin).

Primas H. (1987): Contextual quantum objects and their ontic interpretation. In *Symposium on the Foundations of Modern Physics 1987*, ed. by P. Lathi and P. Mittelstaedt (World Scientific, Singapore), pp. 251–275.

Primas H. (1990a): Zur Quantenmechanik makroskopischer Systeme. In *Wieviele Leben hat Schrödingers Katze?*, ed. by J. Audretsch and K. Mainzer (BI Wissenschaftsverlag, Mannheim), pp. 209–243.

Primas H. (1990b): Realistic interpretation of the quantum theory for individual objects. *La Nuova Critica* **13-14**, pp. 41–72.

Prosser R.T. (1963): On the ideal structure of operator algebras. *Mem. Amer. Math. Soc.* **45**, 1–28.

Raggio G. and Rieckers A. (1983): Coherence and incompatibility in W*-algebraic quantum theory. *Int. J. Theor. Phys.* **22**, 267–291.

Rieckers A. (1990): Condensed Cooper pairs and macroscopic quantum phenomena. In *Large Scale Molecular Systems: Quantum and Stochastic Aspects*, ed. by W. Gans, A. Blumen, and A. Amann (Plenum, New York), pp. 33–76.

Rieckers A. and Ullrich M. (1985): Extended gauge transformations and the physical dynamics in a finite temperature BCS-model. *Acta Phys. Austriaca* **56**, 131–152.

Rieckers A. and Ullrich M. (1986): On the microscopic derivation of the finite-temperature Josephson relations in operator form. *J. Math. Phys.* **27**, 1082–1092.

Roos H. (1980): KMS-condition in a Schrödinger picture of the dynamics. *Physica A* **100**, 183–195.

Rüttimann G.T. (1979): Detectable properties and spectral quantum logics. In *Interpretation and Foundation of Quantum Theory*, ed. by H. Neumann (BI Wissenschaftsverlag, Mannheim), pp. 35–47.

Schrödinger E. (1935): Die gegenwärtige Situation in der Quantenmechanik. *Naturwissenschaften* **23**, 807–812, 823–828, 844–849.

Tesche C.D. (1990): Can a noninvasive measurement of magnetic flux be performed with superconducting circuits? *Phys. Rev. Lett.* **64**, 2358–2361.

Werner R. (1992): Large deviations and mean field quantum systems. In *Quantum Probability and Related Topics VII*, ed. by L. Accardi (World Scientific, Singapore), pp. 349–381.

Zanzinger S. (1993): Coherent superposability of states. In *Symposium on the Foundations of Modern Physics 1993*, ed. by P.J. Lahti, P. Busch, and P. Mittelstaedt (World Scientific, Singapore) pp. 450–457.

CONTEXTUAL BACKGROUND

Karl Gustafson:

THE GEOMETRY OF QUANTUM PROBABILITIES

When the experiments by Aspect et al. (1982) demonstrated beyond any reasonable doubt that Bell's (1964) inequalities are violated by the behavior of certain physical systems, a number of scientists immediately realized the breakthrough in our understanding of quantum mechanics that this result implied. The amazing point was, of course, not so much that the predictions of quantum theory were correct – it was the consequence that the traditional and unquestioned assumptions of generations of scientists about the principle of locality in the physical world were shattered.

Subsequently, many attempts were made to understand this "breakdown of locality". It turned out that different concepts of locality have to be distinguished, that locality has to be contrasted with separability, and far-reaching implications have been discussed for the notion of physical reality. The catchwords of "quantum nonlocality" or "quantum holism" became fashionable (and, no wonder, misused by many fascinated outsiders and popularizers). An informative volume containing different ways to explore and interpret this revolutionary development in contemporary physics has been edited by Selleri (1988); see also Brody (1989).

The nonlocal features of quantum systems sharpened the insight that the essential distinction between quantum and classical systems lies in the non-commutativity of the observables, not in the size of the system. As a consequence, there may be mesoscopic or even macroscopic systems with nonlocal quantum features. In order to discuss these systems, the original irreducible Hilbert space representation of quantum mechanics (which is adequate for simple systems such as studied by Aspect et al.) is generally insufficient since infinitely many degrees of freedom need to be considered.

For this reason it is interesting to look for the form of Bell's inequalities in representations more general than Hilbert spaces. Two corresponding papers were published by Landau (1987) and Baez (1987). Both of them argued in the framework of algebraic quantum mechanics, and they immedi-

diately caught the interest of Hans Primas. While his former student Raggio published a remark on the paper by Baez (Raggio 1988), Primas himself reviewed Landau's paper for the Mathematical Review (Primas 1988). Later he emphasized that Landau's generalization is free from conditions implying any Hilbert space structure (Primas 1996).

Karl Gustafson's interest in Bell-type inequalities is in a sense opposite to this direction of research. He outlines an approach to show that such inequalities are a direct mathematical consequence of the Hilbert space structure, without any additional physical ideas involved. More specifically, he shows that the Bell inequalities are special cases of an operator trigonometry introduced by Gustafson himself about thirty years ago. Wigner's (1970) probabilistic reformulation of Bell's (1964) argument is clarified as a special case of a general operator angle triangle inequality. Accardi and Fedullo's (1982) notion of statistical invariants for quantum probability theories is placed within the operator trigonometric geometrical frame.

References

Accardi L. and Fedullo A. (1982): On the statistical meaning of complex numbers in quantum mechanics. *Lett. Nuovo Cim.* **34**, 161–172.

Aspect A., Dalibard J., and Roger G. (1982): Experimental test of Bell's inequalities using time-varying analyzers. *Phys. Rev. Lett.* **49**, 1804–1807.

Bell J. (1964): On the Einstein Podolsky Rosen paradox. *Physics* **1**, 195–200.

Baez J. (1987): Bell's inequality for C*-algebras. *Lett. Math. Phys.* **13**, 135–136.

Brody T.(1989): The Bell inequality I: Joint Measurability. *Rev. Mex. Fis. Suppl.* **35**, 52–70. Reprinted in *The Philosophy Behind Physics*, ed. by L. de la Peña and P. Hodgson (Springer, Berlin 1993), pp. 205–222.

Landau L. (1987): On the violation of Bell's inequality in quantum theory. *Phys. Lett. A* **120**, 54–56.

Primas H. (1988): Comment on Landau (1987). *Math. Rev.* **88b**, 81014.

Primas H. (1996): Synchronizität und Zufall. *Z. Grenzgeb. Psych.* **38**, 61–91, here: p. 88.

Raggio G.A. (1988): A remark on Bell's inequality and decomposable normal states. *Lett. Math. Phys.* **15**, 27–29.

Selleri F., ed. (1988): *Quantum Mechanics Versus Local Realism* (Plenum, New York).

Wigner E.P. (1970): On hidden variables and quantum mechanical probabilities. *Am. J. Phys.* **38**, 1005–1009.

THE GEOMETRY OF QUANTUM PROBABILITIES

KARL GUSTAFSON
Department of Mathematics, University of Colorado,
Boulder, Colorado 80309-0395, USA
and
International Solvay Institute, Université Libre de Bruxelles,
Campus Plaine 231, Boulevard du Triomphe,
B-1050 Brussels, Belgium

1. Introduction

The EPR paradox of Einstein, Podolsky, and Rosen (1935) was presented as an argument for the need for additional variables in quantum theory to restore causality and locality to that theory. Such a proposition goes to the heart of the controversies about the proper interpretation of quantum mechanics, crystallized nicely in the long debate between Niels Bohr and Albert Einstein, see Bohr (1949). Bell (1964) demonstrated the limitations on locality that any such hidden variable theory might have. Wigner (1970) simplified and made more specific the argument of Bell for a quantum mechanical system of two spin-$\frac{1}{2}$ particles. In the intervening years a huge literature concerning the conceptual difficulties with quantum mechanics has accumulated and we have no intention nor do we make any pretense of dealing with it here. See the collection of articles by Bell (1987) and, for example, the recent exposition of Bohm (1996), or the very recent articles by Goldstein (1998) for just some of the viewpoints held and flavors preferred within the physics community as concerns the proper physical interpretations of quantum mechanics.

Accardi and Fedullo's (1982) justification of the complex numbers in quantum mechanics was based upon a theory of statistical invariants: for a given set of statistical data, with respect to an assumed probabilistic model, statistical invariants are precise mathematical constraints which are necessary and sufficient conditions for the statistical data to admit a description within the assumed probabilistic model. For example, Bayes' formula for conditional probabilities is a statistical invariant for typical Kolmogorov probability models. See Accardi (1988, 1995) for more on the theory of statistical invariants as a part of the quantum probabilistic approach to the foundations of quantum mechanics. There is much of general interest to be learned and discerned there and here we wish only to deal with one cornerstone of that theory.

When perusing the paper by Accardi and Fedullo (1982), we noticed that their Proposition 3, Eq. (19), namely

$$\cos^2 \alpha + \cos^2 \beta + \cos^2 \gamma - 1 \overset{\leq}{=} 2\cos\alpha\cos\beta\cos\gamma , \tag{1}$$

for the angles α, β, γ of a spin model (more details will be given below) in a complex Hilbert space, is precisely the same as our (see, e.g., Gustafson and Rao 1997, Lemma 3.3-1, Eq. (3.3-3))

$$1 - a_1^2 - a_2^2 - a_3^2 + 2a_1 a_2 a_3 \overset{\geq}{=} 0 \tag{2}$$

for the real cosines a_1, a_2, a_3 of the angles between arbitrary unit vectors in a complex Hilbert space (more details will be given below). The angles of inequality (1) are related to transition probability matrices $P(A \mid B)$, $P(B \mid C)$, $P(C \mid A)$ for three observables A, B, C which may take two values. The angles of (2) are related to a triangle inequality for general operator angles within a little known but general operator trigonometry.

Wigner sensed something more general: with respect to his version (Wigner 1970, Eq. (3)) of Bell's inequality, namely

$$\frac{1}{2}\sin^2\frac{1}{2}\theta_{23} + \frac{1}{2}\sin^2\frac{1}{2}\theta_{12} \overset{\geq}{=} \frac{1}{2}\sin^2\frac{1}{2}\theta_{31} , \tag{3}$$

he stated (Wigner 1970, p. 1009):

> The conditions (3), and the conditions obtained therefrom by cyclic interchanges of the indices, have the form of triangular inequalities for three sides $\sin^2\frac{1}{2}\theta_{ik}$.

Accardi also sensed something beyond what he was doing. In his analogy of statistical invariants for models to geometrical invariants for models he states (Accardi 1988, p. 266):

> It is well known how to solve the geometrical problem: There are mathematical constraints on the geometrical data, called geometrical invariants, which provide necessary and sufficient conditions for the given set of data to be describable with a geometrical model. For example, if all the triples of inner angles add up to π, then we can describe them within the usual model of Euclidean space. If their sum is less than π, we try a hyperbolic geometry model; if greater – an elliptic model.

One important outcome of this paper is the providing to Accardi's theory of statistical invariants for models in quantum probability theory its natural underlying geometry: the general operator trigonometry originated by this author about thirty years ago.

It is worth mentioning that the operator trigonometry was relatively neglected until recently and that it is quite likely that the connection between it and quantum probability theory established in this paper will be beneficial to both. As establishing that connection is the main goal of this paper, we must decide with which of the two theories to begin. To resolve the toss-up, we assume that most readers of this paper will be physicists, already familiar with the quantum mechanical considerations, so in Section 2 we recall the essentials of the operator trigonometry. Then in Section 3 we return to the quantum mechanical spin systems of Bell's theory and Accardi's more general theory of quantum mechanical conditional probabilities. There we will be content to establish the assertions made above, with no attempt here at full elaboration.

2. Operator Trigonometry

The operator trigonometry developed in the period 1966 to 1990 comprises Chapter 3 of Gustafson and Rao (1997) and is rather completely described there. Also recommended for more recent developments 1990 to 1996 is Part III of Gustafson (1997a). Additionally the interested reader may consult the recent papers by Gustafson (1994a,b, 1995, 1997b, 1998a,b).

The operator trigonometry began in 1966 with a problem concerning the multiplicative perturbation of the infinitesimal generator A of a contraction semigroup W_t, see Gustafson (1968a,b,c). That led to the question of when the operator product BA of two accretive operators would itself be accretive: $\mathrm{Re}\,\langle BAx, x\rangle \geqq 0$. For two strongly accretive operators A and B a sufficient condition was found:

$$\inf_{\epsilon} \|\epsilon B - I\| \leqq \inf_{x} \frac{\mathrm{Re}\,\langle Ax, x\rangle}{\|Ax\|\|x\|} \,. \tag{4}$$

The right side of (4) was naturally called $\cos A$, or more properly $\cos \phi(A)$, where $\phi(A)$ is the largest angle through which A may turn a vector. The angle $\phi(A)$ was called (Gustafson 1968a) the operator angle. Later (Gustafson 1972) its cosine was called the first antieigenvalue of A. The left side of (4) turns out to be $\sin B$, or more properly $\sin \phi(B)$. This is not obvious but follows by establishing the operator-trigonometric identity

$$\sin^2 B + \cos^2 B = 1 \tag{5}$$

for any strongly accretive bounded operator B on a complex Hilbert space. Thus the condition (4) for an accretive operator product is trigonometric:

$$\sin B \overset{\le}{=} \cos A . \tag{6}$$

A little later Krein (1969) developed an analogous theory of operator angles (he called them deviations). It may be shown that his dev(A) is the same as our $\phi(A)$. The inequality (6) may be seen to be equivalent to a triangle inequality used by Krein,

$$\phi(BA) \overset{\le}{=} \phi(B) + \phi(A) . \tag{7}$$

Geometrically (7) is almost transparent but its proof is less evident (Krein did not provide a proof); see the accounts in Gustafson and Rao (1997) and Gustafson (1996). For more general results than (6) or (7), see Gustafson and Rao (1977).

Our proof of (7) used the Gram matrix

$$G = \begin{bmatrix} \langle x,x \rangle & \langle x,y \rangle & \langle x,z \rangle \\ \langle y,x \rangle & \langle y,y \rangle & \langle y,z \rangle \\ \langle z,x \rangle & \langle z,y \rangle & \langle z,z \rangle \end{bmatrix} \tag{8}$$

for x, y, z any three unit vectors in a Hilbert space. As in the following (7) or its equivalent forms such as (6) will include similar inequalities from the Bell–Wigner–Accardi theory, let us indicate here the proof of (7) from (8), for more details see Gustafson and Rao (1977, 1997). The Gram matrix (8) is positive semidefinite, and definite if and only if the three vectors are linearly independent. Let $\langle x,y \rangle = a_1 + ib_1$, $\langle y,z \rangle = a_2 + ib_2$, $\langle x,z \rangle = a_3 + ib_3$, and define the angles ϕ_{xy}, ϕ_{yz}, ϕ_{xz} in $[0, \pi]$ by $\cos\phi_{xy} = a_1$, $\cos\phi_{yz} = a_2$, $\cos\phi_{xz} = a_3$. To show

$$\phi_{xz} \overset{\le}{=} \phi_{xy} + \phi_{yz} , \tag{9}$$

it suffices to show

$$\cos\phi_{xz} \overset{\ge}{=} \cos(\phi_{xy} + \phi_{yz}) , \tag{10}$$

which by the sum formula for cosines is equivalent to

$$\sqrt{1 - a_1^2} \sqrt{1 - a_2^2} \overset{\ge}{=} a_1 a_2 - a_3 . \tag{11}$$

The desired result (9) follows trivially when the right side of (11) is negative. In the other case we need

$$(1 - a_1^2)(1 - a_2^2) \overset{\ge}{=} (a_1 a_2 - a_3)^2, \tag{12}$$

which is equivalent to (2). But for unit vectors the determinant of the Gram matrix (8) becomes (using complex cancellations)

$$|G| = \begin{vmatrix} 1 & a_1 & a_3 \\ a_1 & 1 & a_2 \\ a_3 & a_2 & 1 \end{vmatrix} = 1 + 2a_1 a_2 a_3 - (a_1^2 + a_2^2 + a_3^2) \gtreqless 0 , \quad (13)$$

which gives (2), hence (10), hence (9). Then (7) follows from (9) by taking suprema of the expression, assuming B to be invertible,

$$\begin{aligned} \phi(BAx, x) &\lesseqgtr \phi(BAx, B^{-1}x) + \phi(B^{-1}x, x) \\ &= \phi(Ax, x) + \phi(Bx, x) . \end{aligned} \quad (14)$$

One more essential of the operator trigonometry will be needed in the next section. Specializing to A a symmetric positive definite matrix with eigenvalues $0 < \lambda_1 < \cdots < \lambda_n$, it is known that

$$\cos A = \frac{2\sqrt{\lambda_1 \lambda_n}}{\lambda_1 + \lambda_2}, \qquad \sin A = \frac{\lambda_n - \lambda_1}{\lambda_n + \lambda_1}, \quad (15)$$

and that the normalized first antieigenvectors are the pair

$$x_{\pm}^1 = \pm \left(\frac{\lambda_n}{\lambda_1 + \lambda_n}\right)^{1/2} x_1 + \left(\frac{\lambda_1}{\lambda_1 + \lambda_n}\right)^{1/2} x_n , \quad (16)$$

where x_1 and x_n denote the normalized eigenvectors corresponding to λ_1 and λ_n, respectively.

3. Quantum Spin Probabilities

Here we will develop further the Bell–Wigner–Accardi spin system quantum probability theory as a special case of the operator trigonometry. We have already established the basic connection between the theories in the introduction. In the preceding section we showed how the crucial quantum spin model inequality (1) is a special instance of the operator trigonometric inequality (2) which is the same as (9) or (10) or (11) or (12). Here we wish to strengthen this connection between the operator trigonometry and the quantum probabilities by focusing almost completely on some details in the paper by Wigner (1970) and its successor by Accardi and Fedullo (1982). As this theory developed from Bell's (1964) original analysis and has been carried much further by Accardi in a series of papers, we shall just refer to this as the Bell–Wigner–Accardi theory, even though there are a multitude of other papers dealing with the same issue raised originally by Bell, see, e.g., Johansen (1997).

Wigner (1970) essentially reduces and simplifies Bell's (1964) general
set-up to the inequality (3) which we gave above. Then to make the point
very clear, he specializes to the case in which the three directions $\omega_1, \omega_2, \omega_3$
in 3 space are coplanar and with ω_2 bisecting the angle between ω_1 and ω_3.
Then $\theta_{12} = \theta_{23} = \theta_{31}/2$ and inequality (3) becomes

$$\sin^2\left(\frac{1}{2}\theta_{12}\right) \gtreqless \frac{1}{2}\sin^2(\theta_{12}) = 2\sin^2\left(\frac{1}{2}\theta_{12}\right)\cos^2\left(\frac{1}{2}\theta_{12}\right), \qquad (17)$$

from which $\cos^2(\frac{1}{2}\theta_{12}) \lesseqgtr 1/2$ and hence $\theta_{31} = 2\theta_{12} \gtreqless \pi$. Thus the condition
(3) necessary for appropriate quantum mechanical spin probabilities for
the hidden variable theories considered by Bell (1964) and Wigner (1970)
is violated for all $\theta_{31} < \pi$. Wigner then asserts (without giving the details)
that the same conclusion may be drawn for all coplanar directions.

Let us now look at this conclusion (17) and its extension to all coplanar
directions from the operator trigonometric perspective. The Gram determinant G of (8) vanishes if and only if the three directions are coplanar,
no matter what their frame of reference. Then we may write the equality
in (13) as follows:

$$(1 - a_1^2) + (1 - a_2^2) - (1 - a_3^2) = 2a_3(a_3 - a_1a_2), \qquad (18)$$

or in the terminology of (3):

$$\begin{aligned}\sin^2\left(\tfrac{1}{2}\theta_{12}\right) + \sin^2\left(\tfrac{1}{2}\theta_{23}\right) - \sin^2\left(\tfrac{1}{2}\theta_{13}\right) = \\ = 2\cos\left(\tfrac{1}{2}\theta_{13}\right)\left[\cos\left(\tfrac{1}{2}\theta_{13}\right) - \cos\left(\tfrac{1}{2}\theta_{12}\right)\cos\left(\tfrac{1}{2}\theta_{23}\right)\right].\end{aligned} \qquad (19)$$

Nonquantum probability violation (3) in the coplanar case is equivalent to
the right side of (19) being nonnegative. Since all half-angles in (19) do not
exceed $\pi/2$, except for the trivial case when $\frac{1}{2}\theta_{13} = \pi/2$, the nonnegativity
of (19) means that of its second factors. By choosing the direction ω_2 to
be the "one in between" among the half-angles, we can without loss of
generality assume that $\frac{1}{2}\theta_{12} + \frac{1}{2}\theta_{23} = \frac{1}{2}\theta_{13}$. The required nonnegativity of
(19) then reduces by the elementary cosine sum formula to

$$\cos\left(\frac{\theta_{12} + \theta_{23}}{2}\right) \gtreqless \frac{1}{2}\left[\cos\left(\frac{\theta_{12} + \theta_{23}}{2}\right) + \cos\left(\frac{\theta_{12} - \theta_{23}}{2}\right)\right], \qquad (20)$$

i.e., $\cos((\theta_{12} + \theta_{23})/2) \gtreqless \cos((\theta_{12} - \theta_{23})/2)$, which is false for positive θ_{23}.
This completes Wigner's argument and is the meaning of coplanar quantum
probability violation (13).

Wigner considers two other configurations for quantum nonprobability
violation, namely (Wigner 1970, from Eqs. (6) and (7), respectively)

$$1 - \frac{1}{2}\left(\sin^2\frac{1}{2}\theta_{12} + \sin^2\frac{1}{2}\theta_{23} + \sin^2\frac{1}{2}\theta_{31}\right) \gtreqless 0 \qquad (21)$$

and

$$\frac{1}{2}\left(\sin^2\frac{1}{2}\theta_{12} + \sin^2\frac{1}{2}\theta_{23} - \sin^2\frac{1}{2}\theta_{31}\right) \geq 0 . \tag{22}$$

As Wigner notes, (22) gives (3), and moreover, the positivity of the three cyclically interchanged versions of (22), plus that of (21), is a necessary and sufficient condition for the possibility to interpret the spin measurements in the ω_i directions on a singlet state in terms of hidden variables.

How do these look from the operator trigonometric perspective, using only the Gram matrix and cosine sums? Using from (13) the general Grammian expression

$$|G| = (1 - a_1^2) + (1 - a_2^2) + (1 - a_3^2) + 2(a_1 a_2 a_3 - 1) , \tag{23}$$

immediately (21) and (22) become, respectively,

$$a_1 a_2 a_3 - \frac{1}{2}|G| \geq 0 \tag{24}$$

and

$$2a_3(a_1 a_2 - a_3) \geq 0 . \tag{25}$$

We note that (23) brings us more quickly to (25) which is the same as (19) and (22) but now for arbitrary directions. Also (23) expresses the positivity of (21) in a more delicate manner (24) which includes the degree of linear independence of the ω_i directions. Thus the operator trigonometry perspective makes more precise both qualitatively and quantitatively the arguments of Wigner (1970) which more naturally followed along the lines of Bell (1964). In particular the fundamental triangle inequality (9) and the Grammian are useful new ingredients in connecting the Bell quantum spin probabilities to the operator trigonometry.

We turn next to Accardi and Fedullo (1982). This was an important paper which not only advanced Wigner's (1970) treatment of the Bell theory but also clarified the statistical meaning of the complex numbers in quantum mechanics, a longstanding question. In addition, Accardi and Fedullo (1982) stressed the notion of statistical invariants to determine whether probability models were Kolmogorovian or not. As we did with Wigner (1970), we wish to look closely at some details in Accardi and Fedullo (1982) while ignoring some larger picture considerations.

Accardi and Fedullo (1982) emphasize conditional probabilities, also called transition probabilities, in contrast to Bell's (1964) arguments with correlations and Wigner's (1970) arguments with configuration combinatorics. Conditional probabilities

$$P(A = a_\alpha \mid B = b_\beta), \quad P(B = b_\beta \mid C = c_\gamma), \quad P(C = c_\gamma \mid A = a_\alpha) \tag{26}$$

are assumed to satisfy symmetry conditions

$$P(A = a_\alpha \mid B = b_\beta) = P(B = b_\beta \mid A = a_\alpha), \quad \text{etc.} \qquad (27)$$

and are said to admit a Kolmogorovian probability model if there exists a probability space $(\Omega, \mathcal{O}, \mu)$ and for each observable A, B, C a measurable partition $A_\alpha, B_\beta, C_\gamma$, of Ω such that for each observable outcome α, β, γ

$$P(A = a_\alpha \mid B = b_\beta) = \frac{\mu(A_\alpha \cap B_\beta)}{\mu(B_\beta)}. \qquad (28)$$

The transition probabilities (26) are said to satisfy a complex Hilbert space probability model if there exists a complex Hilbert space $\mathcal{H}$ such that for each observable A, B, C there exists an orthonormal basis $(\phi_\alpha), (\psi_\beta), (\chi_\gamma)$ such that for each α, β, γ

$$P(A = a_\alpha \mid B = b_\beta) = |\langle \phi_\alpha, \psi_\beta \rangle|^2. \qquad (29)$$

Real Hilbert space versions are also defined the same way. We refer the reader to Accardi and Fedullo (1982) and Accardi (1988, 1995) for more details.

Limiting discussion to three observables taking only two values, the conditional probabilities (26) may be represented by the following transition probability matrices:

$$P = P(A \mid B) = \begin{bmatrix} p & 1-p \\ 1-p & p \end{bmatrix} = \begin{bmatrix} \cos^2(\alpha/2) & \sin^2(\alpha/2) \\ \sin^2(\alpha/2) & \cos^2(\alpha/2) \end{bmatrix}$$

$$Q = P(B \mid C) = \begin{bmatrix} q & 1-q \\ 1-q & q \end{bmatrix} = \begin{bmatrix} \cos^2(\beta/2) & \sin^2(\beta/2) \\ \sin^2(\beta/2) & \cos^2(\beta/2) \end{bmatrix} \qquad (30)$$

$$R = P(C \mid A) = \begin{bmatrix} r & 1-r \\ 1-r & r \end{bmatrix} = \begin{bmatrix} \cos^2(\gamma/2) & \sin^2(\gamma/2) \\ \sin^2(\gamma/2) & \cos^2(\gamma/2) \end{bmatrix}.$$

As in Accardi and Fedullo (1982), we assume for simplicity $0 < p, q, r < 1$, $0 < \alpha, \beta, \gamma < \pi$. Some reasonable probability completeness and positivity assumptions are made, and then it is shown (Accardi and Fedullo 1982) that P, Q, R of (30) admit a Kolmogorovian probability model if and only if the inequality

$$|p + q - 1| \overset{\leq}{} r \overset{\leq}{} 1 - |p - q| \qquad (31)$$

holds.

For conditions for a quantum mechanical spin system to have a complex Hilbert space model existing, the Pauli matrices

$$\sigma_1 = \begin{bmatrix} 0 & 1 \\ 1 & 0 \end{bmatrix}, \quad \sigma_2 = \begin{bmatrix} 0 & -i \\ i & 0 \end{bmatrix}, \quad \sigma_3 = \begin{bmatrix} 1 & 0 \\ 0 & -1 \end{bmatrix} \qquad (32)$$

and spin operators $\sigma \bullet a = \sigma_1 a_1 + \sigma_2 a_2 + \sigma_3 a_3$ for $a = (a_1, a_2, a_3)$ a real 3-vector of norm 1 are considered. A spin model for the transition probabilities (30) is said to exist if there exist three normalized 3 vectors a, b, c such that the orthonormal bases $\psi_\alpha(a), \psi_\beta(b), \psi_\gamma(c)$ realize the matrices P, Q, R of (30) in the sense of (29). In this way the question of the existence of a Hilbert space probability model is reduced to the question of the existence of three norm-1 vectors a, b, c such that

$$\begin{aligned} |\langle \psi_1(a), \psi_1(b)\rangle|^2 &= \cos^2 \theta_{ab}/2, \\ |\langle \psi_1(a), \psi_2(b)\rangle|^2 &= \sin^2 \theta_{ab}/2; \end{aligned} \tag{33}$$

where $\cos \alpha = \cos \theta_{ab}$, $\cos \beta = \cos \theta_{bc}$, $\cos \gamma = \cos \theta_{ac}$ link the angles of (30) to the angles θ between the sought-for directions a, b, c. The setup is the same as in Bell (1964) and more particularly Wigner (1970) but the setting is now framed specifically in terms of competing classical or nonclassical probability models. In particular, it is shown (Accardi and Fedullo 1982, Proposition 3, Corollary 6, Theorem 7) that such vectors a, b, c exist if and only if

$$\cos^2 \alpha + \cos^2 \beta + \cos^2 \gamma - 1 \overset{\leq}{=} 2 \cos \alpha \cos \beta \cos \gamma. \tag{34}$$

Immediately we recognize that (34) is the condition for the nonnegativity of our Gram determinant (13). In other words, (34) is always satisfied for any three normalized direction vectors, with equality holding if and only if those vectors happen to be coplanar. Thus the interest in (34) is only in its use as a sufficient condition. From what we have shown above, those sufficient conditions, e.g., Wigner's (1970), are clearly understandable in terms of the operator trigonometry. In other words, the operator trigonometry provides the natural geometrical model for the statistical invariants of quantum probability theory. As we have said before, that is an emphasized main outcome of this paper.

Accardi and Fedullo (1982) get very close to discovering parts of the operator trigonometry when in their remark on p. 170 they state:

It is easy to verify that because of (11) [the condition $0 < \alpha, \beta, \gamma < \pi$] condition (37) [(34) stated here] is equivalent to the following couple of inequalities

$$|\alpha - \beta| \overset{\leq}{=} \gamma \overset{\leq}{=} \alpha + \beta, \quad \alpha + \beta + \gamma < 2\pi$$

which are necessary and sufficient conditions on α, β, γ to be adjacent angles of a tetrahedron in $\mathbb{R}^3$. This provides a geometrical interpretation for condition (37).

This statement is actually an instance (for real vectors, in three dimensions) of the triangle inequality (9) of the operator trigonometry. However, the general situation, that of a triangle inequality (9) and a related operator trigonometry which exists for arbitrary vectors in any Hilbert space of arbitrary dimension and real or complex scalars, is not glimpsed.

Finally, let us work out some of the details of the operator trigonometry of the transition probability matrices (30). For the operator P and $0 < p < 1$ we find by elementary considerations its two eigenvalues

$$\lambda_1 = 2p - 1 = 2\cos^2(\alpha/2) - 1 = \cos\alpha; \quad \lambda_2 = 1 ; \tag{35}$$

likewise for Q and R. We note in passing that the expression (34) is equivalently a statement about spectra:

$$\lambda_1^2(P) + \lambda_1^2(Q) + \lambda_1^2(R) - 2\lambda_1(P)\lambda_1(Q)\lambda_1(R) \leqq 1 . \tag{36}$$

Similar spectral statements apply to other necessary conditions, e.g., for a quantum probability Hilbert space model (Accardi and Fedullo 1982, Eqs. (34), (35)); we omit the details.

For the operator trigonometry of the matrices P, Q, R, for simplicity let us assume that the underlying bases for P, Q, R have been chosen so that $1 > p, q, r > 1/2$, i.e., we may speak only of the P, Q, R which are symmetric positive definite matrices. Then from (15) we obtain the operator trigonometry of P, Q, R in terms of their spin angles, namely,

$$\cos P = \frac{2\sqrt{\cos\alpha}}{1 + \cos\alpha}, \quad \sin P = \frac{1 - \cos\alpha}{1 + \cos\alpha} . \tag{37}$$

Recall that either of these expressions defines the operator angle $\phi(P)$, now in terms of the quantum probability geometric, e.g., spin orientation, angle α. Second, from (37) we also conveniently see the operator angle $\phi(P)$ in terms of the assigned probabilities p and $1 - p$:

$$\cos P = \frac{\sqrt{2p - 1}}{p}, \quad \sin P = \frac{1 - p}{p} . \tag{38}$$

Less immediately, we know from Gustafson (1997b, Lemma 8.1) that for any symmetric positive definite matrix A we have the square root relation $\sin(A^{1/2}) = \left[\frac{1 - \cos A}{1 + \cos A}\right]^{1/2}$, which for our quantum probabilities gives

$$\sin(P^{1/2}) = \left[\frac{1 - \frac{2\sqrt{\cos\alpha}}{1 + \cos\alpha}}{1 + \frac{2\sqrt{\cos\alpha}}{1 + \cos\alpha}}\right]^{1/2} = \frac{1 - \sqrt{\cos\alpha}}{1 + \sqrt{\cos\alpha}} \tag{39}$$

for the square root transition probability matrix $P^{1/2}$. Of course, this could also be obtained directly from the spectral mapping theorem. From (16) we see that the normalized first antieigenvectors are

$$
\begin{aligned}
x_{\pm}^{1} &= \pm \tfrac{1}{\sqrt{2}} \left[\left(\tfrac{1}{1+\cos\alpha} \right)^{1/2} \begin{bmatrix} 1 \\ -1 \end{bmatrix} + \left(\tfrac{\cos\alpha}{1+\cos\alpha} \right)^{1/2} \begin{bmatrix} 1 \\ 1 \end{bmatrix} \right] \\
&= \pm \tfrac{1}{\sqrt{2}} \left[\left(\tfrac{1}{2p} \right)^{1/2} \begin{bmatrix} 1 \\ -1 \end{bmatrix} + \left(\tfrac{2p-1}{2p} \right)^{1/2} \begin{bmatrix} 1 \\ 1 \end{bmatrix} \right]
\end{aligned}
\tag{40}
$$

in terms of P's spin angles or assigned probabilities. Removing the unnecessary normalizing factors, these antieigenvectors are more easily seen as

$$
\begin{aligned}
x_{+}^{1} &\cong \begin{bmatrix} 1 + \sqrt{\cos\alpha} \\ -1 + 1\sqrt{\cos\alpha} \end{bmatrix} \cong \begin{bmatrix} \sqrt{2p-1}+1 \\ \sqrt{2p-1}-1 \end{bmatrix}, \\
x_{-}^{1} &\cong \begin{bmatrix} -1 + \sqrt{\cos\alpha} \\ 1 + \sqrt{\cos\alpha} \end{bmatrix} \cong \begin{bmatrix} \sqrt{2p-1}-1 \\ \sqrt{2p-1}+1 \end{bmatrix},
\end{aligned}
\tag{41}
$$

in terms of the spin angle or probability. These are the directions that P turns the most; similarly for Q and R.

The operators P, Q, R, due to their special form, commute and thus, for $p, q, r > 1/2$ as we have taken them, all operator products PQ, PR, QR remain positive and of the same bistochastic form. For example,

$$
PQ = \begin{bmatrix} s & 1-s \\ 1-s & s \end{bmatrix}
\tag{42}
$$

with $s = 1 - (p+q-2pq)$; similarly for PR, QR. Thus one can easily work out details of the operator trigonometry for the calculus generated by P, Q, and R. For example,

$$
\sin PQ = \frac{p + q - 2pq}{1 - (p + q - 2pq)}
\tag{43}
$$

tells us the maximum turn angle $\phi(PQ)$ in terms of the contributing probabilities of the transition matrices P and Q, or if preferred, in terms of their spin orientation angles. On the other hand, that P, Q, R commute makes them somewhat less interesting from the viewpoint of the operator trigonometry, which really comes into its own when guaranteeing the positivity of (the real part of) operator products PQ for noncommuting operators P and Q. It is expected that that part of the operator trigonometry will be of interest when treating noncommutative quantum probabilities in the future.

4. Reversibility

It may be asserted as a dictum that underlying all reversible physical theories is somewhere, declared or not, an assumption of detailed balance. In the quantum spin systems discussed here, a detailed balance assumption is implicit in the conditional probability symmetry conditions (27). These imply that the observables A, B, C condition each other independent of order of event occurrence. However, it is well known (Heitler 1954) that in elementary quantum spin systems under time invariance assumptions, detailed balance need not hold when the direction of spin is taken into consideration. Bell (1964) assumed that any hidden variables, which determine the spin components of the first particle, do so in such a way as to be independent from their determination of the spin components of the second particle. This is what he calls the locality assumption. Essentially it assumes that each particle has its own measuring apparatus and that they do not affect each other. So here we discern some possible discrepancies between the detailed balance assumptions (27), the general nonvalidity of such assumptions in elementary spin systems, and the total lack of any simultaneous measuring conditioning in Bell's locality assumption. These will be looked at in more detail elsewhere, as they are not central to this paper. However, we mention them here as a prelude to the point developed below concerning noncommutativity.

It should also be briefly noted here that detailed balance assumptions also may be central to treatments of irreversible processes (Glansdorff and Prigogine 1971). Just as in Bell's (1964) analysis of hidden variables for spin systems, by assuming that the phenomenological coefficients (e.g., thermal diffusion) of mutual interference are symmetric, a reduction in the number of additional variables or coefficients which need be accounted for in irreversible processes is obtained. In other words, the symmetry conditions (27) need not necessarily be tied to reversibility. This general situation will be investigated elsewhere.

These observations lead up to the main point of this section: what happens when the detailed balance conditional probability symmetries (27) are not assumed? Then the transition probability matrices analogous to P, Q, R above need not be symmetric and need not commute. Generally we may expect the noncommutativity to indicate a causality or time ordering, hence irreversibility. However, the operator trigonometry will still apply. In other words, the breaking of detailed balance is an instance of the situation described at the end of the previous section.

5. Conclusion

The main conclusion of this paper is that the Bell–Wigner–Accardi theory of quantum probabilities in spin systems may be placed within the general operator trigonometry developed independently by this author about thirty years ago. It is noted that the use of the Grammian from the operator trigonometry simplifies and clarifies the analysis of Wigner. A general vector and operator triangle inequality from the operator trigonometry clarifies and generalizes the analysis of Accardi. The statistical meaning of the complex numbers in quantum mechanics is seen to be that of the natural geometry of the operator trigonometry.

Acknowledgments: The author appreciates an informal lunch-time discussion about quantum probabilities with I. Antoniou, V. Belavkin, L. Accardi and others at Les Treilles, France, in July 1996.

References

Accardi L. (1988): Foundations of quantum mechanics: a quantum probabilistic approach. In *The Nature of Quantum Paradoxes*, ed. by G. Tarozzi and A. van der Merwe (Kluwer, Dordrecht), pp. 257–323.

Accardi L. (1995): Can mathematics help solving the interpretational problems of quantum theory? *Nuovo Cim.* **110B**, 685–721.

Accardi L. and Fedullo A. (1982): On the statistical meaning of complex numbers in quantum mechanics. *Lett. Nuovo Cim.* **34**, 161–172.

Bell J. (1964): On the Einstein Podolsky Rosen paradox. *Physics* **1**, 195–200.

Bell J. (1987): *Speakable and Unspeakable in Quantum Mechanics* (Cambridge University Press, Cambridge).

Bohm D. (1996): On the role of hidden variables in the fundamental structure of physics. *Found. Phys.* **26**, 719–786.

Bohr N. (1949): Discussion with Einstein on epistemological problems in atomic physics. In *Albert Einstein: Philosopher-Scientist*, ed. by P. Schilpp (Open Court, Evanston), pp. 200–241.

Einstein A., Podolsky B., and Rosen, N. (1935): Can quantum-mechanical description of physical reality be considered complete? *Phys. Rev.* **47**, 777–780.

Glansdorff P. and Prigogine I. (1971): *Thermodynamic Theory of Structure, Stability and Fluctuations* (Wiley, New York).

Goldstein S. (1998): Quantum theory without observers – Parts I, II. *Physics Today* **51**, no. 3, 42–46, no. 4, 38–42.

Gustafson K. (1968a): The angle of an operator and positive operator products. *Bull. Amer. Math. Soc.* **74**, 488–492.

Gustafson K. (1968b): Positive (noncommuting) operator products and semigroups. *Math. Zeit.* **105**, 160–172.

Gustafson K. (1968c): A note on left multiplication of semigroup generators. *Pac. J. Math.* **23**, 463–465.

Gustafson K. (1972): Anti-eigenvalue inequalities in operator theory. In *Inequalities III*, ed. by O. Shisha (Academic Press, New York), pp. 115–119.

Gustafson K. (1994a): Operator trigonometry. *Lin. Multilin. Alg.* **37**, 139–159.

Gustafson K. (1994b): Antieigenvalues. *Lin. Alg. & Applic.* **208/209**, 437–454.

Gustafson K. (1995): Matrix trigonometry. *Lin. Alg. & Applic.* **217**, 117–140.

Gustafson K. (1996): Commentary on *Topics in the Analytic Theory of Matrices*, Section 23: Singular angles of a square matrix. In *Collected Works of Helmut Wielandt, Vol. 2*, ed. by B. Huppert and H. Schneider (deGruyter, Berlin), pp. 356–367.

Gustafson K. (1997a): *Lectures on Computational Fluid Dynamics, Mathematical Physics, and Linear Algebra* (World Scientific, Singapore).

Gustafson K. (1997b): Operator trigonometry of iterative methods. *Num. Lin. Alg. Applic.* **4**, 333–347.

Gustafson K. (1998a): Domain decomposition, operator trigonometry, Robin condition. *Contemporary Mathematics* **218**, 455–460.

Gustafson K. (1998b): Operator trigonometry of wavelet frames. In *Iterative Methods in Scientific Computation*, ed. by J. Wang, M. Allen, B. Chen, and T. Mathew (IMACS, New Brunswick), pp. 161–166.

Gustafson K. and Rao D. (1977): Numerical range and accretivity of operator products. *J. Math. Anal. Appl.* **60**, 693–702.

Gustafson K. and Rao D. (1997): *Numerical Range* (Springer, Berlin).

Heitler W. (1954): *The Quantum Theory of Radiation* (Oxford University Press, London).

Johansen L. (1997): Equivalence between Bell's inequality and a constraint on stochastic field theories for Einstein–Podolsky–Rosen states. *Phys. Rev. A* **56**, 100–107.

Krein M. (1969): Angular localization of the spectrum of a multiplicative integral in a Hilbert space. *Funct. Anal. Appl.* **3**, 89–90.

Wigner E.P. (1970): On hidden variables and quantum mechanical probabilities. *Amer. J. Phys.* **38**, 1005–1009.

CONTEXTUAL BACKGROUND

E.C.G. Sudarshan:
PROBABILITY AND QUANTUM DYNAMICS

Johann von Neumann, the father of Hilbert space quantum mechanics, wrote to Birkhoff in a letter dated November 13, presumably 1935: "I would like to make a confession that may seem immoral: I do not believe absolutely in Hilbert space anymore" (Birkhoff 1961). With this statement, von Neumann indicated a shift in emphasis from states in a Hilbert space to the lattice of all linear closed subspaces of this Hilbert space. In the mid 1930s he and Birkhoff introduced lattice theoretical ideas into quantum theory, thus laying the foundations for the field of quantum logic (Birkhoff and von Neumann 1936).

There is another way, though, to look at von Neumann's statement, even if this might not have been his original intention. Since quite a number of basic problems is encountered when one tries to enforce a rigorous mathematical formulation of quantum mechanics in the usual Hilbert space representation, it is an interesting option to explore the possibility of representing states of a system not by rays in Hilbert space but by distributions in generalized spaces.

Dirac's formulation of quantum theory in terms of bras and kets (Dirac 1930) leads to one such difficulty: the δ-functions required in Dirac's approach are undefined in Hilbert space. A first step to meet this problem was done when Schwartz (1950) developed his theory of distributions with δ-functions as singular limiting cases. A special class of distribution spaces was introduced somewhat later by the Russian mathematician Gel'fand and his collaborators (Gel'fand and Vilenkin 1964): the so-called rigged Hilbert spaces. At about the same time, both J.E. Roberts and A. Bohm utilized rigged Hilbert spaces to make Dirac's formalism mathematically rigorous. By the end of the 1970s, it turned out that some basic physical problems of Hilbert space quantum mechanics, notably in the context of decaying states or resonances, could be clarified in terms of rigged Hilbert spaces. For more details see Bohm and Gadella (1989).

More recently, the rigged Hilbert space approach has been used quite extensively by the Brussels-Austin-group of Prigogine and collaborators. Their special interest in rigged Hilbert spaces was due to the fact that this formalism provides a natural way to derive two semigroups from the basic unitary time evolution of both classical and quantum systems under certain conditions (Antoniou and Prigogine 1993). One of these semigroups describes an evolution toward the future, the other toward the past. Hence it is clear that the rigged Hilbert space formulation does not dispense us from the need of (ad hoc) selecting the proper semigroup. In addition to the rigged Hilbert space approach, other extensions are possible such as, e.g., the Liouville extension (Petrosky and Prigogine 1997).

Sudarshan's contribution describes yet another way to generalize quantum theory in order to cover situations for which the usual Hilbert space approach is insufficient. In his formulation, the states of dynamical systems are identified with distributions which assign numerical values to all dynamical variables. The notion of dynamics is then generalized to mappings (which could include irreversible mappings) of the convex set of states onto themselves. Stochastic quantum dynamics is studied with regard to both the convex set of dynamical maps of density matrices and their generic construction as contractions of extended systems. The analytic continuation of this dynamics to dual analytic spaces is carried out and the dominant metastable modes are identified. The various generalized spaces, often used indiscriminately in the literature, are defined and distinguished. The concept of the age of a decaying system obeying a semigroup is introduced and illustrated.

References

Antoniou I. and Prigogine I. (1993): Intrinsic irreversibility and integrability of dynamics. *Physica A* **192**, 443–464.

Birkhoff G. (1961); Lattices in applied mathematics. In *Proceedings in Pure Mathematics, Vol. 2* (American Mathematical Society, Providence, Rhode Island), pp. 155–184.

Birkhoff G. and Neumann J. von (1936): The logic of quantum mechanics. *Ann. Math.* **37**, 823–843.

Bohm A. and Gadella M. (1989): *Dirac Kets, Gamow Vectors, and Gelfand Triplets. Lecture Notes in Physics, Vol. 348*, ed. by A. Bohm and J.D. Dollard (Springer, Berlin). Triplets (Springer, Berlin).

Dirac P.A.M. (1930): *The Principles of Quantum Mechanics* (Clarendon, Oxford).

Gel'fand I.M. and Vilenkin N.Ya. (1964): *Generalized Functions, Vol. 4* (Academic, New York). Russian original published 1961 in Moscow.

Petrosky T. and Prigogine I. (1997): The Liouville space extension of quantum mechanics. *Adv. Chem. Phys.* **XCIV**, 1–120.

Schwartz L. (1950): *Théorie des Distributions* (Hermann, Paris).

PROBABILITY AND QUANTUM DYNAMICS

E.C.G. SUDARSHAN
Physics Department and Center for Particle Physics,
University of Texas, Austin, TX 78712-1081, USA

1. Introduction

The simplest dynamical system is the point particle characterized by its mass. Its state is specified by its position and momentum. The dynamical law is the description of how these quantities change in time. For a free particle, the momentum remains constant while the position increases in the direction of the momentum. The increase is directly proportional to the elapsed time and inversely proportional to the mass. When such a particle is subjected to a force, the momentum also changes proportional to the force. The instantaneous state may be represented as a point in phase space which moves along a trajectory. For complicated interactions, the trajectories may form an intricate pattern but they do not intersect, and the generic dynamical evolution may be viewed as a mapping of the phase space onto itself.

What happens when the forces acting on the particle are randomly fluctuating? The evolved phase point would also fluctuate; and the only way to specify the final state is to describe the probability for various points in phase space (Bachelier 1900, Einstein 1905). In any particular realization of the motion there would be a final specific phase point but there is no way to predict it; rather the predictable quantity is the probability distribution. This is the quantity that evolves according to a definite law. Since now the final state is a probability distribution in phase space, can we consider a probability distribution over phase space as the generic state (Segal 1947, Haag and Kastler 1964)? The canonical transformations that bring about Hamiltonian time evolution would now be viewed as the evolution of the probability distribution. For a canonical evolution, the probability distribution over the phase space variables behaves as a scalar field:

$$\omega \to \omega'; \quad p \to p'; \quad p'(\omega') = p(\omega). \tag{1.1}$$

The case of a fluctuating force acting on the system is more complicated; it depends on the statistical distribution of the interactions. In particular, a "pure" initial state which is concentrated on a phase point can become a smooth distribution corresponding to a "mixed" statistical state. For a Hamiltonian evolution, the mappings of phase point to phase point as well as of the density distribution are invertible – there is no irreversibility. This way of viewing statistical states of a classical dynamical system (Segal 1947, Haag and Kastler 1964) is in contrast to the popular view that the probability distribution reflects our ignorance of the phase point and that any particle will have a definite position and a definite momentum. While every realization can be a pure state, these pure states do not have a definite evolution. Realizations are pure but the distributions are the entities which evolve according to a well-defined law.

We must also recognize that even the best measurements provide small patches of phase space, and even with a Hamiltonian evolution such a phase patch can spread into a long-tentacled octopus shape. So "almost pure" states can become pretty much mixed up. On the other hand, the probability distributions have a well-defined evolution law, no matter whether the evolution is Hamiltonian or not. Such a description can incorporate irreversible evolutions (Prigogine 1997).

Probability distributions are non-negative measures on phase space and form a convex set (Segal 1947, Haag and Kastler 1964). The distribution

$$p(\omega) = p_1(\omega) \cos^2 \Theta + p_2(\omega) \sin^2 \Theta \qquad (1.2)$$

is an admissible distribution provided $p_1(\omega)$ and $p_2(\omega)$ are admissible distributions. The extremal states are the states whose distributions are concentrated on individual phase points.

Recall that phase space has a symplectic geometry and that, therefore, there is no intrinsic notion of "distance" between two distinct phase space points (Sudarshan et al. 1961, Sudarshan and Mukunda 1974). But there are no guarantees that "nearby" states evolve into "nearby" states, however the notion of "nearness" is defined. The octopus-like phase distribution evolving from a phase patch is evidence of this behavior. But this is not an irreversible evolution, since retracing the final distribution to the initial regular patch is a Hamiltonian evolution, and if we go back further the regular patch would be seen to have evolved from an octopus-like irregular patch. Of course, this does only apply for a reversible Hamiltonian evolution and not for stochastic evolutions.

Though we have commented on Hamiltonian classical systems so far, the concept of distributions applies equally for dynamical systems in a generic sense. For example, if we have a finite set of phase points, only the generic state is a probability vector whose elements sum up to unity. The evolution maps are now stochastic matrices (Ramakrishnan 1959), and the only reversible evolutions are permutations. Hence, there are no continuous reversible evolutions in time for such systems. The generic stochastic evolution is a contraction map with limit states toward which the map converges. Even "pure" states which are concentrated on a specific phase point converge to "mixed" states. Given a generic state of such a system, we could retrace its evolution for a finite time. At that time the inverse image is a statistical state which cannot be traced back any further unless the non-negativity of the elements of the probability vector is violated.

The dynamical evolutions form a convex set. If the map is

$$p(\omega) \rightarrow A\{p(\omega)\} \ ,$$

then

$$A = \cos^2 \Theta A_1 + \sin^2 \Theta A_2 \ ,$$

that is,

$$A\{p(\omega)\} = \cos^2 \Theta A_1\{p(w)\} + \sin^2 \Theta A_2\{p(\omega)\}$$

is also an admissible stochastic evolution. For stochastic matrices this implies that the conditions

$$A_{j,k} \gg 0; \quad \sum A_{j,k} = 1$$

are preserved by such convex combinations. The possibility of a measurement of all dynamical variables to arbitrary accuracy enables us to view the extremal states as probability distributions concentrated on a point in phase space. If this is not true, i.e., if the algebra of dynamical variables is non-Abelian, then the states can no longer be such concentrations (Segal 1947, Haag and Kastler 1964), nor need the distributions be pointwise positive over the phase space. For a canonical quantum system, the Wigner-Moyal distribution (Wigner 1932, Moyal 1949) and the diagonal coherent state distribution (Sudarshan 1963, Mehta and Sudarshan 1965) are not non-negative, nor are the extremal distributions always concentrated at phase points.

For quantum systems with a discrete set of states, the kinematic characterization of a state is given by a non-negative matrix of unit trace. These characterizations form a convex set, the extremal states being projections of rank one. The dynamical evolution must map the corresponding matrix into a matrix of the same set. We can now have reversible non-trivial evolutions corresponding to unitary transformations of this matrix. In addition, there are also the stochastic evolutions, which are irreversible.

2. Quantum Kinematics, Convex Sets of States

The generic state of a quantum system is specified by its density distribution which may be viewed as a linear non-negative normalized number-valued linear functional (Segal 1947, Haag and Kastler 1964; singular linear functionals are used in Sudarshan (1963) and in Mehta and Sudarshan (1965)) on the operators. In a more restrictive form, in which a state is identified as a trace class operator (von Neumann 1955) in the Hilbert space, it has the canonical decomposition

$$\rho = \sum_1^\infty c_n \psi_n \tilde{\psi}_n; \quad c_n > 0; \quad \sum_1^\infty c_n = 1 \tag{2.1}$$

with

$$\tilde{\psi} = \psi^\dagger . \tag{2.2}$$

The set of density distributions can be enlarged provided the dynamical variables for which expectation values are sought are restricted. Conversely, if the set of density distributions is restricted a wider set of dynamical variables may be constructed.

The conditions of positivity and normalization still allow us to form normalized convex combinations:

$$\rho = \rho_1 \cos^2 \theta + \rho_2 \sin^2 \theta . \tag{2.3}$$

They *do not* form a vector space. For several purposes including that of the stochastic dynamics of density distributions, it is advantageous to consider the vector space generated by the density distributions. The additional distributions so obtained may not satisfy either positivity or normalization, or both. We will see that metastable (decaying) "states" are realized by such pseudodensity distributions.

Given the convex set of density distributions, we could seek the boundary elements and, more specifically, the generating extremal elements. The latter are those density distributions whose (normalized) convex combinations generate all density distributions but which themselves have no nontrivial decomposition. Trace class operators in Hilbert space are a compact set under the Hilbert-Schmidt norm and the extremals are one-dimensional projections. For more general definitions of the density distributions, it is necessary to investigate the situation case by case.

If the dynamical variables undergo a unitary transformation, their expectation values change. This is equivalent to a linear transformation on the density distributions:

$$\rho_U(A) = \rho(U A U) . \tag{2.4}$$

If the unitary transformations concerned form a group, then the linear transformations

$$\rho \to \rho_U \tag{2.5}$$

furnish a realization of the same group. On the other hand, if we consider the linear transformations on the density distribution we have new possibilities. This is particularly the case for time evolutions. For simple Hamiltonian systems there is a one-parameter group of transformations on the density distributions, but we have the more general possibility of non-invertible dissipative transformations. Whenever such a dissipation is involved the inverse transformations cannot act on all density distributions. Instead of a time translation group the best we can obtain is a semigroup of dissipative evolutions. In either case we refer to the generator of the group or the semigroup as the Liouvillean.

The generic time evolution is a subject of stochastic dynamics (Sudarshan et al. 1961) and has been systematically studied for decades (Davies 1969, 1970, 1971). The special case of the relaxation of spin systems in an external magnetic field was studied in terms of the Bloch equations half a century ago (Lindblad 1975, Gorini et al. 1976, Gorini et al. 1978). But the convex set of dynamical maps has an intricate structure even for 2×2 density matrices.

One way of arriving at stochastic dynamics is by considering the system as being embedded in a larger system with a time translation group and then contracting out the extraneous degrees of freedom. When one does this one arrives at a subclass of stochastic dynamical maps, namely those of the completely positive type. Conversely, given a completely positive dynamical map we can realize it constructively in terms of the contraction of an extended Hamiltonian time translation (Bloch 1946).

Stochastic dynamics thus involves a *sense of time* and thus a *breaking of time symmetry*. In the contraction procedure the time symmetry breaking is explicit. It has been of continuing interest for more than a century whether the breaking can occur spontaneously, without any explicitly asymmetric procedure and without any restriction on the initial states. On the other hand, if the dynamical laws are time symmetric it is to be expected that the time reversed sequence of states corresponds to a time reversed semigroup. What, then, selects the forward semigroup for time evolution? In other words: what is the ingredient implicit in the dynamics or involved in the choice of physical states that assures the time symmetry breaking second law of thermodynamics? With the increasing recognition of the role of deterministic chaos and the relevance of the Poincaré catastrophe for large Poincaré systems the simple objections of Loschmidt and Zermelo to Boltzmann's H-theorem appear to be less compelling.

This paper is an attempt to clarify this issue by a careful characterization of the various families of states, of the varieties of dual pairs of states and dynamical variables. It appears from this analysis that states exhibiting dissipation and hence breaking time symmetry are to be selected from an extended set of states. There are the time reversed states which exhibit negative dissipation and, hence, are not acceptable as physical states obeying the second law of thermodynamics.

3. Stochastic Quantum Dynamics

For a finite dimensional system, the density distributions are non-negative density matrices of unit trace:

$$\rho \neq 0; \quad \text{Tr}\,\rho = 1; \quad \dim \rho = N \,. \tag{3.1}$$

The extremal generating elements of the convex set of density matrices are projections of rank one:

$$\rho = \Pi_\psi = \psi\tilde{\psi}; \quad \tilde{\psi} = \psi^\dagger \,. \tag{3.2}$$

There is an infinite number of such extremal elements. For a time-independent (Hermitian) Hamiltonian H, the time evolutions are given by a unitary one-parameter group:

$$\rho(t) = \exp\left(-i\mathcal{L}t\right)\rho \equiv \mathrm{e}^{-\mathrm{i}tH}\,\rho\,\mathrm{e}^{\mathrm{i}tH} \,. \tag{3.3}$$

Here, $\mathcal{L}$ is the Liouville superoperator

$$\mathcal{L}\rho = H\rho - \rho H \,. \tag{3.4}$$

A much more general evolution is given by a parametrized map:

$$\rho \to A(t)\rho; \quad \rho_{rs}(t) = \sum_{r's'} A_{rs,r's'}(t)\,\rho_{r's'} \,. \tag{3.5}$$

Since the properties of a density matrix must be preserved by the mapping, we have

$$\sum_r A_{rr,r's'} = \delta_{r's'}; \quad A_{sr,s'r'} = A^*_{rs,r's'} \,. \tag{3.6}$$

If we define the new $N^2 \times N^2$ matrix B with elements

$$B_{rr',ss'} = A_{rs,r's'} \,, \tag{3.7}$$

then the properties of A may be used to deduce

$$B_{ss',rr'} = B^*_{rr',ss'}; \quad \sum_r B_{rs',rs'} = \delta_{r's'} \,. \tag{3.8}$$

If the matrix B is non-negative, we say that the dynamical map $\rho \to A\rho$ is strictly positive. Not all maps need be strictly positive (Sudarshan 1985); the simplest not strictly positive map is the map

$$\rho \to \rho^* . \tag{3.9}$$

In all cases, since B is Hermitian and finite-dimensional it can be diagonalized. For strictly positive maps all eigenvalues are non-negative. In this case we obtain

$$\rho \to \sum_{\mu} V(\mu)\rho V^{\dagger}(\mu); \quad \sum_{\mu} V^{\dagger}(\mu)V(\mu) = 1 , \tag{3.10}$$

where the sum over μ in general runs from 1 to N^2. The dynamical maps themselves form a convex set since

$$B = B_1 \cos^2 \phi + B_2 \sin^2 \phi \tag{3.11}$$

is an acceptable map if B_1 and B_2 are. If both B_1 and B_2 are strictly positive, so is B. The question naturally arises how to find all generating extremal elements of (i) all dynamical maps and of (ii) all strictly positive dynamical maps. The first problem is very complicated and has been done completely only for $N = 2$ (Choi 1972, 1974). For $N > 2$ we know many extremal maps like unitary, antiunitary, and "pin" maps but a complete characterization is still missing.

For strictly positive maps, such a complete characterization is available. The extremal cases may be separated into families of rank R, where $1 \le R \le N$. The case $R = 1$ corresponds to unitary maps while $R = N$ corresponds to the "pin" maps. Moreover, we have a simple construction algorithm for finding all the extremal maps.

The strictly positive maps are obtained by a unitary evolution of an extended system – consisting of the N-dimensional system of interest and an auxiliary R-dimensional system – which is then contracted by taking the partial trace with respect to the auxiliary system. Moreover, it can be shown that any strictly positive map may be displayed as a contraction of a rank one map of an extended system.

In this context we note that these maps can be multiplied by performing them in sequence. The result is again a dynamical map. The dynamical maps therefore form a forward semigroup. In particular, the strictly positive dynamical maps form a forward semigroup. However, these maps are in general not invertible to form a group: they take the set of density matrices into indefinite matrices (of unit trace!).

Given a unitary map we can consider it being generated by a Liouvillean derived from a Hamiltonian. We can then speak of a continuous group with a generator $\mathcal{L}$ and ask for a continuous parameter semigroup for the generic semigroup of completely positive maps. In view of the fact that any such map can be obtained by contraction of a unitary map, we may look for clues to the structure of the generator of a dissipative semigroup (Petrosky and Prigogine 1993, 1988) in such a procedure. By expanding the unitary matrix to second order (where dissipation starts to play a role) we get

$$\rho \to \rho - \mathrm{it}[H, \rho] + \frac{(\mathrm{it})^2}{2!} [H, [H, \rho]] + \cdots . \tag{3.12}$$

Applying it to the extended system and then taking the partial trace, we obtain

$$\rho \to \rho - \mathrm{it}\, [H, \rho] + \mathrm{it}\, \left[L_\alpha, \left[L_\beta^\dagger, \rho\right]\right] . \tag{3.13}$$

It is possible to show that this is the generic generator of a completely positive semigroup. It is noteworthy that for a finite-dimensional system there is no self-adjoint Hamiltonian which could lead to dissipation for an isolated system and for which the Liouvillean eigenvalues are the differences of the Hamiltonian eigenvalues. But if it is coupled to an auxiliary finite-dimensional system the contraction map can exhibit dissipation (Sudarshan 1985). In order to obtain the time evolution as a continuous one-parameter semigroup we may have to take limiting cases of weak coupling and scaling of time (Gorini et al. 1976, 1978).

When the number of dimensions of the vector space becomes denumerably infinite but the Hamiltonian has still a discrete spectrum the situation is not changed dramatically. The only essential change is that there are dynamical maps of arbitrarily high rank which cannot only be obtained from unitary but also from *isometric* operators in the extended space.

4. Liouville Dynamics with Continuous Spectrum

The Liouvillean dynamics of a system with a continuous spectrum furnishes richer possibilities. If ν is a point in the continuous spectrum, $0 < \nu < \infty$, then the density matrix may be parametrized by ν_1, ν_2:

$$\begin{aligned} H\,\rho(\nu_1, \nu_2) &= \nu_1\, \rho(\nu_1, \nu_2) , \\ \rho(\nu_1, \nu_2)H &= \nu_2\, \rho(\nu_1, \nu_2) . \end{aligned} \tag{4.1}$$

Then

$$\mathcal{L}\,\rho(\nu_1, \nu_2) = (\nu_1 - \nu_2)\, \rho\,(\nu_1, \nu_2) , \tag{4.2}$$

and we can relabel the density matrix in the form

$$\rho(\nu; E) \equiv \rho\left(E + \frac{1}{2}\nu,\ E - \frac{1}{2}\nu\right); \quad -2E < \nu < 2E . \tag{4.3}$$

The unitary time evolution is

$$e^{-i\mathcal{L}t}\rho(\nu; E) = e^{-i\nu t}\rho(\nu; E).$$ (4.4)

The trace is invariant under this evolution,

$$\text{Tr}\left(e^{-i\mathcal{L}t}\rho(\nu; E)\right) = \int dE\, \rho(0; E),$$ (4.5)

and the positivity is preserved:

$$\rho(\nu; E) \geq 0 \rightarrow e^{-i\mathcal{L}t}\rho(\nu; E) \geq 0.$$ (4.6)

This is equivalent to the statement

$$\int_{-\infty}^{\infty} \int_{|v|}^{\infty} \rho(\nu; E)\, f^*\left(E + \frac{1}{2}\nu\right) f\left(E - \frac{1}{2}\nu\right) dE\, d\nu \geq 0.$$ (4.7)

For any energy E or any finite range of energies $0 < E < E_0$ the time dependent density matrix is an entire function of t and always obeys a group rather than a semigroup. But under suitable conditions, the survival probability

$$P(t) = \text{Tr}\left(\rho\, e^{i\mathcal{L}t}\rho\right) \equiv \int\int \rho(-\nu; E)e^{-i\nu t}\, \rho(\nu; E)\, d\nu\, dE$$

$$= \int\int \rho^{\dagger}(\nu; E)\, e^{-i\nu t}\, \rho(\nu; E)\, d\nu\, dE$$ (4.8)

may exhibit appropriate exponential behavior. Clearly, $P(t)$ is real and bounded by unity.

Since the density distribution ρ may be expressed in the form

$$\rho(\nu_1, \nu_2) = \sum_{\alpha} r_{\alpha}\psi_{\alpha}(\nu_1)\psi_{\alpha}^{\dagger}(\nu_2)$$ (4.9)

with $0 \leq \nu_1, \nu_2$ and $0 < r_{\alpha} < 1$, it follows that

$$\left(e^{-i\mathcal{L}t}\rho\right)(\nu_1, \nu_2) = \sum_{\alpha} r_{\alpha} e^{-i(\nu_1 - \nu_2)t}\, \psi_{\alpha}(\nu_1)\psi_{\alpha}^{\dagger}(\nu_2)$$

$$= \sum_{\alpha} r_{\alpha}\, \psi_{\alpha}(\nu_1)\, e^{-i\nu_1 t}\left(\psi_{\alpha}(\nu_2)e^{-i\nu_2 t}\right)^{\dagger}.$$ (4.10)

Hence, the survival probability has the decomposition

$$P(t) = \sum_{\alpha} r_{\alpha}^2\, P_{\alpha}(t); \quad P_{\alpha}(t) = \left|\int_0^{\infty} \psi_{\alpha}^{\dagger}(\nu)\, e^{-i\nu t}\, \psi_{\alpha}(\nu)\right|^2,$$ (4.11)

so that $P_\alpha(t)$ are the absolute values squared of functions of t analytic in the lower half plane. Then, we have by the Paley-Wiener theorem:

$$\left| \int \frac{\log P_\alpha(t)}{1 + t^2} \, dt \right| < \infty . \tag{4.12}$$

This is not possible if the convex sum of those functions decreases exponentially with t for $t > 0$. This is a slight generalization of a result derived four decades ago by Khalfin (1958).

At this stage it will be useful to classify the kinds of density distributions that we may consider, their analytic continuations, and the extension of the set of density distributions. Before going into this, we note that, given any set of dynamical variables, we may consider the density distributions as their duals. If we consider too large a class of distributions we restrict the set of observables and vice versa. Finally, if we consider analytic continuations of the dynamical variables, the density distributions themselves should be analytically continued initially (Kapur and Peierls 1938, Hu 1948, Sudarshan et al. 1978, Parravicini et al. 1980). The dual correspondence should be maintained.

5. Varieties of Statistical State Spaces

Given the density distribution $\rho(\nu; E)$, the time evolution can be displayed as

$$\left(e^{-i\mathcal{L}t} \rho \right) (\nu; E) = e^{-i\nu t} \rho(\nu; E) , \tag{5.1}$$

and the survival probability can be written in the form

$$P(t) = \int_0^\infty dE \int_{-2E}^{2E} d\nu \, e^{-i\nu t} \rho^*(\nu; E) \, \rho(\nu; E) . \tag{5.2}$$

The integration over the finite segment $-2E < \nu < 2E$ may be deformed to run along some path in the complex plane provided the function $\rho(\nu; E)$ is analytic in ν in a suitable domain in which the new open contour C from $-2E$ to $2E$ lies:

$$P(t) = \int dE \, P(t, E) ,$$
$$P(t, E) = \int_C \rho^*(z^*, E) \, e^{-izt} \rho(z, E) \, dz . \tag{5.3}$$

We can now define various spaces associated with density distributions $\rho(z, E)$. Let us start with noting that the density distributions constitute a convex set, not a vector space. We can, however, relax the positivity condition and define the vector space spanned by the density distributions.

We distinguish the following spaces:

1. the space $\mathcal{B}$ of integrable distributions where the ν integration extends over the bounded range $-2E < \nu < 2E$;
2. the space $\mathcal{C}$ of square integrable (and integrable) distributions (this is the analog of the classical Koopman phase space densities);
3. the space $\mathcal{D}$ of distributions which are boundary values of functions analytic in a domain providing the analytic continuations for complex contours in the variables ν and E (where the E integration extends from 0 to ∞);
4. the space $\mathcal{E}$ of distributions where the variable E extends from $-\infty$ to $+\infty$;
5. the space $\mathcal{F}$ of distributions analytic in a half plane for ν (except for an essential singularity at infinity) and suitably analytic in the variable E, so that we can develop a forward semigroup in time;
6. the space $\mathcal{A}$ of distributions analytic in a half plane for ν and suitably analytic in the variable E.

It is clear that these spaces are different. The space $\mathcal{C}$ contains the space $\mathcal{B}$, the space $\mathcal{A}$ is contained in the space $\mathcal{F}$, and the space $\mathcal{B}$ is contained in the space $\mathcal{E}$. As sets, $\mathcal{D}$ and $\mathcal{B}$ are dense in each other but there are elements in $\mathcal{B}$ which have no counterpart in $\mathcal{D}$.

For an isolated system the total energy E is bounded from below, but *for a system which is open to dynamical interaction with other systems this may not be an essential requirement.* It is only under this provision that the spaces $\mathcal{E}$, $\mathcal{F}$, and $\mathcal{A}$ are relevant.

A piecewise analytic function or any general measure which belongs to the space can be arbitrarily closely approximated by boundary values of analytic functions. Similarly a distribution along a complex contour in $\mathcal{C}$ can be approximated arbitrarily closely by functions in $\mathcal{B}$ though there is no one-to-one correspondence between the vectors (Sudarshan 1992). A specially interesting case is the complex delta distribution which assigns, to a function representing a vector in the dual space which is analytic in a domain containing the particular complex point, the value of the function at the complex point. There is *no vector in $\mathcal{B}$ or in $\mathcal{C}$* which corresponds to this vector, but in $\mathcal{E}$ there is such a vector. This vector in $\mathcal{E}$ would be appropriate for describing the simplest metastable excitations (see next section).

If we take a physical state in $\mathcal{D}$ and analytically continue it, we can consider it as a function along a complex contour together with one or more isolated poles (or, more generally, branch cuts). The pole terms control the behavior of the survival probability *but they are always accompanied by a background integral.* This background integral is essential: it reproduces the correct behavior at short (Zeno) times and long (Khalfin) times. An isolated pole by itself would not have a corresponding state in the space $\mathcal{D}$ of physical states.

In the space $\mathcal{F}$ the situation is quite different. There exist states in $\mathcal{F}$ which correspond precisely to a discrete complex point (or points). For a single complex point these are the familiar Gamow-Breit-Wigner states with a unique exponential dependence of the survival probability. These correspond to unique vectors in $\mathcal{F}$. The correspondence between $\mathcal{F}$ and its analytic continuation is a correspondence of complete spaces, not merely dense sets.

A special subset of these functions is analytic in an entire half plane. Such functions constitute the Hardy class functions with many interesting properties and are often taken to represent *nascent* metastable states. But the Hardy class property is *not* preserved by time evolution since

$$\rho(\nu; E) \rightarrow e^{-i\nu t}\rho(\nu; E) \tag{5.4}$$

has an *essential (exponential) singularity at infinity*. So after any finite time has elapsed, a nascent state evolves into a non-Hardy class function. These non-Hardy class functions are labeled by the index of exponential growth at infinity.

6. Need for Extended Space: Breaking of Time Symmetry

If the states having a purely exponential survival probability are to be included as natural ("physical") states, the spectrum of energies E has to be extended from $0 < E < \infty$ to $-\infty < E < \infty$. This leads to the spaces $\mathcal{E}$ and $\mathcal{F}$ of functions which are the boundary values of analytic functions, analytic in a half plane except, perhaps, for an exponential type singularity at infinity. The extension of the energy spectrum from $0 < E < \infty$ to $-\infty < E < \infty$ is equivalent to lifting the restriction

$$-2E < \nu < 2E \tag{6.1}$$

and allowing the ν integration to run from $-\infty$ to $+\infty$. So if the survival amplitude

$$P(t) = \int_{-\infty}^{\infty} \int_{-\infty}^{\infty} \rho(\nu; E) \, e^{-i\nu t} \, \rho^*(\nu; E) \, d\nu \, dE \tag{6.2}$$

is calculated for positive and negative times, we will get two distinct functions. For $t > 0$, $e^{-i\nu t}$ is a convergence factor for the lower half plane and an exponential increase for the upper half plane. Hence,

$$P(t) = \int_{-\infty}^{\infty} dE \int_{-\infty}^{\infty} d\nu \, e^{-i\nu t}\rho(\nu; E) \, \rho^*(\nu^*; E) \tag{6.3}$$

can be considered as a closed contour integration where the integration in the lower half plane is closed by an infinite semicircle whose contribution vanishes. The result is, then,

$$P(t) = \int_{-\infty}^{\infty} dE \cdot 2\pi i \sum_{\substack{\text{Residues in the} \\ \text{upper half plane}}} \rho(z, E) \rho^*(z^*, E) \, e^{-izt} , \qquad (6.4)$$

while for $t < 0$:

$$P(t) = \int_{-\infty}^{\infty} dE \cdot 2\pi i \sum_{\substack{\text{Residues in the} \\ \text{upper half plane}}} \rho(z, E) \, \rho^*(z^*, E) \, e^{-iz^*t} . \qquad (6.5)$$

In particular, if there are no poles in the upper half plane then

$$P(t) = \sum a_j \, e^{-\gamma_j t}, \qquad t \leq 0 . \qquad (6.6)$$

For the density functions in the extended space $\mathcal{F}$, with no poles in the lower half plane but poles in the upper half plane, $P(t)$ is exponentially decreasing with $|t|$ for the past $(t < 0)$.

These two classes of functions are disjoint except for the constant function, but functions constant in ν lead to an unphysical survival "probability"

$$P(t) = \delta(t) . \qquad (6.7)$$

In the extended space $\mathcal{F}$ there are thus two disjoint sets of states. The forward evolving states with

$$1 \geq P(t) = \text{Tr} \left(\rho(t) \rho(0) \right) > 0, \quad t > 0 , \qquad (6.8)$$

are the states consistent with the second law of thermodynamics. The second set is a time reversed set of backward regressing states with

$$1 \geq P(t) > 0, \quad t < 0 . \qquad (6.9)$$

These states are *not* suitable for a system that obeys the second law. The choice of physical states as forward evolving *is the breaking of time symmetry*. It is not dependent upon objective information or the act of isolated measurements but it is a property of thermodynamically adapted states and is picked automatically and universally. Open systems must have this time symmetry breaking if the second law of thermodynamics is regarded to be generally valid.

We now consider in detail the correspondence between the states of open systems in the space $\mathcal{F}$ and the states of closed systems in the spaces $\mathcal{C}$ and $\mathcal{D}$. Given any element of $\mathcal{F}$, we can restrict it to the domain

$$-2E < \nu < 2E, \quad 0 < E < \infty , \qquad (6.10)$$

and this yields an element of $\mathcal{D}$. However, given an element of $\mathcal{D}$, we cannot automatically extend it to $\mathcal{F}$ since analytic continuation to the negative real axis may not be possible.

Despite this *there is a natural splitting of any vector in $\mathcal{C}$ or in $\mathcal{D}$* into two vectors in $\mathcal{F}$ with different domains of analytic continuation. Given the function $f(\nu; E)$ for a vector in $\mathcal{C}$ which vanishes outside the range $-2E < \nu < 2E$, $E > 0$, we define (Sudarshan 1992)

$$g(\nu; E) = \frac{1}{\pi \mathrm{i}} \int_{-2E}^{2E} d\nu' \, \frac{f(\nu'; E)}{\nu - \nu' + \mathrm{i}\epsilon} \, . \tag{6.11}$$

This integral, if it exists, defines a function for all values of ν and is analytic in the lower half plane. Hence, it is a suitable member of $\mathcal{F}$ appropriate for describing an open system with forward (dissipative) evolution. A companion state with backward (dissipative) evolution is given by

$$h(\nu; E) = \frac{1}{\pi \mathrm{i}} \int_{-2E}^{2E} \frac{f(\nu'; E)}{\nu' - \nu - \mathrm{i}\epsilon} \, . \tag{6.12}$$

Clearly,

$$g(\nu; E) + h(\nu; E) = f(\nu; E) \, , \tag{6.13}$$

and hence

$$h(\nu; E) = -g(\nu; E) \quad \text{for} \quad |\nu| > 2E \, . \tag{6.14}$$

The functions $g(\nu; E)$ obtained here belong to the space $\mathcal{A}$ of functions analytic in the upper half plane, more restrictive than the space $\mathcal{F}$ admitting essential singularities of exponential type at infinity. This class of functions are the Hardy class functions. They are sometimes used to describe the corresponding states (Rosenblum and Rovyak 1985, Bohm and Gadella 1989).

If $f(\nu; E)$ behaves like $\exp(-\mathrm{i}\nu t)$ at infinity for $t > 0$, the definition of $g(\nu; E)$ remains unaltered, except that it will also behave like $\exp(-\mathrm{i}\nu t)$ at infinity. These types of behavior are therefore quite appropriate to describe forward dissipative evolution. The corresponding functions are not in the Hardy space $\mathcal{A}$ but in the space $\mathcal{F}$ discussed above.

The behavior of the density distribution is automatic with temporal evolution. Given $\rho(\nu; E)$ at time $t = 0$, the density distribution function at time t becomes

$$\rho_t(\nu; E) = \rho_o(\nu; E) \, \mathrm{e}^{-\mathrm{i}\nu t} \tag{6.15}$$

which belongs to the space $\mathcal{F}$ but not to the Hardy space $\mathcal{A}$. Each such state is labeled by a parameter τ characterizing the rate of exponential growth at infinity (see next section for more details). This parameter increases linearly with time evolution and may therefore be called the age of the state. The nascent states introduced before correspond to states of age zero.

7. Dynamical Processes and Dissipative Evolution

Our discussion has been focussed on generic systems so far. We have not yet talked about interactions, scattering, and explicit dissipative evolutions. Let us now address a generic system with a "total Hamiltonian" H which may be written

$$H = H_C + V \ , \tag{7.1}$$

where H_C is *isospectral* with H and is a simple structure, say a collection of "free Hamiltonians". If the states $|E, r\rangle$ are a set of (ideal) eigenstates of H_C with degeneracy label r, and if the states $|E, r\rangle\rangle$ are a corresponding set of (ideal) eigenstates of H, then there would be, by definition of the isospectral property of H and H_C, a one-to-one correspondence of states with the same degeneracies. Apart from normalization this correspondence may be written as:

$$
\begin{aligned}
Z^{-1/2}|E\rangle\rangle &= |E\rangle + (E - H_C + i\epsilon)^{-1} V |E\rangle\rangle \\
&= \left[1 - (E - H_C + i\epsilon)^{-1} V\right]^{-1} |E\rangle \ ,
\end{aligned} \tag{7.2}
$$

where Z is the wavefunction renormalization constant (Sudarshan et al. 1994). We shall omit this factor in what follows. Equation (7.2) can be formally expanded in a perturbation series

$$|E\rangle\rangle = |E\rangle + \sum_1^\infty \{G_C(E) V\}^n |E\rangle; \quad G_C(E) = (E - H_C + i\epsilon)^{-1} \ . \tag{7.3}$$

This corresponds to the "in" state appropriate for the initial state of a scattering process. Choosing the energy denominators with $-i\epsilon$ furnishes the "out" states. Both the solution and the perturbation expansion can be extended from the Hilbert spaces to the analytically continued spaces (Sudarshan et al. 1994). No substantial change is needed if the spectrum condition is relaxed to include arbitrarily large negative energy continua.

Rather than discussing the problem of the correspondence between the (ideal) eigenstates of H and H_C we could do the same in terms of (ideal) density distributions in relation to the Liouville operators $\mathcal{L}$ and $\mathcal{L}_C$:

$$
\begin{aligned}
\mathcal{L} R(\nu; E) &= \nu R(\nu; E), \\
\mathcal{L}_C R_C(\nu; E) &= \nu R_C(\nu; E) \ .
\end{aligned} \tag{7.4}
$$

Here, $R(\nu; E)$ and $R_C(\nu; E)$ are the (ideal) density distributions

$$
\begin{aligned}
R_{rs}(\nu; E) &= |E + \frac{1}{2}\nu, r\rangle\rangle\langle\langle E - \frac{1}{2}\nu, s|, \\
R_{c_{rs}}(\nu; E) &= |E + \frac{1}{2}\nu, r\rangle\,\langle E - \frac{1}{2}\nu, s|.
\end{aligned}
\tag{7.5}
$$

They are related by

$$
\begin{aligned}
R_{\text{in}}(\nu; E) &= \left\{1 - G_C\left(E + \frac{1}{2}\nu + i\epsilon\right)V\right\}^{-1} \times \\
&\quad \times\; R_C(\nu; E)\left\{1 - G_C^\dagger\left(E - \frac{1}{2}\nu + i\epsilon\right)V\right\}^{-1}, \\
R_{\text{out}}(\nu; E) &= \left\{1 - G_C\left(E + \frac{1}{2}\nu - i\epsilon\right)V\right\}^{-1} \times \\
&\quad \times\; R_C(\nu; E)\left\{1 - G_C^\dagger\left(E - \frac{1}{2}\nu - i\epsilon\right)V\right\}^{-1}.
\end{aligned}
\tag{7.6}
$$

The scattering probability for ideal states is

$$
\begin{aligned}
P_{\text{scatt}} &= \text{Tr}\left\{R_{\text{out}_{rs}}^\dagger(\nu, E)\,R_{\text{in}_{r's'}}(\nu', E')\right\} = \\
&= \langle\langle E - \frac{1}{2}\nu, s, \text{out}\,|\,E' - \frac{1}{2}\nu', s', \text{in}\rangle\rangle\,\langle\langle E' + \frac{1}{2}\nu', r', \text{in}\,|\,E + \frac{1}{2}\nu, r, \text{out}\rangle\rangle.
\end{aligned}
\tag{7.7}
$$

But

$$
\langle\langle E - \frac{1}{2}\nu, s, \text{out}\,|\,E' - \frac{1}{2}\nu', s', \text{in}\rangle\rangle = \delta\left(E - E' - \frac{1}{2}\nu + \frac{1}{2}\nu'\right)S_{rs}\left(E - \frac{1}{2}\nu\right)
\tag{7.8}
$$

is the scattering matrix. So the scattering probability is given by

$$
P_{\text{scatt}} = \delta(E - E')\,\delta(\nu - \nu')\,S_{rs}\left(E - \frac{1}{2}\nu\right)S_{r's'}^\dagger\left(E + \frac{1}{2}\nu\right).
\tag{7.9}
$$

With the proper understanding of the adjoints and duals, these considerations do not only apply to the real spectrum representations but also to analytic continuations. Of course, as long as one deals with the real spectrum representation, no metastabilities per se occur in the scattering probabilities. Rather, the resonances manifest themselves by characteristic "resonant shapes" of the probability distributions. If we want to consider the role of metastable states and the scattering of metastable excitations we should consider the analytic continuations which would uncover the resonant states as members of complete sets of states.

The (unnormalized) states $|E\rangle\rangle$ can be normalized by suitable state sensitive multiplicative changes. When this is done we denote

$$|E, \text{in}\rangle\rangle = \Omega_{\text{in}}\,|E\rangle; \quad |E, r; \text{in}\rangle\rangle = \Omega_{\text{in},r,s}|E, s\rangle \qquad (7.10)$$

with $\Omega_{\text{in},r,s}$ as a *unitary* operator. Then, if F_C is any invariant for the Hamiltonian H_C,

$$[F_C, H_C] = 0\,, \qquad (7.11)$$

then there is an invariant of H given by

$$F = \Omega F_C \Omega^\dagger, \quad [F, H] = 0\,, \qquad (7.12)$$

by virtue of

$$H = \Omega H_C \Omega^\dagger. \qquad (7.13)$$

But there is no guarantee that, if the matrix elements of F_C are smooth non-singular functions of E, the matrix elements of F are non-singular functions. In cases of non-trivial scattering the matrix elements of F will definitely be singular functions. F and F_C are unitarily equivalent and are constants of motion for H and H_C, respectively. If H and H_C share some symmetry properties, the corresponding operators are regular constants of motion for both H_C and H; this is analogous to the traditional constants of motion for the total Hamiltonian in classical dynamics. But there are additional constants of motion.

Let us now consider the time evolution. For the Hilbert space, the time evolution is the exponential of an imaginary multiple of a Hermitian Hamiltonian and, as such, it is unitary (norm preserving) no matter whether the energy spectrum is bounded from below or not. When we generalize to dual spaces, there is no longer a norm for the state. We must rather consider the invariance of the scalar product bilinear in the vector of the two dual spaces. If $\psi, \tilde{\phi}$ are such vectors, we have

$$\tilde{\phi}\psi \to \tilde{\phi}(t)\,\psi(t) = \tilde{\phi}\,e^{iHt}\,e^{-iHt}\psi = \tilde{\phi}\psi\,. \qquad (7.14)$$

If H has complex eigenvalues for ψ, there are ϕ with the *same* complex eigenvalues, and therefore the product of the two remains constant (Sudarshan et al. 1978, Sudarshan and Chiu 1993). But it is no longer true that $\psi(t)$ has the same "length" as $\psi(0)$: the "length" of a vector is not defined in dual spaces. But if there were complex eigenvalues of H, then it is clear that $\psi(t)$ can be a complex multiple of $\psi(0)$. But $\tilde{\phi}(t)$ would be the inverse multiple of $\tilde{\phi}(0)$.

Similar considerations apply for the spaces spanned by density distributions. In this case, there is always an invariant state with normalized trace, and all the other states are pseudodensities with trace zero. The evolution is "unitary", that is, it preserves scalar products between duals.

A measure of this scale change is provided by the survival amplitude

$$A(t) = \tilde{\phi}(0)\,\psi(t) \equiv \tilde{\phi}\,\mathrm{e}^{-\mathrm{i}Ht}\psi \,. \tag{7.15}$$

As ψ changes, so does $A(t)$. In particular, if the state ψ is dominated by a complex pole at z, then the survival amplitude has the dependence

$$A(t) = A(0)\,\mathrm{e}^{-\mathrm{i}zt} \,. \tag{7.16}$$

Whenever $\mathrm{Im}\,z < 0$ for $t > 0$, then we have:

$$|A(t)| < |A(0)| \,. \tag{7.17}$$

Thus, *in this sense*, the complex energy state is a *decaying* state.

From the vectors $\psi, \tilde{\phi}$ in the dual spaces we can construct pseudodensity distributions

$$\rho = \psi_1 \psi_2^{\dagger}, \quad \sigma = \tilde{\phi}_1^{\dagger} \tilde{\phi}_2 \,, \tag{7.18}$$

which generate dual spaces (Segal 1947, Haag and Kastler 1964). With these we can calculate the survival probability

$$
\begin{aligned}
P(t) &= \mathrm{Tr}(\sigma\rho(t)) = \mathrm{Tr}\left(\sigma\,\mathrm{e}^{-\mathrm{i}Ht}\rho\,\mathrm{e}^{\mathrm{i}Ht}\right) \\
&= \left(\tilde{\phi}\,\mathrm{e}^{-\mathrm{i}Ht}\psi\right)\left(\psi^{\dagger}\mathrm{e}^{\mathrm{i}Ht}\tilde{\phi}^{\dagger}\right) \\
&= |A(t)|^2 \,.
\end{aligned}
\tag{7.19}
$$

Therefore, if we know the survival amplitude, the survival probability can be computed.

When the states contain a superposition of eigenvectors of H, the behavior is given by:

$$P(t) = \left|\sum \tilde{\phi}_E \psi_E \,\mathrm{e}^{-\mathrm{i}Et}\right|^2 \longrightarrow \left|\int \tilde{\phi}(E)\psi(E)\,\mathrm{e}^{-\mathrm{i}Et}dE\right|^2 \,. \tag{7.20}$$

As long as $\psi(E)$ and $\tilde{\phi}(E)$ are boundary values of functions analytic in the lower half plane except for poles (or "short" branch cuts), we can evaluate the integral over E by closing the contour with an infinite semicircle in the lower half plane (for $t > 0$). If there is only one pole in the lower half plane, the entire survival amplitude A is as if there were only one complex "energy" point eigenvalue contributing to the integral. More generally, A will be the superposition of several such "energies" and possibly an integral over them.

Let us consider the single complex eigenvalue in detail, even though the discrete complex eigenvalue must be accompanied by associated branch cuts. The state vector

$$\psi_0(E) = N_0 \, (E - z)^{-1}; \quad N_0^2 = \frac{z^* - z}{2\pi i} , \tag{7.21}$$

and its dual

$$\tilde{\phi}_0(E) = N_0 \, (E - z^*)^{-1} \tag{7.22}$$

give the survival amplitude for $t > 0$:

$$
\begin{aligned}
A_0(t) &= \int \tilde{\phi}_0(E) \, e^{iEt} \phi_0(E) dE \\
&= N_0^2 \int \frac{e^{-iEt}}{(E - z)(E - z^*)} dE \\
&= \frac{-2\pi i}{z - z*} N_0^2 \, e^{-izt} = e^{-izt} .
\end{aligned}
\tag{7.23}
$$

In this case, the analytic continuation of the wave function vanishes as z^{-1} at infinity. For $t < 0$ we get similarly:

$$A_0(t) = e^{iz^* t} . \tag{7.24}$$

The survival amplitude as a function of t is therefore the join of two *distinct* analytic functions, one for $t > 0$ and another one for $t < 0$.

Now consider the state

$$
\begin{aligned}
\psi_\tau(E) &= N_0(E - z)^{-1} e^{-iz t} \\
\tilde{\phi}_\tau(E) &= N_0(E - z^*)^{-1} e^{-iz^* t}
\end{aligned}
\tag{7.25}
$$

obtained by a multiplicative transformation. Then the survival amplitude is

$$
\begin{aligned}
A_\tau(t) &= N_0^2 \int (E - Z)^{-1} (E - Z^*)^{-1} e^{i(z - z^*)\tau} \, e^{-izt} dE \\
&= N_0^2 \frac{-2\pi i}{z - z^*} e^{-izt}
\end{aligned}
\tag{7.26}
$$

which may be written

$$A_\tau(t) = A_0(t + \tau) . \tag{7.27}$$

In other words: the state ψ_τ may be thought of as having been created at time $t = \tau$. If τ is positive, we extrapolate, *for these states*, the semigroup for negative values of t such that

$$t + \tau > 0 . \tag{7.28}$$

This quantity τ may be called the "age" of the state in the *extended* space (Ramakrishnan 1959, Sudarshan 1992).

Having defined the age and the survival amplitude for the states we can define the age and survival probability for density distributions. Analytic density distributions in the space $\mathcal{F}$ can be chosen so that we can define the forward semigroup on them. But after the time evolution for any finite time is considered, the states are no longer in $\mathcal{F}$ but are in $\mathcal{E}$. If we denote a state in $\mathcal{F}$ at $\tau = 0$ by $\rho_0(\nu; E)$, then

$$\rho_\tau(\nu; E) = e^{-i\nu\tau} \rho_0(\nu; E) \qquad (7.29)$$

is not in the space $\mathcal{F}$ but remains in the space $\mathcal{E}$. For them, the forward semigroup can be extrapolated to *negative* values of t such that $t + \tau > 0$. These are the metastable states with age τ.

In the preceding discussions we have labeled the density distribution $\rho(\nu; E)$ with the labels appropriate for the total energy and total Liouvillean. In many cases, however, we have a comparison Hamiltonian H_C and an interaction V such that H_C is isospectral with H and

$$H = H_C + V . \qquad (7.30)$$

Then we could have an alternative labeling of the states by ν_C, E_C appropriate to

$$H_C \rho = \left(E_C + \frac{1}{2}\nu_C\right)\rho ,$$
$$\rho H_C = \left(E_C - \frac{1}{2}\nu_C\right)\rho . \qquad (7.31)$$

To avoid confusion we use the symbol R for the density distribution labeled by ν_C, E_C so that

$$R(\nu_C; E_C) \equiv \rho(\nu; E), \qquad (7.32)$$

with ν_C having the same range as ν and E_C the same range as E. Then

$$R(\nu; E) = \left(1 - G_C\left(E + \frac{1}{2}\nu\right)V\right)^{-1} R_0 \left(\left(1 - G_C\left(E - \frac{1}{2}\nu\right)V\right)^\dagger\right)^{-1}. \qquad (7.33)$$

It may be that the creation of the state is most simply described in the comparison Hamiltonian representation R_C. Then the dependence on the variables ν, E is governed by the wave matrix factors preceding and following R_C. In the special case of the Dirac-Friedrichs-Lee model of a discrete (metastable) state coupled to a continuum, the resonant complex pole plus background dependence is immediately realized if the initial state is the discrete state of the comparison Hamiltonian with the discrete energy level chosen to prevent instability, and then by continuing in the "mass" parameter of the model. The time dependence of the survival amplitude and survival probability have been studied extensively in the literature (see, e.g., Chiu et al. 1997, Chiu and Sudarshan 1990, Sudarshan et al. 1978).

More generally, the wave matrix

$$\Omega(E) = (1 - G(E)\,V)^{-1} \tag{7.34}$$

has an *analytic dependence* on E. As a consequence, if $R_C(\nu; E)$ is a simple function of E and ν, then $R(\nu; E)$ will be *analytic* in both ν and E. The singularities of the wave matrix in the complex variable of energy reappears in the survival probability. While both the scattering amplitude and wave matrix depend on both the total Hamiltonian H and the comparison Hamiltonian H_C, it is known that only the singularities of the wave matrix appear in the survival amplitude. The redundant poles of the scattering amplitude, if there are any, do not contribute. Anyway, the *survival amplitude depends on both the total Hamiltonian H and the comparison Hamiltonian H_C*.

Another point to be noted is that when we consider the time evolution, despite the fact that the (norm)^2 of the state is not defined directly, we still can talk of affine scale; that is, whether the state gets multiplied by a number e^{izt}. Such states *do not* exist in $\mathcal{C}$ but they exist in the extended space $\mathcal{E}$. They may be realized along the real axis but could equally well be identified as complex discrete energy states. As a consequence, while the product of a state and its dual is invariant under time evolution, the survival amplitude depends on time. For the special state corresponding to a discrete complex pole, the dependence is purely exponential.

References

Bachelier L. (1900): Theorie de speculation. *Annal. Scientifique de la Ecole Normale* **16**, 21–102.

Bloch F. (1946): Nuclear induction. *Phys. Rev.* **70**, 460–474.

Bohm A. and Gadella M. (1989): *Dirac Kets, Gamow Vectors and Gelfand Triplets. Lecture Notes in Physics, Vol. 348*, ed. by A. Bohm and J.D. Dollard (Springer, Berlin).

Chiu C.B. and Sudarshan E.C.G. (1990): Decay and evolution of the neutral kaon. *Phys. Rev. D* **42**, 3712–3723.

Chiu C.B., Sudarshan E.C.G., and Bhamathi G. (1997): Unstable systems in generalized quantum theory. *Advances in Chemical Physics* **99**, 121–210. See also references given there.

Choi M.D. (1972): Positive linear maps on C* algebras. *Illinois J. Math.* **18**, 565–574.

Choi M.D. (1974): A Schwarz inequality for positive linear maps on C* algebras. *Can. J. Math.* **24**, 520–529.

Davies E.B. (1969): Quantum stochastic processes. *Commun. Math. Phys.* **15**, 277–304.

Davies E.B. (1970): Quantum stochastic processes II. *Commun. Math. Phys.* **19**, 83–105.

Davies E.B. (1971): Quantum stochastic processes III. *Commun. Math. Phys.* **23**, 51–70.

Einstein A. (1905): Zur Elektrodynamik bewegter Körper. *Annalen der Physik* **17**, 891–921.

Gorini V., Kossakowski A., and Sudarshan E.C.G. (1976): Completely positive dynamical semigroups of N-level systems. *J. Math. Phys.* **17**, 821–825.

Gorini V., Frigerio A., Verri F., Kossakowski A., and Sudarshan E.C.G. (1978): Properties of quantum Markovian master equations. *Rep. Math. Phys.* **13**, 149–173.

Haag R. and Kastler D. (1964): An algebraic approach to quantum field theory. *J. Math. Phys.* **5**, 848–861.

Hu N. (1948): On the application of Heisenberg's theory of S-matrix to the problems of resonance scattering and reactions in nuclear physics. *Phys. Rev.* **74**, 131–140.

Kapur P.L. and Peierls R. (1938): The dispersion formula for nuclear reactions. *Proc. Roy. Soc. Lond. A* **166**, 277–295.

Khalfin L. (1958): Contribution to the decay theory of a quasi-stationary state. *Soviet Physics JETP* **6**, 1053–1063.

Lindblad G. (1975): Completely positive maps and entropy inequalities. *Commun. Math. Phys.* **40**, 147–151.

Mehta C.L. and Sudarshan E.C.G. (1965): Relation between quantum and semiclassical description of optical coherence. *Phys. Rev. B* **138**, 274–280.

Moyal J.E. (1949): Quantum mechanics as a statistical theory. *Proc. Cambridge Phil. Soc.* **45**, 99–124.

Neumann J. von (1955): *Mathematical Foundations of Quantum Mechanics* (Princeton University Press, Princeton).

Parravicini G., Gorini V., and Sudarshan E.C.G. (1980): Resonances, scattering theory, and rigged Hilbert spaces. *J. Math. Phys.* **21**, 2208–2226.

Petrosky T. and Prigogine I. (1988): Poincaré theorem and unitary transformations for classical and quantum systems. *Physica A*, **147**, 439–460.

Petrosky T. and Prigogine I. (1993): Poincaré resonances and the limits of trajectory dynamics. *Proc. Nat. Acad. Sci. USA*, **90**, 9393–9397.

Prigogine I. (1997): *The End of Certainty* (Free Press, New York).

Ramakrishnan A. (1959): Probability and stochastic processes. In *Handbuch der Physik, Vol. III/2*, ed. by S. Flügge (Springer, Berlin).

Rosenblum M. and Rovyak J. (1985): *Hardy Classes and Operator Theory* (Oxford University Press, Oxford).

Segal I.E. (1947): Postulates for general quantum mechanics. *Ann. Math.* **48**, 930–948.

Sudarshan E.C.G. (1963): Equivalence of semiclassical and quantum mechanical descriptions of statistical light beams. *Phys. Rev. Lett.* **10**, 277–279.

Sudarshan E.C.G. (1985): Quantum measurements and dynamical maps. In *From SU(3) to Gravity*, ed. by E. Gotsman and G. Tauber (Cambridge University Press, Cambridge), p. 33.

Sudarshan E.C.G. (1992): Quantum dynamics, metastable states, and contractive semigroups. *Phys. Rev A* **46**, 37–48.

Sudarshan E.C.G., Mathews P.M., and Rau J. (1961): Stochastic dynamics of quantum mechanical systems. *Phys. Rev.* **121**, 920–924.

Sudarshan E.C.G. and Mukunda N. (1974): *Classical Dynamics* (Wiley, New York).

Sudarshan E.C.G., Chiu C.B., and Gorini V. (1978): Decaying states as complex eigenvectors in generalized quantum mechanics. *Phys. Rev. D* **18**, 2914–2929.

Sudarshan E.C.G. and Chiu C.B. (1993): Analytic continuation of quantum systems and their temporal evolution. *Phys. Rev. D* **47**, 2602–2614.

Sudarshan E.C.G., Chiu C.B., and Bhamathi G. (1994): Perturbation theory on generalized quantum mechanical systems. *Physica A* **202**, 540–552.

Wigner E.P. (1932): On the quantum correction for thermodynamic equilibrium. *Phys. Rev.* **40**, 749–759.

CONTEXTUAL BACKGROUND

Dennis H. Rouvray:
PERSPECTIVES ON THE MOLECULAR MICROWORLD

Throughout his entire scientific life Primas kept careful track of mathematical literature. In 1955 he discovered that in some remote place the distinguished mathematician Collatz had published some seminal work (Collatz and Sinogowitz 1948) on an eigenvalue problem in connection with graphs. Primas realized that the (2-parameter) Hückel-Hamiltonian is nothing else than the incidence matrix of the π-electron system. In his characteristic way he concluded that Hückel theory is "only" graph theory and is, in fact, free of quantum mechanics. (As far as we know, the Hückel-Hamiltonian was never rigorously derived from quantum mechanics. It is a heuristic invention, remotely inspired by the Heitler-London theory of the chemical bond.)

Primas' contribution was ignored for about twenty years until the Yugoslavian school boosted the graph theory program (a prominent exponent being N. Trinajstić). In many of these papers Primas was mentioned as the early father of this program. Ironically, they attempted to go into a direction opposite to that suggested by Primas: they justified the application of graph theory to organic chemistry by its relation to Hückel theory, whose poor quantum mechanical status was not generally known among organic chemists at that time. Primas, however, never accepted this kind of argument, and there are people who recall quite unpleasant seminar discussions with protagonists of the chemical graph theory research program.

In his contribution, Rouvray puts forward a different, more optimistic point of view. He proposes that the graph theoretical approach be taken seriously. He is at pains, on the one hand, to deduce genuine chemical concepts from this approach, and, on the other hand, to emphasize the compatibility of chemical graph theory with basic quantum mechanical concepts such as the Born-Oppenheimer-approximation. Rouvray combines this presentation with a glance at the open problems of theoretical chemistry as a whole, from his own point of view. This makes it interesting to compare his standpoint with the view on theoretical chemistry which Hans Primas used to give his students at the end of the third semester as recorded in the textbook by Primas and Müller-Herold (1990):

"In our opinion, the more fundamental problems of theoretical chemistry lie neither with calculators nor with mathematical problems but with conceptual questions. The problem can be viewed from the standpoint of the scientist, the philosopher, or the engineer. These standpoints are different, but they are all legitimate. ... So we have to ask ourselves:

- *What are we trying to explain? (from the viewpoint of the scientist)*

- *What are we trying to understand? (from the viewpoint of the philosopher)*

- *What are we trying to master? (from the viewpoint of the engineer)*

Depending on whether one wishes to develop theoretical chemistry along the lines of scientific practice, philosophical perception, or in engineering terms, one will arrive at a different definition of priorities. Independently of any future programs, it is good to remind ourselves of everything we do not understand at the moment and all the things we cannot do. On the basis of the principles of quantum mechanics, some examples of the things we do not know are:

- *what is the state of matter at zero Kelvin temperature (solid? ferromagnetic? suprafluid?),*

- *how many aggregate states of matter are there,*

- *how can a liquid be theoretically characterized,*

- *what is the precise connection between molecules and chemically pure substances (prize question: what exactly does liquid water have to do with the individual molecule H_2O)?.*

These and similar problems are closely connected with the following much more fundamental question: *to what extent is quantum mechanics really the fundamental theory for all chemical and molecular-biological phenomena?* That is a philosophical question and is hence avoided by many scientists and engineers. In fact, researchers who claim that they 'do not want to bother with philosophy' often take as their tacit starting point an unreflected and thus poor philosophy; as a result 'their prejudices lead them to irrelevant questions' (Heisenberg 1976).

The assertion that molecules exist in an absolute sense cannot be reconciled with the basic principles of quantum mechanics. From the standpoint of quantum theory, the object of scientific investigations is not an undivided reality but rather abstract patterns. In classical physics, we can divide the world up into objects and perceive the correlations between the objects by means of reciprocal actions. In quantum mechanics, there are no objects a priori. The familiar division of the world into objects is not a fundamental characteristic of objective reality but an abstraction produced by our intellect in order to think in terms of categories.

The abstractions necessary for the concept of molecules are basically different from those that lead to the concept of substances and to chemical thermodynamics. In other words, the quantum theory of molecules and chemical thermodynamics are mutually exclusive forms of observation, yet they do not contradict each other. According to the context, one form of observation may be convenient, the other inconvenient, without the classification of 'true' or 'false' being appropriate. A form of observation is never true; it may be correct if it is appropriate, expedient and in conformity with a clearly specified class of experiments.

The molecular description of matter is compatible with quantum mechanics and – as far as we know – empirically correct, but it comprises only s small fraction of the essence of matter. We need various basically different descriptions and complementary standpoints if the full reality is to become perceptible. As the choice of viewpoint is not laid down for us by quantum mechanics, a constructive mind can exploit this freedom, create a new reality, and thus become the fabricator mundi, in the style of Leonardo da Vinci.

Thus a topical and urgent task for theoretical chemistry is to develop the various complementary points of view, to elucidate the reciprocal dependencies of complementary descriptions, and to find novel descriptions."

References

Collatz L . and Sinogowitz U. (1948): Spektren der Graphen (unpublished), referred to in W. Süess: *Reine Mathematik, Teil 2*, p. 251; Vol. 2 of the German edition of "FIAT Review", Wiesbaden 1948.

Heisenberg W. (1976): Was ist ein Elementarteilchen? *Naturwissenschaften* **63**, 1–7.

Primas H. and Müller-Herold U. (1990): *Elementare Quantenchemie* (Teubner, Stuttgart), Chap. 6.5: Offene Probleme und Ausblick.

PERSPECTIVES ON THE MOLECULAR MICROWORLD

DENNIS H. ROUVRAY
Department of Chemistry, University of Georgia,
Athens, Georgia 30602-2556, USA

One thing I have learned in a long life:
that all our science,
measured against reality,
is primitive and childlike –
and yet it is the most precious thing we have.

(Albert Einstein 1951)

1. General Introduction

By the time that the present millennium expires at the end of the year 2000, quantum theory will have been with us for some 75 years. From the moment of its inception, quantum theory provoked an immense furor and throughout its entire existence has continued to exert a highly disruptive influence on science as a whole. Quantum theory may truly be said to be the *enfant terrible* of science! What is it about this theory that causes so much heat to be generated? Broadly speaking, it is the friction between the ideas of quantum theory and more traditional views of our world that have evolved over the past three centuries in particular. Quantum theory calls upon us to abandon several of the most cherished concepts of classical physics and to start thinking in strange and unaccustomed ways. To begin with, we are required to relinquish any notion that the entities constituting the microworld of atoms and molecules behave like classical particles. Exactly what these entities are is unknown at the present time. Many workers have suggested that such entities can no longer be viewed as substantive particles, even though they possess a rest mass. They are to be characterized solely in terms of wave functions that describe the vibrations of supposedly empty space. But the wave functions employed are in general complex, i.e., they include the square root of minus one, which means that to obtain anything approaching solidity from them we have to take the square of the amplitude of the modulus of the wave function.[1] This computation yields a quantity that gives us a measure of the probability of the particle being located at a given point in space at a given time.[2] As Heisenberg summed up the situation (Heisenberg 1958, p. 180): "in quantum theory all the classical concepts, when applied to the atom, ... are correlated with statistical expectations; only in rare cases may the expectation become the equivalent of certainty."

Quantum theory is also unique in being the first theory in modern times to adopt a holistic view of the world. A simple definition of a holistic world is one that cannot be divided up into nontrivial subsystems in a manner such that the states of the subsystems determine the state of the world (Primas 1983, p. 324). Within the context of quantum theory it is thus not feasible to excise some subsystem from the world and to examine it in isolation from the rest of the world. It can be shown that such an excised portion of the world would always be inextricably interconnected with the remainder of the world via Einstein-Podolsky-Rosen-type correlations (Einstein et al. 1935). This fact casts serious doubt on whether it is theoretically possible to study molecules in isolation from their environment and even whether we can define molecules in any chemically relevant way by making use of quantum theory. A careful analysis of the issues involved led Primas to formulate a number of general guidelines. He concluded (Primas 1983, p. 253) that "a reasonable and consistent unified language for theoretical chemistry" would need to be found and that it would have to be one that "choose[s] its regulative principles in such a manner that the fair requirements of a moderate realism are fulfilled". Underlying these conclusions was the recognition that "[t]here are no entities in our world which have observable attributes independently of any abstraction. Observable phenomena are created by abstracting from some Einstein-Podolsky-Rosen correlations. Without such an abstraction there are no phenomena."

The proposed embarkation point for any foray we may wish to make into the molecular microworld is thus a theory that is still contentious in many respects. Why we should want to make use of this particular vehicle for our excursions into the molecular realm is however easily answered. Quantum theory has proved time and again to be a very reliable means of arriving at our goal of predicting the behavior of the microworld. Indeed, it has been so successful in modeling physical and chemical phenomena that it (along

---

[1]Note added by the editors: This statement could be misleading. Professor Rouvray is, of course, well aware that quantum mechanics can be formulated in real Hilbert spaces as well. There is, however, no direct connection between physical "solidity" on the one hand and the number field of quantum mechanical Hilbert spaces on the other.

[2]Note added by the editors: More precisely, it is the probability of a particle being located *by a measurement of the first kind* at a given point in space at a given time.

with relativity theory) has been hailed (see, for instance, Jammer 1966, p. 23) as one of the two major breakthroughs achieved by twentieth century science. The sole justification, however, for the now almost universal acceptance of quantum theory is its undisputed ability to make predictions that accord with experimental findings. In many instances the concordance between the two has been astonishing. For example, use of the Dirac equation has enabled us to predict with remarkable accuracy the fine structure of the hydrogen spectrum (Milonni 1994, p. 328). Similarly, this equation can yield the deviations from the Thomson cross section that occur in the Compton scattering of light by electrons (Milonni 1994, p. 328). One of its most phenomenal successes has been predictions of the magnetic moment of the electron (Penrose 1997, p. 51). In Dirac units, this moment is predicted to be 1.001159652(46) whereas the observed value is 1.0011596521(93) – an accuracy close to one part in a hundred billion (10^{11}). The overall picture is equally satisfactory: there is currently no known disagreement between the predictions of quantum theory and any unequivocally established experimental result.

2. Chemists and Physicists

What has been the response of the scientific community to the unruly child that is quantum theory? As might be expected with something so awkward and troublesome, the response has been very varied. It has ranged from warm embrace to hesitant tolerance. Let us consider the physicists first. Quotes from three major physicists of the twentieth century reveal just how dramatically opinions differ on the merits and significance of quantum theory. Paul Dirac, a lifelong proponent of the theory, proclaimed as early as 1929 that (Dirac 1929) "[t]he general theory of quantum mechanics is now almost complete. ... The underlying physical laws necessary for the mathematical theory of a large part of physics and the whole of chemistry are thus completely known, and the difficulty is only that the exact application of these laws leads to equations much too complicated to be soluble." Contrasting sharply with this enthusiastic support are the doubts and misgivings of Albert Einstein, a lifelong pessimist about the value of the theory, who had to admit that (Einstein 1926) "[q]uantum mechanics is very impressive" but who went on to confess that "an inner voice tells me that it is not the real thing. ... In any case I am convinced that He [God] does not play dice".[3] The third physicist we quote, Richard Feynman, observed in somewhat jocular vein that, whether one be a supporter or a detractor,

---

[3]Originally in German: "Die Quantenmechanik ist sehr achtung-gebietend. Aber eine innere Stimme sagt mir, dass das doch nicht der wahre Jakob ist. ... Jedenfalls bin ich überzeugt, dass der [Herrgott] nicht würfelt."

the theory is basically unintelligible. At the beginning of one of his noted lecture series, he warned his audience that (Feynman 1985, p. 9f) "the way we have to describe Nature is generally incomprehensible to us. You're not going to be able to understand it ... my physics students don't understand it either. That is because I don't understand it. Nobody does."

The attempts of quantum theoreticians to elucidate the nature of matter have been characterized by several different approaches. From the outset, three kinds of mathematics were employed for the description of matter: Schrödinger (1926) used wave mechanics and viewed matter in terms of wave patterns, Heisenberg (1925) used matrix mechanics which led to a discrete representation of reality, and Dirac (1925) used so-called q-number mechanics (based on the noncommutative behavior of the parameters x (position) and p (momentum), i.e., on the fact that $xp \neq px$) which gave no specific interpretation of the nature of matter. This early work, often referred to now as pioneer quantum mechanics (Primas 1983, p. 345), immediately provoked the question whether three different interpretations of physical reality were being presented. Surprisingly, however, all three of these approaches yielded identical predictions in most cases, though the Dirac approach became the one of choice whenever relativistic effects play a significant role because it is the only one of the three that is relativistically correct, i.e., its basic equation is Lorentz-Fitzgerald invariant (Pyykkö 1986, p. 153). In more recent times, most of the ideas of pioneer quantum mechanics have been modified, elaborated or incorporated into more all-embracing theories. Some of the newer thinking that has emerged over the past three decades has been so astonishing that it would appear to have come straight out of the annals of science fiction! We may include in this category the indefinable holomovement of Bohm (1980), the many-worlds interpretation of Everett (1957), and the bootstrap theory of particles due to Chew (1966) which claims that nature cannot be reduced to any fundamental entities at all.

While such exotic ideas now appear to be the mainstay of modern theoretical physics, the chemical community at large has remained aloof and has generally adopted a highly skeptical attitude toward the more bizarre claims and theories advocated by physicists. As far as most chemists are concerned, quantum theory means essentially pioneer quantum theory; anything else is simply left aside for the physicists to argue about. Accordingly, the time-independent Schrödinger equation is usually taken to be the last word on the subject, though occasional use is made of the Dirac equation. Chemists have thus imposed an artificial limit on the extent to which they are prepared to admit quantum ideas into their domain. Why is there such great reluctance on the part of chemists to embrace the more cutting-edge ideas of modern quantum theory? Could it be that chemists believe there is nothing left to explore – that they tacitly accept the provocative claim of Dirac (1929) that the Dirac equation contains all of chemistry? It is unlikely that such thinking motivates chemists since the latter have always been rather uneasy about the excessive mathematization of their discipline. While it is widely acknowledged that mathematics is a useful tool, the feeling is that its application should not be overdone, especially if it appears to be reducing reality to evanescent matter waves or even removing fundamental entities altogether!

3. Theories and Models

The situation in which chemists find themselves has been clearly recognized by Primas, who commented (see, e.g., Rouvray 1995a) that "everyone (including the theoreticians) believes in some kind of realism. It is disturbing and inconsistent to expect that a scientist accepts the existence of an external reality in the world of everyday experience and in his laboratory but not if he is working on his theories." Generations of chemists have been nurtured on a picture of reality that is solid and substantial. The first ball-and-stick models of molecules were constructed by Dalton around the year 1810 and these have been widely used, especially for pedagogical purposes, ever since (Rouvray 1997). Moreover, as early as 1808, Wollaston understood that (Wollaston 1808) "when our views are sufficiently extended to enable us to reason with precision concerning the proportions of elementary atoms, we shall ... be obliged to acquire a geometrical conception of their relative arrangement in all the three dimensions of solid extension". Some of his depictions (Wollaston 1813) of three-dimensional molecular species are reproduced in Figure 1.

It was during the period 1845–1875, however, when molecules really began to take shape. This period marks the rise of structure theory in chemistry, a time when the structures of many small molecules such as those of methane and benzene were being worked out largely on the basis of chemical intuition (Benfey 1964). The idea that to a considerable extent the properties of a molecule are determined by the geometric arrangement of its atoms in space took firm hold (Benfey 1981). Structure theory has continued to be an indispensable theoretical foundation for chemistry ever since. Indeed, Frankland commented that (see Russell 1996, p. 276) "[c]hemistry owes its progress from empiricism to exact science entirely to Strukturchemie". Lewis went even further and claimed that (Lewis 1923, p. 1) "no generalization of science, even if we include those capable of exact mathematical statement, has ever achieved a greater success in assembling in a simple way a multitude of heterogeneous observations than this group of ideas".

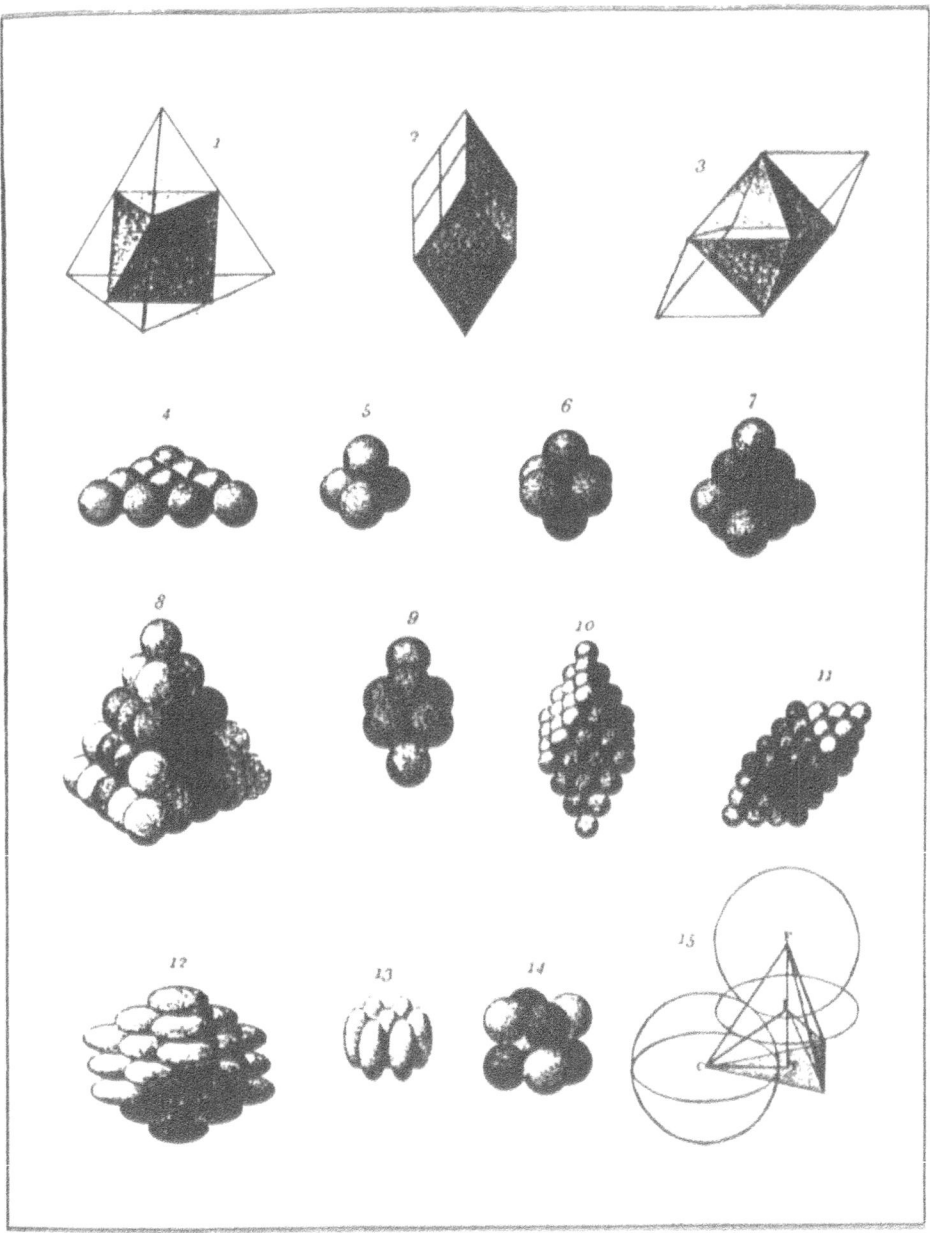

Figure 1.   One of the earliest depictions of molecules as three-dimensional structures, due to William Wollaston and dating from the year 1813. Reproduced from Wollaston (1813).

From these early beginnings, the use of models in the chemical domain has continued to gather momentum down to the present day. In particular, the advent of the computer age and the now widespread use of computer facilities has rendered the construction and manipulation of molecular models in real time on screen a commonplace (Doucet and Weber 1996). Are chemists justified in placing such reliance in their models? On the face of it, the stance that they adopt appears entirely reasonable. It is understandable that chemists are loath to relinquish a part of their heritage that has proven to be so highly efficacious in enabling them to get to grips with molecular reality. It is probably no exaggeration to assert that the majority of chemists feel viscerally about molecules and that this feeling is borne of years of hands-on experience in dealing with chemical substances. Molecules begin to assume characters, one might almost say personalities, and this engenders an affectionate response on the part of chemists. Such feelings arise, however, only because molecules are perceived to be real and tangible. It would be out of the question to entertain the same kind of feelings toward matter waves! The alien and arcane nature of matter waves was brought into sharp focus by Coulson who observed that (Coulson 1955) "the tangible, the real, the solid, is explained by the intangible, the unreal, the purely mental. Yet that is what we chemists are always doing, wave-mechanically or otherwise." Are chemists convinced that what they are alleged to be doing all the time is a valid exercise? If they were, it would hardly have been necessary for Pauling to remind them that (Pauling 1985) "[c]hemistry is a quantum phenomenon, or, rather, a great collection of quantum phenomena". Nor would a recent editorial in a drug design journal (Counts 1991) have bluntly stated that chemists do not believe in wave functions.

Chemists are inveterate doubters and, along with Cassirer, demand to know if (Cassirer 1964, p. 300) "all this is anything other and anything more than a strange phantasmagoria? How can the multitude of models and their continual refinement help us when none of them can make any claim to actual or ultimate truth?"[4] A strategy that suggests itself in such a vexing situation is to take a hard look at all the available experimental evidence. Rather surprisingly, there is a substantial body of evidence, much of it quite recent, that supports the stance of most chemists. Whereas it has proved possible to image the inhabitants of the microcosm in particulate form, it has never been possible to get a direct glimpse of matter waves. From the earliest researches into the nature of electrons, it has been the case that individual electrons impinge on phosphorescent screens as particles and not as waves (Thomson 1937, p. 335). In the case of atoms, a variety of exper-

---

[4]Originally in German: "Ist all dies etwas anderes und etwas mehr als eine seltsame Phantasmagorie? Was hilft uns die Fülle der Bilder und ihre ständige Verfeinerung, wenn keines von ihnen Anspruch auf wirkliche, auf endgültige Wahrheit hat?"

imental techniques, including field ion microscopy, X-ray crystallography and scanning tunneling microscopy, have revealed that atoms behave as point particles (Müller 1965; Ball 1994, p. 83). More recent work, which has generated considerable excitement and eventually led to the award of the 1989 Nobel Prize in physics (Pool 1989), involved the isolation of single particles in Penning magnetic traps. To date it has been possible to trap a number of different species; these have included an electron, a positron, various individual atoms and ions, and even antimatter atoms (Dehmelt 1988). All of these appeared to behave as tiny spheres, the radius of the electron proving to have a value that was less than 10^{-20} cm. Particles as large as atoms could even be viewed with the unaided eye provided they were stimulated with laser light. The appearance of a barium atom observed by von Baeyer was described as (von Baeyer 1992, p. 98) "not the calm, serene globe she appeared to be but a seething dynamic little lady, flinging photons in all directions like a whirling dervish".

4. Geometry and Topology

Chemistry has developed into an unashamedly naive and realist science, the philosopher Bradley (1955) reminding us that to "Avogadro and Canniz-zaro, as to Couper and Kekulé, the molecules and atoms considered ... were real objects; they were thought of in the same way as one thinks of tables and chairs". Even today, the majority of chemists regard molecules as real and isolable entities characterized by a more or less fixed geometry. Criticism of such ideas is seldom heard, and when it is, it tends to be surprisingly muted. One can certainly sense the hesitancy felt by Primas in approaching this theme. After asking (Primas 1983, p. 292): "Are there molecules?" he goes on to declare that "[t]his seems to be a silly question. ... The doctrine of the existence of molecules is so widely received and so respectable that we may have some difficulty to discuss it seriously." However, within the rather constricted confines of pioneer quantum mechanics, a molecule cannot be assigned a structure for the simple reason that the notion of a fixed geometry or a molecular shape is an alien one. Pioneer quantum mechanics is quite unable to offer a correct or a consistent description of structure at the molecular level (Woolley 1978). Molecular structure is a classical concept that has to be introduced by making use of an approach put forward by Born and Oppenheimer (1927). Structure is effectively created by regarding the atomic nuclei as classical particles, i.e., by assuming that the nuclei are fixed in their positions and that only the electrons are smeared out in the form of matter waves. In the Born-Oppenheimer limit of nuclei with infinite mass, the notion of molecular structure, of course, becomes a genuinely classical concept.

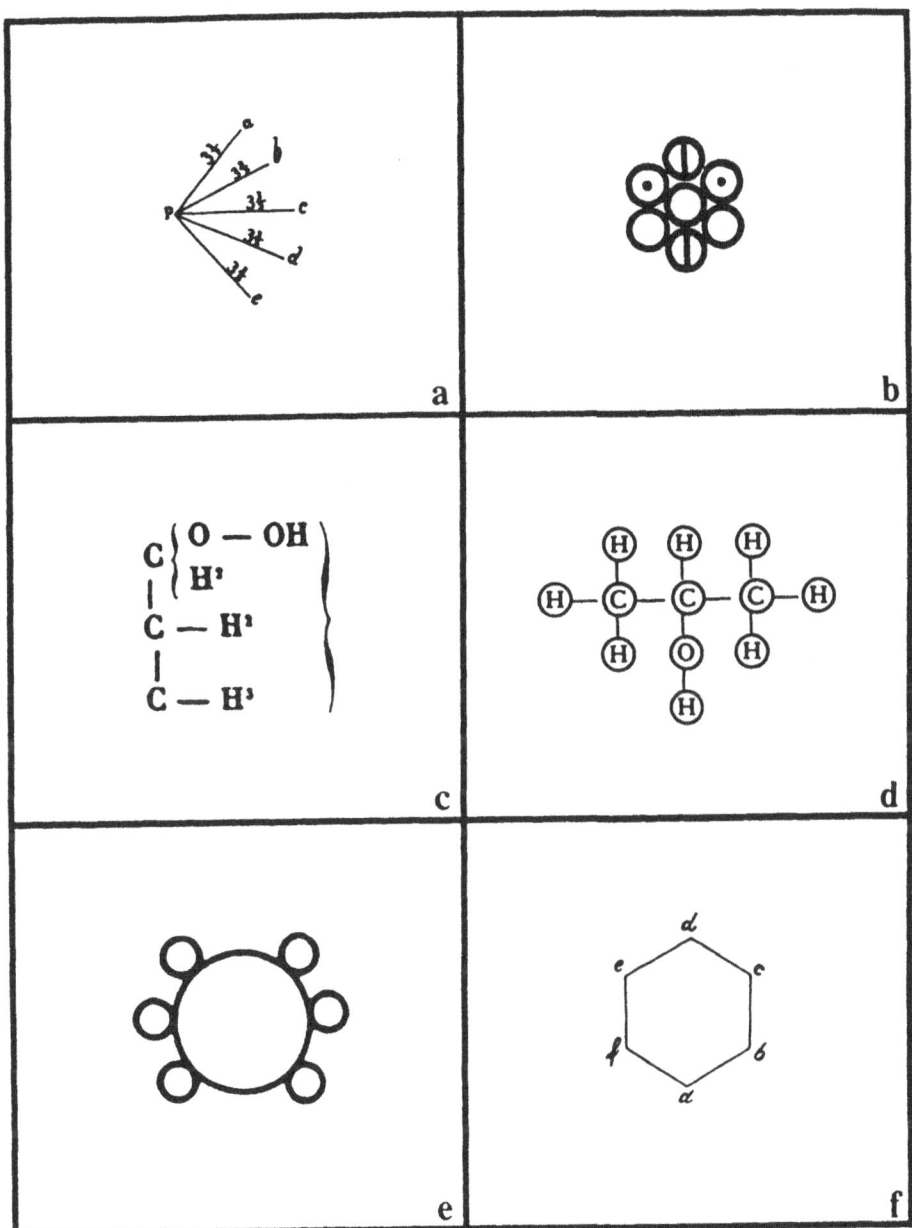

Figure 2.   Several of the early attempts at the graphical representation of molecules. Reproduced in the boxes are (a) an affinity diagram, due to William Higgins, dating from 1789; (b) an atomic combination, due to John Dalton, dating from 1808; (c) a constitutional formula, due to Archibald Couper, dating from 1858; (d) a constitutional formula due to Alexander Crum Brown, dating from 1861; (e) a representation of benzene, due to Joseph Loschmidt, dating from 1861; and (f) another representation of benzene, due to August Kekulé, dating from 1866.

Of other possible approaches that might be adopted to deal with the concept of molecular structure, none would be capable of offering us more than a caricature of individual species rather than a complete description. In the words of Primas, the fact remains that (Primas 1983, p. 107) "[i]n every observation and in every experiment we have to leave out of consideration an unlimited number of effects as irrelevant ... Different viewpoints lead to inequivalent descriptions of nature." One caricature of molecular reality that has continued to fascinate chemists stems from the intense interest chemists have in the linkages that hold molecules together. In Figure 2 we illustrate some of the earliest attempts to depict these linkages. Such linkages were eventually given the name chemical bonds by Frankland (1866).

The precise nature of the chemical bond has been a topic of ongoing debate within the chemical community for some 200 years, though it has to be admitted that the concept is still not generally well understood. In fact, the vast majority of chemists tend to take the existence of chemical bonds for granted and leave others to figure out the exact nature of the beast. Representation of the chemical bond by a straight line, as originally envisaged by Couper (1858), makes it tempting to simplify matters by omitting any reference to weak intra-molecular bonding or to the interactions that connect a molecule to its environment. Inevitably the spotlight then falls on the remaining comparatively strong bonds within the molecule. This kind of approach is one that has been followed by chemists for well over a century. Perhaps surprisingly, yielding to this temptation brings significant insights into the way in which molecular structure determines molecular behavior.

The idea of considering molecules as abstract networks of atoms and bonds was pioneered (see Rouvray 1989) by the mathematicians Arthur Cayley and James Sylvester in the mid-1870s. Sylvester (1878a) introduced the term 'chemicograph' – which he soon shortened to 'graph' – to describe the molecular networks. The term actually derived from the chemists' 'graphic notation', an expression in common use at the time to denote a structural formula. A number of the chemical graphs investigated by Sylvester (1878b) are reproduced in Figure 3. Cayley (1874) was the first to use chemical graphs in which the atoms were represented by bullet points. With his graphs he demonstrated how it was possible to enumerate the members in several different homologous series, including the alcohols and the alkanes (Cayley 1875). Cayley's graphs, like most graphs used to represent chemical species, were planar graphs, i.e. they could be drawn in the plane without any crossing of the lines. The neglect of the geometry of the species in this kind of representation might at first sight seem to be a serious drawback or even to render graphs inappropriate for use in

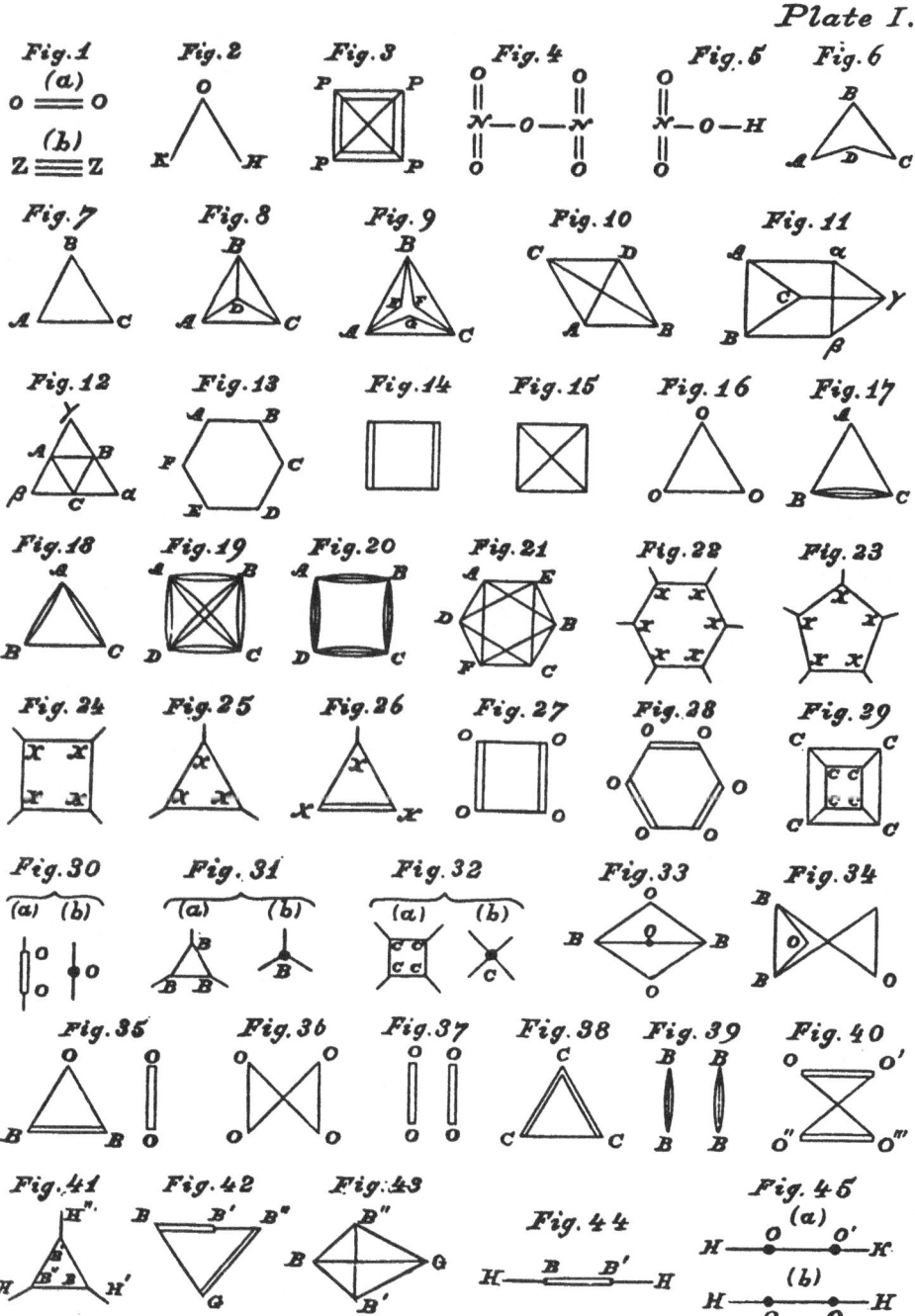

Figure 3. A depiction of several of the chemical graphs studied by the mathematician James Sylvester. This work dates from 1878. Reproduced from Sylvester (1878b, p. 82).

the chemical context. This, however, has proven not to be the case: graphs are very useful and a multitude of applications have been found for them in chemistry (Mallion and Rouvray 1990). A simple depiction of the basic links holding a molecule together – frequently referred to nowadays as its topology – is enough to yield a great deal of information about the species.

5. Eigenvalues and Energies

Let us consider, for instance, the kinds of information that can be obtained about the energy levels of various molecular species from an examination of their chemical graphs. Species having graphs that are bipartite, i.e., graphs whose vertices can be colored consistently with two colors, will always have their eigenvalues paired into plus-minus pairs of the same value. Expressed in more chemical language, it may be said that molecules which are at-ternant will have their energy levels paired. This result, which had been widely adumbrated before its formal publication as a theorem in 1940, is known to chemists as the Coulson-Rushbrooke pairing theorem (Coulson and Rushbrooke 1940). Since 1940, the theorem has found numerous appli-cations in chemistry and has been extended in a variety of different ways. For example, there are now dozens of theorems in existence that relate the colorability of molecular graphs to their energy level spectra. Some of the species that have been embraced by such results include a wide variety of hydrocarbons, including systems that are cyclic and noncyclic, saturated and unsaturated, möbius and nonmöbius, and of high-spin and low-spin; organic ferromagnets; positive and negative hydrocarbon ions; magnetic media; crystals and metals (Mallion and Rouvray 1990). These results re-tain their validity whatever level of approximation is employed in solving the relevant Schrödinger equations. They may thus be said to have (Coulson et al. 1978, p. 159) "an underlying mathematical significance outside the context of, not merely the ... model, but quantum chemistry and, indeed, chemistry itself".

It became increasingly clear during the 1950s just what the signifi-cance of these results was. Independent investigations by Ruedenberg and his co-workers (Ruedenberg 1954, 1955; Ham and Ruedenberg 1958a,b) in the United States and by Günthard and Primas (1956) in Switzerland succeeded in establishing an intimate connection between the molecular graph of a species and the approximate energy levels of the corresponding molecule. This was achieved by first converting the molecular graph into its vertex adjacency matrix[5] and then showing that the eigenvalues of this matrix were identical to the energy levels obtained from solution of the ap-proximate Hückel equation for the species under investigation. The actual

---

[5]For graph-theoretical terminology see, for instance, Wilson (1972).

relationship established was that the vertex adjacency matrix of any conjugated organic species is isomorphic to its Hückel molecular orbital matrix.[6] This implies that the vertex adjacency matrix eigenvalues are identical to the energy eigenvalues obtainable from the Hückel matrix, provided due allowance is made for a scaling factor. At the time, this result was novel in that it revealed the equivalence of the Hückel approach and methods based on graph theory for the study of molecular species. Moreover, this result provided a rational foundation for a host of different methods that were later developed to calculate the properties of molecular species. Almost invariably, these methods have been based on analysis of the molecular connectivity (also known as the molecular topology) rather than the geometry of the species in question. The methods have embraced atom additivity schemes (Rouvray 1991, p. 22), bond transferability methods (Klein 1986), and the use of atomic volumes (McGowan and Mellors 1986, p. 24) and topological indices (Rouvray 1987). The role played by Primas in this early work was thus not only that of an advocate of (Primas 1983, p. 253) "the fair requirement of a moderate realism" in modeling molecular reality. He is also to be viewed as a pioneer who was prepared to explore any route that might have possibilities for bringing about a consistent description of the natural world.

Following on from this pioneering work, an upsurge of interest in chemical graphs meant that they rose to ever greater prominence within the chemical domain (Rouvray 1991, p. 17). Of course, they would not have retained their popularity unless there were additional supporting evidence that they were able to satisfactorily model reality. At the fundamental level such evidence was introduced by Mezey and his co-workers (Mezey 1988, 1993; Mezey and Maruani 1990) in the late 1980s and early 1990s. The approach they adopted was to consider molecules as totally fuzzy objects, i.e., to regard the Born-Oppenheimer approximation as invalid. Molecules were accordingly no longer viewed as semiclassical entities with their nuclei occupying fixed positions in space.

On such a basis it was concluded that molecular structure could reasonably be defined in terms of an equivalence class of molecular graphs (Mezey and Maruani 1990). Consequently, every fuzzy, vibrating and rotating molecular entity may be associated with a mathematical structure of known and well-defined connectivity (Wilson 1972). This can be done because in general the motions of the atomic nuclei in a molecule leave the chemical graph intact. It would thus appear to be justifiable in most cases to interpret molecules as topological objects and to explain their behavior by means of the invariant structure of the molecular graph (Rouvray

---

[6]For an introduction to Hückel theory see, for instance, Heilbronner and Bock (1976).

1995b). At least, this avoids a treatment of molecules viewed as consisting of a mixture of classical and quantum entities. This approach has been
considerably buttressed in recent years by the work of Villaveces and Daza
C. (1990, 1997) who have specified in detail the conditions under which
a topological equivalence may be established between molecular structure
characterizations that rely on the Born-Oppenheimer approximation, e.g.,
that of Bader (1994) and those that do not, e.g., that of Mezey (1993). In
particular, these latter workers were able to demonstrate that the category
of stable chemical structures could always be defined in terms of certain
rigorously specified topological spaces (Villavaces and Daza C. 1997).

6. Concluding Comments

As we have indicated, chemists in the generality do not naturally take to
quantum theory. It is regarded by them as alien in that it has been forced
upon them by physicists and it is also viewed as arcane in that quantum
theory, at least on the face of it, would appear to have small relevance
to chemistry. Chemists still hanker after solid and substantial atoms and
are quite repelled by the notion of etherial matter waves. What is to be
done? Chemists will simply have to learn to live with quantum theory,
because, in whatever ways quantum theory may eventually evolve, it seems
highly unlikely that the theory will be completely overthrown or disappear
from the scene. It manifestly contains too many elements of truth to be
overturned and supplanted by some radical new theory. Indeed, far too
much reliable experimental evidence has already been amassed for that to
occur! In the thoughts of Einstein (1936), "[t]here is no doubt that quantum
mechanics has seized hold of a beautiful element of truth, and that it will
be a test stone for any future theoretical basis in that it must be deducible
as a limiting case from that basis". Developments to date would certainly
seem to confirm this prediction, for, although quantum theory has evolved
into some unexpected directions, in general it can be claimed that all of
its newer forms can be reduced to pioneer quantum theory in the limit.
Some of the newer directions have included the many attempts that have
been made to incorporate quantum theory into relativity theory (Sachs
1986, Prugovečki 1995), endeavors to express fundamental quantum notions
in terms of the fractal nature of spacetime (Naber 1988, Nottale 1993),
and several proposals that the underlying logic used in quantum theory be
changed to something more compatible with microreality, possibly to fuzzy
logic (Prugovečki 1984, Pitowsky 1989).

   At present there is no evident pretender waiting in the wings to replace quantum theory. All that we have to work with are the numerous
approaches that have been made to extend, to elaborate and to incorpo-

rate quantum theory into some wider framework. Whatever the eventual outcome, however, allowance will have to be made for the fact that to a good approximation atoms behave as spheres and a good deal of experimental chemistry can be accounted for on this basis. Clearly, it will also be necessary to bridge the gap that exists between the chemist's notion of the molecule as a highly structured and autonomous entity comprised of spherical atoms and whatever model is ultimately propagated by quantum theoreticians. Some accommodation has already been made. As we have pointed out, Mezey (1993) has demonstrated that the molecular graph remains a useful concept even when the Born-Oppenheimer approximation is abandoned and molecules are treated as completely fuzzy objects. It is thus still reasonable to think of molecules as comprised of atoms, and molecules can even be considered to be possessed of a definite topology provided that weaker interactions are neglected. Recovery of the molecular geometry is, however, more difficult and it would seem that the best that can be achieved in this regard is to view molecules as having time-averaged geometries. Not surprisingly, there is now a strong trend to view molecules as topological entities rather than geometric ones (Rouvray 1995b). The current situation in which we find ourselves has been perceptively summed up by Primas (1987) who has commented that "[i]f we accept quantum mechanics as a correct theory of reality, then neither atoms nor molecules exist in an absolute sense ... However, we can describe matter in terms of these entities provided we are willing to abstract from some holistic entanglements ... Of course, other kinds of abstractions are feasible and lead to different descriptions, but abstractions are unavoidable."

References

Bader R.F.W. (1994): *Atoms in Molecules: A Quantum Theory* (Clarendon Press, Oxford), chap. 3, p. 53.

Baeyer H.C. von (1992): *Taming the Atom: The Emergence of the Visible Microworld* (Random House, New York).

Ball P. (1994): *Designing the Molecular World: Chemistry at the Frontier* (Princeton University Press, Princeton).

Benfey O.T. (1964): *From Vital Force to Structural Formulas* (Houghton Mifflin, New York 1964). Reprinted by the Beckman Center for the History of Chemistry, Philadelphia (1992).

Benfey O.T., ed. (1981): *Classics in the Theory of Chemical Combination* (Kreiger, Malabar FL).

Bohm D. (1980): *Wholeness and the Implicate Order* (Routledge and Kegan Paul, London), p. 150.

Born M. and Oppenheimer J.R. (1927): Zur Quantentheorie der Molekeln. *Ann. Phys.* **84**, 457–484.

Bradley J. (1955): On the operational interpretation of classical chemistry. *Brit. J. Phil. Sci.* **6**, 32–42.

Cassirer E. (1964): *Zur modernen Physik* (Wissenschaftliche Buchgesellschaft, Darmstadt).

Cayley A. (1874): On the mathematical theory of isomers. *Phil. Mag.* **47**, 444–447.
Cayley A. (1875): On the analytical forms called trees, with applications to the theory of chemical combinations. *Brit. Assoc. Adv. Sci.* **45**, 257–305.
Chew G.F. (1966): *The Analytic S-Matrix* (Benjamin, New York).
Coulson C.A. (1955): Contributions of wave mechanics to chemistry. *J. Chem. Soc.*, 2096–2084.
Coulson C.A. and Rushbrooke G.S. (1940): Note on the method of molecular orbitals. *Proc. Cambridge Phil. Soc.* **36**, 193–203.
Coulson C.A., O'Leary B., and Mallion R.B. (1978): *Hückel Theory for Organic Chemists* (Academic Press, London), app. D.
Counts R.W. (1991): Do you believe in wavefunctions? *J. Comput.-Aided Mol. Design* **5**, 167–168.
Couper A.S. (1858): Sur une nouvelle théorie chimique. *Ann. Chim. Phys.* **53**(3), 469–489.
Dehmelt H. (1988): A single atomic particle forever floating at rest in free space: a new value for electron radius. *Physica Scripta* **T22**, 102–110.
Dirac P.A.M. (1925): The fundamental equations of quantum mechanics. *Proc. Roy. Soc. London A* **109**, 642–653.
Dirac P.A.M. (1929): Quantum mechanics of many-electron systems. *Proc. Roy. Soc. London A* **123**, 714–733.
Doucet J.P. and Weber J. (1996): *Computer-Aided Molecular Design* (Academic Press, London).
Einstein A. (1926): Letter to Max Born dated Dec. 4, 1926. In *Albert Einstein – Hedwig and Max Born: Briefwechsel 1916–1955*, ed. by M. Born (Nymphenburger Verlagshandlung, München), p.129. English translations are to be found in *The Born-Einstein Letters*, transl. by I. Born (Macmillan, London 1971), p. 91, and in A. Pais: *Subtle is the Lord: the Science and Life of Albert Einstein* (Clarendon Press, Oxford 1982), p. 443.
Einstein A. (1936): Physics and reality. *J. Franklin Inst.* **221**, 313–347.
Einstein A. (1951): Letter to Hans Mühsam dated July 9, 1951. Einstein Archive 36–610.
Einstein A., Podolsky B., and Rosen N. (1935): Can quantum-mechanical description of physical reality be considered complete? *Phys. Rev.* **47**, 777–780.
Everett H. (1957): "Relative state" formulation of quantum mechanics. *Rev. Mod. Phys.* **29**, 454–462.
Feynman R.P. (1985): *QED: The Strange Theory of Light and Matter* (Princeton University Press, Princeton).
Frankland E. (1866): Contributions to the notation of organic and inorganic compounds. *J. Chem. Soc.* **19**, 372–394.
Günthard Hs.H. and Primas H. (1956): Zusammenhang von Graphentheorie und MO-Theorie von Molekeln mit Systemen konjugierter Bindungen. *Helv. Chim. Acta* **39**, 1645–1653.
Ham N.S. and Ruedenberg K. (1958a): Energy levels, atom populations, bond populations in the LCAO MO model and in the FE MO model. A quantitative analysis. *J. Chem. Phys.* **29**, 1199–1214.
Ham N.S. and Ruedenberg K. (1958b): Mobile bond orders in conjugated systems. *J. Chem. Phys.* **29**, 1215–1229.
Heilbronner E. and Bock H. (1976): *The HMO Model and its Applications, 1. Basis and Manipulation* (Wiley, London).
Heisenberg W. (1925): Über quantentheoretische Umdeutung kinematischer und mechanischer Beziehungen. *Z. Phys.* **33**, 879–893.
Heisenberg W. (1958): *Physics and Philosophy: A Revolution in Modern Science* (Harper and Row, New York).
Jammer M. (1966): *The Conceptual Development of Quantum Mechanics* (McGraw Hill, New York).

Klein D.J. (1986): Chemical graph-theoretic cluster expansion. In: *Mathematics and Computational Concepts in Chemistry*, ed. by N. Trinajstić (Horwood, Chichester UK), chap. 16, p. 171.

Lewis G.N. (1923): *Valence and the Structure of Atoms and Molecules* (Chemical Catalog Company, New York).

Mallion R.B. and Rouvray D.H. (1990): The golden jubilee of the Coulson-Rushbrooke pairing theorem. *J. Math. Chem.* **5**, 1–21.

McGowan J.C. and Mellors A. (1986): *Molecular Volumes in Chemistry and Biology: Applications Including Partitioning and Toxicity* (Horwood, Chichester UK), chap. 2.

Mezey P.G. (1988): From geometrical molecules to topological molecules: a quantum mechanical view. In *Molecules in Physics, Chemistry and Biology, Vol. II*, ed. by J. Maruani (Reidel, Dordrecht), pp. 61–81.

Mezey P.G. (1993): *Shape in Chemistry: An Introduction to Molecular Shape and Topology* (VCH, New York), p. 22.

Mezey P.G. and Maruani J. (1990): The concept of 'syntopy'. A continuous extension of the symmetry concept for quasi-symmetric structures using fuzzy-set theory. *Mol. Phys.* **69**, 97–113.

Milonni P.W. (1994): *The Quantum Vacuum: An Introduction to Quantum Electrodynamics* (Academic Press, Boston).

Müller E.W. (1965): Field ion microscopy. *Science* **149**, 591–601.

Naber G.L. (1988): *Spacetime and Singularities: An Introduction* (Cambridge University Press, Cambridge).

Nottale L. (1993): *Fractal Space Time and Microphysics: Towards a Theory of Scale Relativity* (World Scientific, Singapore).

Pauling L. (1985): Why modern chemistry is quantum chemistry. *New Scientist* **108** (7. Nov.), 54–55.

Penrose R. (1997): *The Large, the Small and the Human Mind* (Cambridge University Press, Cambridge).

Pitowsky I. (1989): *Quantum Probability – Quantum Logic* (Springer, Berlin).

Pool R. (1989): Basic measurements lead to Physics Nobel. *Science* **246**, 327–328.

Primas H. (1983): *Chemistry, Quantum Mechanics and Reductionism* (Springer, Berlin), 2nd edition.

Primas H. (1987): Contextual quantum objects and their ontic interpretation. In *Symposium on the Foundations of Modern Physics 1987. The Copenhagen Interpretation 60 Years after the Como Lecture*, ed. by P. Lahti and P. Mittelstaedt (World Scientific, Singapore), pp. 251–275, here: p. 270.

Prugovečki E. (1984): *Stochastic Quantum Mechanics and Quantum Spacetime* (Reidel, Dordrecht).

Prugovečki E. (1995): *Principles of Quantum General Relativity* (World Scientific, Singapore).

Pyykkö P. (1986): *Relativistic Theory of Atoms and Molecules* (Springer, Berlin), chap. 9.

Rouvray D.H. (1987): The modeling of chemical phenomena using topological indices *J. Comput. Chem.* **8**, 470–480.

Rouvray D.H. (1989): The pioneering contributions of Cayley and Sylvester to the mathematical description of chemical structure. *J. Mol. Struct.(Theochem)* **185**, 1–14.

Rouvray D.H. (1991): The origins of chemical graph theory. In *Chemical Graph Theory: Introduction and Fundamentals*, ed. by D. Bonchev and D.H. Rouvray (Gordon and Breach, London), chap. 1, pp. 1–39.

Rouvray D.H. (1995a): John Dalton: the world's first stereochemist. *Endeavour (New Series)* **19**, 52–57.

Rouvray D.H. (1995b): A rationale for the topological approach to chemistry. *J. Mol. Struct.(Theochem)* **336**, 101–114.

Rouvray D.H. (1997): Do molecular models accurately reflect reality? *Chem. & Ind.* **15**, 587–590.

Ruedenberg K. (1954): Free-electron network model for conjugated systems. V. Energies and electron distributions in the FE MO model and in the LCAO MO model. *J. Chem. Phys.* **22**, 1878–1894.

Ruedenberg K. (1955): Errata: Free-electron network model for conjugated systems. V. Energies and electron distributions in the FE MO model and in the LCAO MO model. *J. Chem. Phys.* **23**, 401.

Russell C.A. (1996): *Edward Frankland: Chemistry, Controversy and Conspiracy in Victorian England* (Cambridge University Press, Cambridge).

Sachs M. (1986): *Quantum Mechanics from General Relativity: An Approximation for a Theory of Inertia* (Reidel, Dordrecht).

Schrödinger E. (1926): Über das Verhältnis der Heisenberg-Born-Jordanschen Quantenmechanik zu der meinen. *Ann. Phys.* **79**, 734–756.

Sylvester J.J. (1878a): Chemistry and algebra. *Nature* **17**, 284.

Sylvester J.J. (1878b): On an application of the new atomic theory to the graphical representation of the invariants and covariants of binary quantics, — with three appendices. *Am. J. Math.* **1**, 64–128.

Thomson Sir J.J. (1937): *Recollections and Reflections* (Macmillan, New York).

Villaveces J.L. and Daza C.E.E. (1990): On the topological approach to the concept of chemical structure. *Int. J. Quantum Chem., Quantum Chem. Symp.* **24**, 97–106.

Villaveces J.L. and Daza C.E.E. (1997): The concept of chemical structure. In *Concepts in Chemistry: A Contemporary Challenge*, ed. by D.H. Rouvray (Research Studies Press, Taunton UK), chap. 4, pp. 101–132.

Wilson R.J. (1972): *Introduction to Graph Theory* (Oliver and Boyd, Edinburgh).

Wollaston W.H. (1808): V. On super-acid and sub-acid salts. *Phil. Trans. Roy. Soc. London* **98**, 96–102.

Wollaston W.H. (1813): Dr. Wollaston on the elementary particles. *Phil. Trans. Roy. Soc. London* **103**, 62–63.

Woolley R.G. (1978): Must a molecule have a shape? *J. Amer. Chem. Soc.* **100**, 1073–1078.

CONTEXTUAL BACKGROUND

Charles P. Enz:

THE SCIENCE OF MATTER: FASCINATION AND LIMITS

To those working alongside him, it often came as a surprise how quickly and radically Hans Primas could change his mind, even in matters that were important to him. His occasional comment in such situations: "Why should I worry over what I was on about yesterday?" In the context of research, he was unruffled by possible contradictions that might crop up in his own conduct: "After all, I'm not Boolean." On the other hand, in questions of central importance there were always positions that he never encroached on. For instance, the following views of his on the relation of reality and quantum mechanics are remarkably constant:

- The essential new aspect of quantum mechanics is the noncommutativity of observables, not the microscopic size of the system considered.
- Planck's constant h is also relevant and nonzero in classical physics. Classical mechanics is not given as the limit $h \to 0$ of a quantum mechanical theory. Classical features are described by the center (containing those observables which commute with all other observables) of the overall algebra of observables.
- Holistic quantum effects do also arise for macroscopic systems, and the numerical smallness of h does not necessarily imply the contrary.
- There is no instantaneous "collapse of the wave function". One should derive a generalized quantum theory in which the state reduction can be described by a continuous dynamics without quantum jumps. Stochastic "Schrödinger equations", derived from the coupling of a small system to an environment, e.g., the electromagnetic or gravitational field, play a prominent role in this respect.
- Quantum mechanics does not contradict a realistic view of the world. Realism and non-locality are compatible, but the relationship between them is highly nontrivial.

In addition to these formal and conceptual topics, Primas' work shows a remarkable predilection for engineering-type formal computations not necessarily related to mathematical theorems about fundamental quantum theory. For instance, he studied the Husimi and Wigner distributions for quantum systems in various disguises, and he discussed stochastic quantum dynamics at a technically advanced level. This "Fermian" way of looking at quantum theory may be traced back to his work on magnetic resonance. "Fermi calculates like an engineer" is a sentence of Wolfgang Pauli's much quoted by Primas.

In his essay, Charles Enz combines conceptual discussions with such engineering-type estimates. Using the wave aspect of matter in the small, Planck's constant h is shown to give rise to a limit of divisibility. Combined with the limit of gravitational instability one is led to the Planck mass as an upper limit of the particle spectrum. Then a selection of experiments is discussed with the aim to show that a confined electron may be seen in one case as a free point-particle and in another as a wave, and that strong interaction may even give rise to quasi-particles with fractional charge. These facts are taken to demonstrate the ontological reality of the wavefunction. Finally it is argued that the existence of limits of predictability in biology and technology may imply serious threats to mankind and that, therefore, scientists should act as engaged observers guided by responsibility and compassion.

THE SCIENCE OF MATTER: FASCINATION AND LIMITS

CHARLES P. ENZ

Département de Physique Théorique, Université de Genève,
24 quai Ernest Ansermet, CH-1211 Genève 4, Switzerland

The ability to reduce everything to simple fundamental laws does not imply the ability to start from those laws and reconstruct the universe. In fact, the more the elementary-particle physicists tell us about the nature of the fundamental laws, the less relevance they seem to have to the very real problems of the rest of science, much less to those of society.

(P.W. Anderson 1972)

1. Limit of Divisibility

The usual understanding of the ancient Greek atoms proposed by *Leucippus* and his disciple *Democritus* some 2440 years ago is that of indivisible objects of various forms and sizes that one may stick together like LEGO-pieces in an unlimited number of ways to form the material bodies. From this picture two conclusions are usually drawn: Some philosophers declare the Greek atomists to be the first materialists (Gex 1949), and some scientists say that "quantum mechanics also put an *end to atomism*" (Primas 1994). Personally I find both conclusions unfortunate. To me the *Greek atomists* were the first *physicists*. Contrary to the *geometers*, the *Pythagoreans*, *Aristotle* and, above all, *Descartes* who saw the world as a continuum, these physicists saw it in the light of the polarity between the *full* and the *void*. For them dividing a body meant to intercalate a void between two fulls which were one before. Since this process cannot be repeated indefinitely they realized that it is necessary for describing the material world to postulate the existence of a *limit of divisibility*. This, it seems to me, is the essence of the Greek atoms, and not the LEGO-folklore. Pauli (1994a, p. 140) characterizes this aspect as "a way out of the difficulties of the relation between unity and multiplicity". And he also says: "It would not be correct to designate these thinkers as materialists in the modern sense."

Of course, the weakness of the atomic doctrine was that the size of the atoms was completely unknown, so that the opponents, above all Descartes, had an easy game to demolish it in declaring matter to be nothing more than *res extensa*. That, however, was before *Planck*, who at the turn of this century discovered the *quantum of action h*, a universal constant of fundamental importance, as we shall see all along this paper. Consider now the limit of divisibility in this new light:

Successive division of a material body will reveal that beneath a certain size (radius) a, the *imprecision* Δx in the position of the body grows so large that division becomes impossible because $\Delta x > a$. This happens when the *quantum nature* or *wave aspect* (wavelength λ, light velocity c) starts to appear, that is, approximately, when $\Delta x = \lambda$. But the wave aspect is linked to the *particle aspect* (mass m, energy E) of the body through Planck's constant: $E = hc/\lambda$. With the rest energy $E = mc^2$ one thus finds that

$$\lambda(m) = \frac{h}{mc} \ . \tag{1}$$

It means that for a given mass m subdivision becomes impossible for a size $a < \lambda(m)$.

Physically important is the distinction between the two types of matter that we know, namely *chemical* and *nuclear*. We may express the sizes a_{ch} and a_{nu} of chemical and nuclear bodies in terms of their respective *characteristic density*. A simple approximative way is obtained by introducing the *radius of the hydrogen atom* a_H, the *radius of the proton* a_p and the *hydrogen mass* which is much the same as the *proton mass* m_p. In terms of these quantities one has

$$\frac{a_{ch}(m)}{a_H} = \frac{a_{nu}(m)}{a_p} = \left(\frac{m}{m_p}\right)^{1/3} \ . \tag{2}$$

The limit of divisibility $\lambda(m) = a(m)$ then determines the *limiting masses* m_{ch} and m_{nu} for *chemical* and *nuclear* matter, respectively, below which division ceases to be possible:

$$m_{ch} = m_p \left(\frac{\lambda_p}{a_H}\right)^{3/4} ; \quad m_{nu} = m_p \left(\frac{\lambda_p}{a_p}\right)^{3/4} , \tag{3}$$

where $\lambda_p = \lambda(m_p)$. One finds $m_{ch} = 0.65 \ m_e$, m_e being the *electron mass*, and $m_{nu} = 1.01 \ m_p$ (Enz 1986). This shows that *chemical matter* (atoms) is divisible which is what chemistry is all about. But it also shows that in *nuclear matter* the proton and, more generally, the *hadrons* (proton, neutron, pion, etc.) are at the *limit of divisibility*. This statement may surprise in view of the experimental fact that quarks may be detected inside the proton. However, according to the idea of *asymptotic freedom* the quarks confined inside a hadron behave as quasi-free only in the limit of high energy. But the uncertainty relation implies that this can happen only in the limit of short time intervals. Taking a longer view the quarks are extended over the whole hadron, and division is impossible.

Therefore, unless entirely new physics is discovered, the simple qualitative picture of a limit of divisibility presented above may serve as a physical definition of *elementary particles* as juxtaposed to the usual group-theoretical definition (see, e.g., Primas 1981, p. 300). Elementary particles in this sense were discussed by the 21-year old Pauli in Part V of his famous *Encyklopädie Article* (Pauli 1958) where he noticed that in classical relativistic theory there is no way to make them stable, and suspected that some kind of *discreteness* should be introduced.

2. Limit of Stability

In this *physical* definition of elementary particles, as in the group-theoretical one, mass is a *classical observable* (Primas 1981, pp. 271–301). This fact implies a second criterion, namely *gravitational stability*. A body of mass m, here assumed to be homogeneous, becomes unstable against body-antibody pair-production when its gravitational energy $3Gm^2/5a$ exceeds $3mc^2$ where G is Newton's gravitational constant. This leads to a limiting radius of $a = Gm/5c^2$. More precisely, the instability is due to a singularity of the space-time metric at the *Schwarzschild radius*

$$\sigma(m) = \frac{2Gm}{c^2} \, , \tag{4}$$

as determined by general relativity. As a consequence, $\sigma(m)$ becomes an *event horizon* which separates the *world inside* from the *world outside*. The body therefore is called a *black hole*. For nuclear matter this instability determined by $a_{\mathrm{nu}}(m) = \sigma(m)$ happens when m is several times the solar mass (Misner et al. 1973).

From Eqs. (1), (2), and (4) we see that, in a logarithmic plot, $\log \lambda(m)$, $\log \sigma(m)$ and $\log a_{\mathrm{ch}}(m)$, $\log a_{\mathrm{nu}}(m)$ as functions of $\log m$ are straight lines with slopes -1, $+1$ and $1/3$, respectively, as shown in Fig. 2 of Enz (1986). The curves $\lambda(m)$ and $\sigma(m)$ intersect at a value $m = \sqrt{hc/2G}$ which, up to a factor $\sqrt{\pi}$, is the *Planck mass*

$$m_{\mathrm{Pl}} = \sqrt{\frac{hc}{2\pi G}} = 1.3 \times 10^{19} m_p = 0.021 \text{ mg} \, . \tag{5}$$

The meaning of the Planck mass therefore is that of an *upper limit for the mass spectrum* of physical elementary particles. The *ultimate limit of divisibility* is thus given by the *Planck scale*

$$a_{Pl} = \lambda(m_{Pl}) = 0.78 \times 10^{-19} a_p = 1.02 \times 10^{-32} \text{cm} . \tag{6}$$

At this scale even the notions of space and time become obsolete. To this Planck point corresponds a *limiting mass density* ρ_{Pl} which, expressed in units of the nuclear density $\rho_{nu} \simeq 1.8 \times 10^{14}$ g/cm^3, is

$$\frac{\rho_{Pl}}{\rho_{nu}} = \frac{m_{Pl}}{m_p}\left(\frac{a_{Pl}}{a_p}\right)^{-3} \simeq 2.7 \times 10^{76} . \tag{7}$$

From here up to the most diluted matter in the universe the region above the curves $\lambda(m)$ and $\sigma(m)$ comprises all that is in the material world.

3. Synchronistic Happenings

Today the expression *correlation* has acquired, in addition to the traditional meaning as an effect of interaction between particles and/or fields, a reading that emerged in recent years out of the abundant discussion of the *Einstein-Podolsky-Rosen* (EPR) paper of 1935 (Einstein et al. 1935). Einstein, Podolsky, and Rosen realized that, in principle, division of a quantum system does not result in independent *elements of reality* but leaves "synchronicities" between the parts at all times after division. "In principle" here means: unless *environmental influences* wash out these effects.

This *environmental averaging* happens, e.g., if I cut a piece of wire in two, since I may walk away happily with one part of the wire, without the wavefunctions of the conduction electrons in the other part preventing me from doing so – and this in spite of the fact that before division the wavefunctions of all the conduction electrons were extended over the uncut wire! (Miraculously, all the electrons managed to escape from being cut in two!) Without this averaging I would be obliged, as Schrödinger with his cat, to describe the cut wire by a *coherent superposition* of product states of both parts, 1 and 2, in which each electron may be assigned to either part,

$$\Psi_{12} = \sum_\lambda c_\lambda \psi_{1\lambda} \psi_{2\lambda} . \tag{8}$$

This means that the two cut parts would be *entangled* and I could not have walked away so happily. This averaging, or *decoherence*, is a typical effect of correlations in its traditional meaning. EPR-correlations, on the other hand, are characterized by just such an *entangled wavefunction* Ψ_{12}.

In Eq. (8) the parameter λ must, of course, assume more than one value. In the simplest case of two-valuedness, λ may represent, e.g., the two polarizations of a *laser-beam* consisting of photons of two colors ω and $\omega_o - \omega$

produced in an appropriate crystal as non-linear response from original photons of color ω_o. Perpendicular polarizations are obtained by exploiting the *birefringence* of the crystal such that the photons of one color form the *ordinary ray* and the photons of the other color the *extraordinary ray*. The division of this system is produced by a non-polarizing semi-transparent beam splitter which sends the split beams into two analyzers. There the angles Θ_1 and Θ_2 formed by the polarizations with two arbitrary transverse directions are measured by counting the number N of *coincidences* per unit time between the photons of the two beams. The result obtained is that *the counting rate only depends on the difference $\phi = \Theta_2 - \Theta_1$, that is, $N = N(\phi)$*. Hence, in determining the polarization of one of the split beams, that of the other beam is known simultaneously, although both are spatially separated! This experiment gives the largest and cleanest EPR-correlation I know, the quantitative aspect being governed by a *Bell inequality* (Kiess et al. 1993).

This example also shows that entangled states may be the result of interactions in the process of their formation – in the example the non-linear response in the crystal – so that the distinction between the two types of correlation seems to be more a question of procedure than of principle. But while EPR-correlations are the result of a non-trivial *composite wavefunction* (8), to which corresponds, as we have seen, a quite precise *reality* in terms of an experimental procedure, correlations due to interactions give rise to even more surprising situations.

4. Tricky Electrons

The spectacular discovery of the new "high-temperature" superconductors by *Alex Müller* and *Georg Bednorz* in 1986, for which the authors were awarded the Nobel Prize the following year already, opened up an entirely new class of *strongly correlated systems* of condensed matter. This new class of systems is characterized not only by strongly correlated, i.e., interacting electrons, but also by a *low dimensionality*, $d = 2$ *or* 1. Concerning this second feature the new superconductors are no exception. Indeed, the essential physics of these substances happens in layers formed by two-dimensional CuO_2-lattices where, in the metallic state, the electrons, or rather the holes, move and form pairs (Enz 1996a). Today such a *two-dimensional electron gas* (2DEG) may be manufactured artificially at the interface between layers of the semiconductors $GaAs$ on one side and $AlGaAs$ on the other. In a lithographically applied multi-electrode arrangement it is possible to control the *surface density* n of the electrons by a *gate* and to determine the resistance ρ_{xx} in the x-direction of the 2DEG by measuring voltage and current at two electrodes placed in this direction.

The 2DEG acquires quite spectacular properties when a *transverse magnetic field H* is applied. Again Planck's constant enters the scene, this time in the form of the *flux quantum* $\phi = hc/e$ where e is the *elementary charge* carried by the electrons. It is natural to measure the magnetic field in units of $n\phi$ as $H = n\phi/\nu$, where ν is a pure number called the *filling factor*. The magnetic field gives rise to the *quantum Hall effect*: a current in the x-direction induces a Hall voltage in the perpendicular y-direction which may be measured as a *Hall resistance* ρ_{xy}. While the Hall effect in the bulk is known since the end of last century (Hall 1879), the quantum nature of the effect which exists only in two dimensions, was discovered only in 1980 by *Klaus von Klitzing* and collaborators (Klitzing et al. 1980). Klitzing received the Nobel Prize for this discovery in 1985.

The quantum nature of the Hall effect reveals itself as *plateaus* in the otherwise linear relationship

$$\rho_{xy}(H) = \frac{R}{\nu} = R\frac{H}{n\phi} . \tag{9}$$

In this equation, $R = h/e^2 = 25.8$ kOhm is another manifestation of Planck's constant, here appearing as a *universal resistance*. For *low magnetic field* where $\nu > 1$ the plateaus show up at integer values $\nu = 1, 2, ...$; this is the so-called *integer quantum Hall effect* (IQHE). At *high magnetic field*, $\nu < 1$, which is the domain of the *fractional quantum Hall effect* (FQHE), the plateaus appear at certain fractional rational values of the filling factor ν. At the plateaus the *longitudinal resistance* ρ_{xx} drops to zero, indicating that the 2DEG behaves as a fluid of free charged *quasi-particles*. While in the IQHE these quasi-particles are almost free, i.e. uncorrelated, electrons, the electrons in the FQHE are strongly correlated and the quasi-particles show *fractional charges*, e.g. $e/3$ (Goldman and Su 1995; see also Dorozhkin et al. 1995). At a certain field H_c the longitudinal resistance ρ_{xx} seems to have the universal value R, independent of the density n and, at still higher fields, the 2DEG seems to undergo a phase transition to an insulator.

The understanding of this strongly correlated system in the FQHE still poses some problems, although the basic framework was laid down by *Robert Laughlin* already in 1983 (Laughlin 1983). Based on the experimental fact that no plateaus are observed for $\nu = p/q$ with p an odd and q an even number and on the assumption that the 2DEG forms an *incompressible fluid*, Laughlin constructed a ground state or *condensate wavefunction* Ψ_k for $q = 2k + 1$ as analytic function of the complex position variables $z_r = x_r + iy_r$ of all the electrons, labeled by the index r, such that Ψ_k vanishes as δ^{2k+1} when the distance $\delta = |z_r - z_s|$ between any electrons r

and s goes to zero. He found that

$$\Psi_k = \{\prod_{r<s}(z_r - z_s)^{2k+1}\} \exp(-\frac{1}{4}\sum_l |z_l|^2) . \tag{10}$$

And he generated an elementary excitation or *quasi-particle* at z_0 "by piercing the fluid at z_0 with an infinitely thin solenoid and passing through it a flux quantum $[\phi]$ adiabatically". In this way he found that these excitations have charge $e/(2k + 1)$ (Laughlin 1983).

This construction of the wavefunction of a strongly correlated system shows that any attempt of applying individual *particle-wave duality* to the electrons is doomed but that reality is again described by a *composite wavefunction* (10) and the excited states following from it. The interpretation of this complex situation found a surprisingly simple solution in 1989 when *Jaindra Jain* took up Laughlin's idea of "an infinitely thin solenoid" carrying one flux quantum ϕ. He proposed a model of local *composite fermions* consisting of localized electrons each of which has $2k$ flux quanta attached to it. These flux quanta have the effect to reduce the field $H = \phi n/\nu$ to an *effective field*

$$H_{\text{eff}} = \phi n\left(\frac{1}{\nu} - 2k\right) = \frac{\phi n}{\nu_{\text{eff}}} \tag{11}$$

which, for $\nu = 1/(2k + 1)$, yields $\nu_{\text{eff}} = 1$ (Jain 1989, 1990). Jain's trick therefore reduces the complex situation of the FQHE to the relatively simple case of a IQHE in an effective field H_{eff}. But the trick can do more: for a field corresponding to the value $\nu = 1/2$, for which no plateau is observed, the composite fermions with $k = 1$ see a field $H_{\text{eff}} = 0$. It has indeed been observed that in a field (11) given by a value of ν slightly off 1/2 these composite fermions move on quasi-classical orbits corresponding to this value of H_{eff} (Willett et al. 1993, Kang et al. 1993, Goldman et al. 1994). More recently, however, investigations have indicated that Jain's composite fermion model is too crude to explain finer details of the FQHE.

In *one dimension*, even without external fields, correlated electrons dissociate into a spin-less *charge liquid* and a neutral *spin liquid*, both moving quite independently from each other. Although this so-called *Luttinger liquid* description has not yet been seen experimentally, possible candidates are the edge excitations in the quantum Hall effect. Theoretical efforts are made to generalize this description to two dimensions. In fact, it had long been suspected by *Philip Anderson* (Nobel Prize 1977) that the 2-dimensional hole liquid in the CuO_2-layers of the *high-temperature superconductors* dissociates into a charge liquid and a spin liquid, the quasi-particles of which he called *holons* and *spinons*, respectively (Enz 1996a). Much effort is spent to find manifestations of a spin liquid in these new superconductors.

5. Master of One Electron

These examples show that under *strong correlation* and in *low dimensionality* electrons have a chameleon-like ability of disguisement. And one may wonder whether *free* electrons exist at all. The answer is an unqualified yes, as the following experiment shows: *Hans Dehmelt* and *Wolfgang Paul* (whom Wolfgang Pauli liked to call "my real part") had developed wall-less "boxes" which keep charged particles trapped by an arrangement of crossed electric and magnetic fields and were awarded the Nobel Prize of 1989 for this work. In one experiment Dehmelt stored a single electron in such a *Penning trap* for ten months (Dehmelt 1990).

The electron oscillates in the trap and makes random quantum jumps which are caused by the coupling of the electron's charge to the ever present *electromagnetic vacuum fluctuations*. These fluctuations constitute the most important *non-classical environment* of any charged particle, giving rise to *renormalization effects* which are the main subject of *quantum electrodynamics*. The *elementary quantum of charge e*, however, is not renormalized. This fact is due to *gauge invariance* and, technically, is the result of a so-called *Ward identity* (see, e.g., Enz 1992, Section 12 and p. 224). It guarantees that the fundamental quantities depending on e, the fine-structure constant $\alpha = 2\pi e^2/hc \simeq 1/137$, the flux quantum ϕ and the universal resistance R, have values that are *independent of the environment*. This is not true, however, for the electron's *magnetic moment*. And, indeed, Dehmelt found in the experiment described above that this magnetic moment is not one Bohr magneton but 1.001159652188(4). This is one of the most accurately determined renormalization effects (Dehmelt 1990). However, the effect is small, and Dehmelt's electron may be considered a *free particle* for all practical purposes.

Using the plausible formula $\mu - 1 = \pi a/\lambda(m)$ connecting magnetic moment μ, radius a, and mass m for spin $1/2$ particles (fermions), Dehmelt bravely extrapolated from the proton and even heavier fermions down to the electron, thereby determining an upper bound for the *electron radius*: $a_e < 10^{-20}$ cm $\simeq 10^{-7}$ a_p. From this result one may safely conclude that for all practical purposes the free electron is a *point particle*. Curiously, in spite of this Dehmelt (1990) speculated about an infinite hierarchy of compositions of the electron!

6. Seeing the Wavefunction

In contrast to the strongly correlated appearances of the electron described in Sect. 4 above the particle-wave duality does make sense for free electrons. A spectacular demonstration of wave ripples of a free electron confined on a copper surface within a fence of 48 iron atoms positioned on a circle with a diameter of 143 Å was made by *Donald Eigler* and his group at IBM's Almaden Research Center in California (Crommie et al. 1993). The tool for both, positioning the iron atoms and measuring the topography of the trapped electron as a function of the radial distance r from the center of the circle, was a *scanning tunneling microscope* (STM), the invention of which by *Gerd Binnig* and *Heini Rohrer* of the IBM Zurich laboratory at Rüschlikon was rewarded by the Noble Prize of 1986.

Topographic measurements with an STM on a metallic surface are made by moving a pointer ending in a one-atom wide tip over the surface, keeping the *tunneling current* through the tip constant by varying the height of the tip over the surface. The adjustable parameter in this procedure is the *bias voltage* V between the pointer and the metallic surface. The circle of 48 *Fe* atoms called a "quantum corral" was the smallest that could be realized against the mutual repulsion between the iron atoms. But it was sufficiently dense to keep an electron inside the fence during a long enough time for its *wavefunction* $\psi(r, \varphi)$ to be well-defined, φ here being the polar angle. The electron is kept on the copper surface by the work function on the vacuum side and by the band gap of the bulk electrons on the metal side.

What is measured in the experiment is the *local density of states* at fixed bias voltage V and neglecting spin,

$$D(V,r) = \sum_{n=1}^{\infty} \sum_{l=0}^{\infty} \delta(E_{n,l} - E_F - V)|\psi_{n,l}(r)|^2 \qquad (12)$$

where $E_{n,l}$ and $\psi_{n,l}(r)$ $e^{il\varphi}$ are the angular momentum *eigenvalues* and *eigenfunctions*, respectively, and E_F is the *Fermi energy*. In practice the *Dirac* δ-function in Eq. (12) acquires finite width and height. Since the eigenfunctions $\psi_{n,l}$ are normalized, integration of $D(V,r)$ over the surface of the circle yields the usual expression for the density of states (see, e.g., Eq. (10.4) of Enz 1992). A calculation of the $E_{n,l}$ for the value 0.38 m_e of the *effective mass* of a surface electron shows that, for sufficiently small bias voltage V, $D(V,r)$ is a linear combination of only three terms, $|\psi_{5,0}|^2$, $|\psi_{4,2}|^2$, and $|\psi_{2,7}|^2$ (see Eq. (12) above and Fig. 4 of Crommie et al. 1993). This calculated local density of states is in excellent agreement with the result of the topographical measurement made at $V = 0.01$ Volt in ultra-high vacuum and at a temperature of 4 K. The latter is shown in the spectacular picture on the cover of the November 1993 issue of *Physics Today*.

Is it also possible to determine the phase ϕ of the wavefunction $\psi = |\psi|e^{i\phi}$? The answer is no, if ψ describes one electron or, more generally, a fixed number N of identical particles. Indeed, ϕ and N are approximately *canonically conjugate variables* and hence satisfy an *uncertainty relation* $\Delta N \Delta \phi > 1$. But since for a fixed number of particles $\Delta N = 0$, this means that ϕ is totally undetermined. In order to have $\Delta N \neq 0$, the system must be coupled to a reservoir of particles. Such a reservoir is automatically present in a *superconductor* or a *superfluid* in the form of non-super particles. Thus *phase differences* are indeed measurable, i.e., become *quantum mechanical observables*, e.g., in a *Josephson junction* (see, e.g., Section 27 of Enz 1992).

7. Reality and the Wavefunction

"Science is a systematic refinement of the concepts of everyday life revealing a deeper and ... not directly visible reality behind the everyday reality of colored, noisy things" (Pauli 1994b, p. 28). In this very open-minded definition of reality by Pauli the emphasis is on the *concepts*. Pauli here follows *Einstein*'s earlier view expressed, e.g., in his obituary for Ernst Mach, in which Einstein says that concepts as well as the points of view by which they are associated to objects make sense only insofar as these objects can be exhibited (Einstein 1916, p. 102). The examples discussed above suggest that, while the objects may be identified in a positivist manner with quantum mechanical observables, the central concept is the *complex wavefunction* "revealing a deeper and not directly visible reality".

Heisenberg reports in relation with his discovery of the uncertainty relations in 1927 that Einstein had expressed a somewhat different view, namely that "only theory decides what can be measured" (translated from Heisenberg 1969, p. 111). This introduces the *observer* in an active role: he chooses the system he wants to study and determines its wavefunction which, ideally, tells him what he can and cannot measure. Of course, there exist *abstract generalizations of quantum mechanics*, in which "the adjective 'quantum' no longer refers to Planck's quantum of action but to the existence of incompatible properties" (Primas 1981, p. 245). But such generalizations, in which the word "wavefunction" has a rather contemptible sound (it does not figure in the *Index* of Primas 1981!), miss the essence of the quantum message, namely the ultimate *wave nature of matter*, by which matter acquires a quality resembling much more the *spiritual realm* rather than Descartes' *res extensa*.

It is this analogy that inspired *Robert Jahn* and *Brenda Dunne* to undertake their psychokinesis experiments and to model *consciousness* in terms of "'probability-of-experience' waves in some generalized 'consciousness space/time' domain, akin to the 'probability-of-observation' waves that satisfy the Schrödinger equation ..." (Jahn and Dunne 1986, p. 737). A similar motivation lies behind *Hans Primas'* assumption of "holistic correlations between psyche and matter" which he models in close analogy with EPR-correlations. However, the effects are so subtle that "in daily life Einstein-

Podolsky-Rosen correlations are only rarely observed" (Primas 1996, p. 83). The reason, obviously, is the numerical smallness of the quantum of action expressed in anthropomorphic units: $h = 6.62 \times 10^{-27}$ erg sec. But in spite of its smallness, h is of fundamental importance, and one would think that the analog of h should be at least of qualitative importance also in Primas' psycho-physical correlations. Of course, the above examples should both be understood "only as a pragmatic procedure in the modeling. It is not claimed that quantum mechanics is directly involved. Beyond a mere analogy, therefore, little may be said" (translated from Mahler 1996, p. 105).

Psycho-physical parallelism has played an important role in Pauli's thinking, as is witnessed in his essays (Enz and Meyenn 1994) and in his extended correspondence. The following passage from a letter of Pauli to *Markus Fierz* of 12 August 1948 is an impressive example (Letter [971] in Meyenn 1993, pp. 559-560; translation taken from Enz 1996b, p. 74):

> "The layman usually thinks that when he says 'reality' he is speaking of something obviously known; while it seems to me just to be the most important and exceedingly difficult task of our time to work for the establishment of a new idea of reality. This is also what I mean when I always emphasize that science and religion <u>must</u> have something in common. (I do <u>not</u> mean 'religion within physics' nor also 'physics within religion' – since both would be onesided – but inclusion of both into a whole.) What appears to me to contain the new idea of reality I would like to tentatively call: the <u>idea of the reality of the symbol</u>. A symbol is, on the one hand, a product of the efforts of man and, on the other hand, a sign for an objective order in the universe of which man is only part. It owns something of the old notion of God and also something of the old notion of object. (An example within physics: 'the atom'. The primary qualities of space-filling are obviously lost. If it were not a symbol, how could it be at the same time wave and also particle?)."

8. Limits of Predictability

As indicated at the beginning of Sect. 1, the *wave aspect* of matter disappears when the body becomes too heavy. Divisibility, however is not the criterion since, indeed, *atomic matter waves* have been observed (Schmiedmayer et al. 1995) and are used today in ultra-short-wave spectroscopy. Very recently, the wave aspect of entire atoms has been demonstrated in spectacular experiments realizing Bose condensation and even an atom laser (see, e.g., Goss Levi 1997). Thus, thanks to quantum mechanics, nuclear and even chemical matter has lost its "materialistic" quality, expressed as "LEGO-folklore" above. However, in modern science this message does not reach very far since already the large molecules making up living systems are, at least *in isolation*, quite LEGO-like, the "atoms" of biology being the "indivisible" *genes*.

Schrödinger, in his famous book *What is Life?*, says in the *Epilogue on Determinism and Free Will*: "To the physicists I wish to emphasize that in my opinion, and contrary to the opinion upheld in some quarters, *quantum indeterminacy* plays no biologically relevant role in them [the space-time events in the body of a living being] ..." (Schrödinger 1967). Indeed, the *hereditary stability* of biological species is quite remarkable. The bonds have sufficiently *low uncertainties*, i.e., the bonding electron will not escape, neither by thermal fluctuations over the bonding potential (unless heated to near the boiling point) nor by quantum tunneling through it. This, of course, may be understood in terms of a *Darwinian selection of the fittest* among the bio-molecules.

Exposed to a *cause from the outside*, however, be it in the form of a *radiation* or of a *chemical agent*, there may result a *mutation in the inside* of these "biological atoms", the genes. Contrary to the *random mutations* of strict neo-Darwinism, these *induced mutations* have such an immediate reality in modern life as the radiation deaths and leukemias in the wake of the *Chernobyl* disaster or the lung cancers of smokers. I wish to evade the rather murky debate of what Pauli in his phantasy *The Piano Lesson* (Pauli 1995) calls "blind, i.e. purposeless chance" (§ 37) or "chance ... [that] sometimes fluctuates systematically" (§ 28) which received an admirably lucid analysis by Müller-Herold (1995). Instead I wish to discuss situations in which the very notion of probability, and hence *predictability*, becomes meaningless, thereby posing serious threats to mankind.

Consider, for example the problem with *low doses* of an ionizing radiation, a chemical agent, or also a low-frequency electromagnetic field. Concerns about the risks of low-level *ionizing radiation* go back to the beginning of the atomic age. Indeed, in a fairly recent study one reads (Upton 1991): "Although the effects of large doses of radiation are well documented, the choice of the appropriate dose-response model for use in estimating the health hazards of small doses remains controversial. The notion that there might be no threshold for certain biological effects of radiation dates from the 1940s, when experiments in genetics suggested that the frequency of mutations varied in proportion to the dose of X-rays, without any threshold." Studies of the A-bomb survivors of *Hiroshima* and *Nagasaki* have played a big role in determining this dose-response relation. Better known are the quite analogous controversies that exist in relation with the effects of *smoking* and of *low-frequency electromagnetic fields* (for the latter see, e.g., Goodwin 1997).

However, the model alluded to in the above quotation, namely that for a sufficiently small *dose d* the *probability p* is proportional to *d*, does not allow an extrapolation to arbitrarily small values of *d*. The reason is that, mathematically, the probability is defined as number of hits H per number of trials N, $p = H/N$, taken for large N. But small doses means small N, so that below a certain limiting dose deviations from the above proportionality cannot be excluded. In his essay *Probability and Physics*, Pauli describes this situation as follows: "... it is necessary somewhere or other to include a rule for the attitude in practice of the human observer, or in particular the scientist, which takes account of the subjective factor as well, namely that the realization, even on a single occasion, of a very unlikely event is regarded from a certain point on as impossible in practice. Theoretically it must be conceded that there is still a chance, different from zero, of error; but in practice actual decisions are arrived at in this way, ..." (Pauli 1994c, p. 45).

In the above problem of low doses this means that, "in practice" a *threshold model*

$$p(d) = \begin{cases} 0 & \text{if } d < d_o \\ \alpha d & \text{if } d > d_o \end{cases} \qquad (13)$$

is often adopted, and low doses mean a small threshold value d_o. But such a model has serious consequences since, depending on the value of d_o, working in radiation-contaminated areas may not be covered by special *health protection*, as long as the dose is less than the declared value of d_o! Ultimately, of course, the solution of the correct dependence $p(d)$ must be sought on the *molecular level*. There an astonishing mechanism exists, namely the *self-repair* of genetic faults caused by erroneous copying but also by external actions. In addition, the molecules in the cells are under constant aggression of oxidation which gives rise to their deterioration. This threat is taken care of by the *proliferation of cell division* since in each division the concentration of faulty molecules is halved, and cells damaged beyond repair are discarded. But the proliferation capacity diminishes with age, which is the reason why cancer caused by external actions may appear with a delay of years or decades. This shows that, seen from the molecular level, the model (13) is much too crude; in the least, some time-dependence must be assigned to d_o.

A parameter which was not mentioned in the above discussion is the *time of exposition* to the external action, τ. Thus, in a *steady-state situation* characterized by an *intensity* or concentration I of the external action the dose is $d = I\tau$. It then follows from Eq. (13) that, for $\tau > d_o/I$, the probability is $p(\tau) = f\tau$, where $f = \alpha I$ is the *probable frequency* of events. Such steady-state situations are typical for the technologies of production and of transportation where p takes the meaning of integrated *probability of failure*. There a typical problem is *metal fatigue* which concerns, e.g., airplane wings, hulls of super-tankers, or containment vessels of nuclear reactors.

It is an amazing fact that the *metallic bond*, which is due to the conduction electrons, is so strong that *rigidity* over macroscopic distances is achieved. But looking into the *microscopic mechanism* of metal fatigue one quickly notices that a linear relationship of $p(\tau)$ does not correspond to reality neither. What happens is that *microscopic cracks* (dislocations), which are always present in the crystallites forming the metal, migrate and merge, eventually giving rise to *macroscopic fissures*. This migration may be temporarily blocked by impurities (carbon atoms in steel). But after a sufficiently long time, the *critical time* τ_c, failure is unavoidable. This means that the probable frequency f practically jumps from a very low to a very high value at τ_c. To predict τ_c, however, is practically impossible.

The *nuclear industry* determines "probabilities" by calculating the probable frequency of the most likely "scenarios". For the French fast reactor *Superphénix* this frequency is estimated, for the highest level of *gravity* of the accident, at 1 in 100 000 years (Tangui 1989). The unfoundedness of this procedure has been shown in a critical and detailed analysis of the methods of evaluation in a report commissioned by the *Association pour l'Appel de Genève* in Geneva (Benecke and Reimann 1989). Ultimately, however, the single most important and patently incalculable factor that limits predictability is *human failure*, as demonstrate the nuclear accidents, as well as many avoided accidents, from *Three Mile Island* to *Chernobyl*. Thus humanity has to live with increasing threats caused by the ambition of materialistic (in the above sense) technocrats obsessed with *size* and *power*, while the insurance companies insist on a ceiling in the sums to be covered, thus leaving the ordinary citizen not only to suffer the consequences of *mega-accidents* but also to pay for them.

9. The Engaged Observer

What did Pauli have in mind when he protested in a letter to Bohr of 15 February 1955 that Bohr (1985) had introduced the notion of the *detached observer* in his lecture at the occasion of the bicentennial of Columbia University? A *rational* answer was given by Pauli in a radio address at the same occasion (Pauli 1994b, p. 33): "In the new pattern of thought we do not assume any longer the *detached observer*, ... but an observer who by his indeterminable effects (Einwirkungen) creates a new situation, theoretically described as a new state of the observed system. In this way every observation is a singling out of a particular factual result (eines realen Einzelereignisses), here and now, from the theoretical possibilities, thereby making obvious the discontinuous aspect of the physical phenomena."

In an earlier paper (Enz 1991; see also Enz 1995, Section 4) I expressed the view that with his idea of a more "engaged observer" Pauli may have had in mind a future field theory "which makes fields without test bodies not only physically but also logically impossible" (Pauli 1994b, p. 34). While this is certainly a valid point of view, there is also an *irrational* reading of Pauli's strange idea. In his *Lecture to the Foreign People* he expresses the opinion that, compared to the older view of a detached observer, "contemporary physics arrived at a new type of interpretation of nature" which "very strongly emphasizes the element of freedom in the happenings of nature" (translated from Pauli 1995, § 32). *Rationally* it is the freedom of planning the experiment which also implies a sacrifice of some complementary information. But for Pauli the detached observer becomes a symbol of a scientific attitude in which the *rational mind*, i.e., *res cogitans*, confronts matter in the form of *res extensa* in a totally *conscious act*. And this attitude must eventually be overcome by taking notice of the *unconscious* and the *irrational*. In the new scientific attitude the words "freedom" and "sacrifice" acquire a *moral quality* implying *responsibility* and *compassion*.

This message has been put forward by Primas in many essays (e.g., Primas 1992, 1995) which often express rather impatiently that "we need a new natural science". However, I think that impatience is not the answer. It would be rather unfair, for example, to criticize Pauli for not having given his "lecture to the foreign people" (see Pauli 1995). For, as Pauli says in the *Piano Lesson*, "First I would have to explain to them in other terms what 'piano' and 'playing the piano' mean ..." (translated from Pauli 1995, § 45). Pauli's motto was "the still older is always the new". It was the consequence of his belief in the action of the *archetypes* which most of the time are latent, potential, but periodically actualize, constellate, giving rise to cultural and personal *epochs*. It was this belief that made Pauli answer the announcement of the belated discovery of the neutrino by "everything comes to him who knows how to wait" (Enz 1995, p. 25).

Complementary to this externalized reading of Pauli's hesitation with *The Lecture to the Foreign People* there is, however, the *subjective significance* to Pauli which was emphasized by *Marie-Louise von Franz* (Franz 1995), namely that Pauli was not ready to draw the consequences of *The Piano Lesson* for his own spiritual life, particularly the aspect characterized above as *compassion* or, what von Franz calls the *feeling*. (Note, however, that "feeling" also has negative aspects!) Indeed, at the end of *The Piano Lesson* Pauli returns "into the customary time and the customary space of everyday life". And, already outside, he notices that "I had put on hat and coat" (translated from § 71 and § 72 of Pauli 1995), the attributes of the illustrious professor of theoretical physics.

But wait! Hadn't he just given the lecture to the foreign people inside and is now stepping outside satisfied, happily reaching for hat and coat? And this is so real that the Lady is moved to say to Pauli: "You have made me a baby, I believe" (§ 44 of Pauli 1995). Therefore, be an engaged observer, you may see things in complementary ways. And it may require your full responsibility and compassion.

Acknowledgments: I am grateful to the late Marie-Louise von Franz for a pertinent comment concerning the last part of this text.

References

Anderson P.W. (1972): More is different. *Science* **177**, 393–396.

Atmanspacher H., Primas H., and Wertenschlag-Birkhäuser E., eds. (1995): *Der Pauli-Jung-Dialog und seine Bedeutung für die moderne Wissenschaft* (Springer, Berlin). In the *Preface* of this reference the editors recall that the theme of the meeting that gave rise to this publication had been *Das Irrationale in den Naturwissenschaften: Wolfgang Paulis Begegnung mit dem Geist der Materie.* Unfortunately this change of title may misguide the expectations of the reader, as is evident from the review by M. Eckert in *Physikalische Blätter* **51**, 1105 (1995).

Benecke J. and Reimann M. (1989): *Das Gefahrenpotential des Superphénix* (Sollner Institut, München).

Bohr N. (1985): *Atomphysik und menschliche Erkenntnis* (Vieweg, Braunschweig), pp. 83 and 85. The passage of Pauli's letter of 15 February 1955 defining the deta⌣hed observer appears as preamble to his essay *Probability and Physics*, Pauli (1994c), p. 43.

Crommie M.F., Lutz C.P., and Eigler D.M. (1993): Confinement of electrons to quantum corrals on a metal surface. *Science* **262**, 218–220.

Dehmelt H. (1990): Experiments with an isolated subatomic particle at rest. *Rev. Mod. Phys.* **62**, 525–530.

Dorozhkin S.I., Haug R.J., Klitzing K. von, and Ploog K. (1995): Experimental determination of the quasiparticle charge and the energy gap in the fractional quantum Hall effect. *Phys. Rev. B* **51**, 14729–14732.

Einstein A. (1916): Ernst Mach. *Physikalische Zeitschrift* **17**, 101–104.

Einstein A., Podolsky B., and Rosen N. (1935): Can quantum-mechanical description of physical reality be considered complete? *Phys. Rev.* **47**, 777–780.

Enz C.P. (1986): Le rôle de l'espace et le problème de localisation en physique moderne, vus en particulier par Wolfgang Pauli. *Archives des Sciences* (Geneva) **39**, 185–200. Reprinted in *Epistemologie* **10**, 187–205 (1987).

Enz C.P. (1991): Quantum theory in the light of modern experiments. In *Advances in Scientific Philosophy. Essays in Honour of Paul Weingartner on the Occasion of the 60th Anniversary of his Birth*, ed. by G. Schurz and G.J.W. Dorn (Rodopi, Amsterdam/Atlanta), pp. 191–201.

Enz C.P. (1992): *A Course on Many-Body Theory Applied to Solid-State Physics* (World Scientific, Singapore).

Enz C.P. (1995): Rationales und Irrationales im Leben Wolfgang Paulis. In Atmanspacher et al. (1995), pp. 21–32.

Enz C.P. (1996a): Review of the physics of high-temperature superconductors. In *From Quantum Mechanics to Technology*, ed. by Z. Petru, J. Przystawa, and K. Rapcewicz (Springer, Berlin), pp. 143–160.

Enz C.P. (1996b): The wavefunction of correlated quantum systems as objects of reality. In *Vastakohtien todellisuus. Juhlakirja professori K.V. Laurikaisen 80-vuotispäivänä*, ed. by U. Ketvel et al. (Helsinki University Press, Helsinki), pp. 61–76.

Enz C.P. and Meyenn K. von, eds. (1994): *Wolfgang Pauli. Writings on Physics and Philosophy* (Springer, Berlin).

Franz M.-L. von (1995): Reflexionen zum "Ring i". In Atmanspacher et al. (1995), pp. 331–332.

Gex M. (1949): *Einführung in die Philosophie* (Francke, Bern). See also *Encyclopaedia Universalis, Vol. 11* (Paris, 1985), p. 875

Goldman V.J., Su B., and Jain J.K. (1994): Detection of composite fermions by magnetic focusing. *Phys. Rev. Lett.* **72**, 2065–2068.

Goldman V.J. and Su B. (1995): Resonant tunneling in the quantum Hall regime: measurement of fractional charge. *Science* **267**, 1010–1012.

Goodwin I. (1997): Research council panel tries to end controversy linking EMF's with cancer and other health disorders. *Physics Today* **50**, No. 1, 49–50.

Goss Levi B. (1997): Bose condensates are coherent inside and outside an atom trap. *Physics Today* **50**, No. 3, 17–18.

Hall E.H. (1879): On a new action of the magnet on electric currents. *American Journal of Mathematics* **2**, 287–292.

Heisenberg W. (1969): *Der Teil und das Ganze* (Piper, München).

Jahn R.G. and Dunne B.J. (1986): On the quantum mechanics of consciousness, with application to anomalous phenomena. *Found. Phys,* **16**, 721–772.

Jain J.K. (1989): Composite-fermion approach for the fractional quantum Hall effect. *Phys. Rev. Lett.* **63**, 199–202.

Jain J.K. (1990): Theory of the fractional quantum Hall effect. *Phys. Rev. B* **41**, 7653–7665.

Kang W., Stormer H.L., Pfeiffer L.N., Baldwin K.W., and West K.W. (1993): How real are composite fermions? *Phys. Rev. Lett.* **71**, 3850–3853.

Kiess T.E., Shih Y.H., Sergienko A.V., and Alley C.O. (1993): Einstein-Podolsky-Rosen-Bohm experiment using pairs of light quanta produced by type-II parametric downconversion. *Phys. Rev. Lett.* **71**, 3893–3897.

Klitzing K. von, Dorda G., and Pepper M. (1980): New method for high-accuracy determination of the fine-structure constant based on quantized Hall resistance. *Phys. Rev. Lett.* **45**, 494–497.

Laughlin R.B. (1983): Anomalous quantum Hall effect: an incompressible quantum fluid with fractionally charged excitations. *Phys. Rev. Lett.* **50**, 1395–1398.

Mahler G. (1996): Was heisst 'nicht klassisch'? Quantentheorie – und darüber hinaus. *Zeitschrift für Parapsychologie und Grenzgebiete der Psychologie* **38**, 92–107.

Meyenn K. von, ed. (1993): *Wolfgang Pauli. Scientific Correspondence with Bohr, Einstein, Heisenberg, a.o., Volume III: 1940-1949* (Springer, Berlin).

Misner C.W., Thorne K.S., and Wheeler J.A. (1973): *Gravitation* (Freeman, San Francisco), Section 23.6.

Müller-Herold U. (1995): Vom Sinn im Zufall: Überlegungen zu Wolfgang Paulis 'Vorlesung an die fremden Leute'. In Atmanspacher et al. (1995), pp. 159–177.

Pauli W. (1958): *Theory of Relativity. With Supplementary Notes by the Author* (Pergamon Press, London). Translated from the German by G. Field.

Pauli W. (1994a): Science and Western thought. In *Wolfgang Pauli. Writings on Physics and Philosophy*, ed. by C.P. Enz and K. von Meyenn (Springer, Berlin), pp. 137–148.

Pauli W. (1994b): Matter. In *Wolfgang Pauli. Writings on Physics and Philosophy*, ed. by C.P. Enz and K. von Meyenn (Springer, Berlin), pp. 27–34.

Pauli W. (1994c): Probability and physics. In *Wolfgang Pauli. Writings on Physics and Philosophy*, ed. by C.P. Enz and K. von Meyenn (Springer, Berlin), pp. 43–48.

Pauli W. (1995): Die Klavierstunde. Eine aktive Phantasie über das Unbewusste. Frl. Dr. Marie-Louise von Franz in Freundschaft gewidmet. In Atmanspacher et al. (1995), pp. 317–330.

Primas H. (1981): *Chemistry, Quantum Mechanics and Reductionism* (Springer, Berlin).

Primas H. (1992): Umdenken in der Naturwissenschaft. *Gaia* **1**, 5–15.

Primas H. (1994): Realism and quantum mechanics. In *Proceedings of the International Congress of Logic, Methodology, and Philosophy of Science*, ed. by D. Prawitz, B. Skyrms, and D. Westerståhl (Elsevier, Amsterdam), pp. 609–631.

Primas H. (1995): Über dunkle Aspekte der Naturwissenschaft. In Atmanspacher et al. (1995), pp. 205–238.

Primas H. (1996): Synchronizität und Zufall. *Zeitschrift für Parapsychologie und Grenzgebiete der Psychologie* **38**, 61–91.

Schmiedmayer J., Chapman M.S., Ekstrom C.R., Hammond T.D., Wehinger S., and Pritchard D.E. (1995): Index of refraction of various gases for sodium matter waves. *Phys. Rev. Lett.* **74**, 1043–1047.

Schrödinger E. (1967): *What is Life? Mind and Matter* (Cambridge University Press, Cambridge), p. 92. Second edition 1983.

Tangui P. (1989): *Rapport de synthèse. La sureté nucléaire à EDF à fin 1989* (Inspection générale pour la sureté nucléaire, Paris), unpublished.

Upton A.C. (1991): Health effects of low-level ionizing radiation. *Physics Today* **44**, No. 8, 34–39.

Willett R.L., Ruel R.R., West K.W., and Pfeiffer L.N. (1993): Experimental demonstration of a Fermi surface at one-half filling of the lowest Landau level. *Phys. Rev. Lett.* **71**, 3846–3849.

CONTEXTUAL BACKGROUND

Abner Shimony:
HOLISM

Holism is regarded as the counter-concept to atomism or particularism. In the context of biology, for example, holism is the doctrine that the parts of an organic whole exhibit characteristics which do not exist if these parts are considered separately. In the historical-philosophical context, Spinoza advocated a material holism according to which there is just one single substance; everything that exists is a mode of this one substance.

In contemporary epistemology there are three different forms of holism. The first is epistemic holism which says that it is not possible to confirm or disconfirm any sentence taken in isolation. The confirmation or disconfirmation of a sentence is always relative to a whole network of sentences (Duhem-Quine-thesis). Second, there is semantic holism which is the thesis that any belief has a meaning only relative to other beliefs within a coherent system of beliefs. Finally, holism of justification asserts that a sentence can only be justified by other sentences. A sentence is justified if and only if it coheres with other sentences in a whole system of sentences.

A special position is held by quantum theoretical holism. It is the only one not to work on the assumption that a whole is made up of given parts which then stand in some form of relation – however problematic – to that whole. What it does do is to take as its starting point the fact that the decomposition of a whole into indivisible (so-called elementary) components is not unique. In other words, different decompositions are due to different kinds of components. According to Primas, one can no longer say that a quantum whole "consists" of these components, but rather that it can be "divided up" into such components. It is only the decomposition which "generates" the more elementary components by means of an experimental intervention.

The most significant manifestation of quantum mechanical holism are the so-called Einstein-Podolsky-Rosen correlations between the various possible sub-systems of a quantum whole. More precisely, quantum mechanical holism manifests itself in a violation of Bell's inequalities (Bell 1964), which is to be seen as a quantitative expression of the Einstein-Podolsky-Rosen correlations. The violation of Bell's inequalities was proven in a series of key experiments conducted by the group around Alain Aspect from 1978 to 1982 in the Institut d'Optique Théorique in Orsay (Université de Paris Sud); see, e.g., Aspect (1982).

As stressed by Primas, this holistic character of the material world creates a new problem: the holistic correlation between a measuring apparatus and an object of observation. Since not only the object but also the measuring apparatus must fundamentally be described by quantum mechanics, holistic correlations between apparatus and object are to be expected. For the measurement, abstractions must be made from these Einstein-Podolsky-Rosen correlations. These abstractions have been metaphorically denoted as Heisenberg cuts (Primas 1993). Without Heisenberg cuts, fundamental quantum physics cannot be linked to experimental physics. Every Heisenberg cut suppresses certain Einstein-Podolsky-Rosen correlations, so that corresponding measurement results can only cover partial aspects of material reality.

Primas made this dilemma the starting point for his proposal of a distinction between endophysics and exophysics. Endophysics is the theory of an undivided world with all Einstein-Podolsky-Rosen correlations between possible sub-systems. In this world there are no observable facts. Exophysics is the theory of a world with Heisenberg cuts. In this world there are experimental facts, but no universal meaning can be attributed to them as they are relative to a proper choice of a Heisenberg cut.

This dichotomy between endo- and exo-quantum mechanics is both acknowledged and criticized in Shimony's contribution. The decisive point of his criticism lies in the fact that Primas sees himself forced to introduce concepts with unclarified ontological status in order to make his distinction between endo- and exophysics: cuts, contexts, experimental outcomes, appearances etc. As a way out of this, Shimony proposes a so-called "phenomenological principle" as a possible back-up for the link between endophysics and exophysics.

References

Aspect A., Dalibard J., and Roger G. (1982): Experimental test of Bell's inequalities using time-varying analyzers. *Phys. Rev. Lett.* **49**, 1804–1807.

Bell J.S. (1964): On the Einstein Podolsky Rosen paradox. *Physics* **1**, 195–200.

Primas H. (1993): The Cartesian cut, the Heisenberg cut, and disentangled observers. In *Symposia on the Foundations of Modern Physics 1992. The Copenhagen Interpretation and Wolfgang Pauli*, ed. by K.V. Laurikainen and C. Montonen (Singapore, World Scientific), pp. 245–269.

HOLISM

ABNER SHIMONY
Departments of Philosophy and Physics,
Boston University, Boston MA 02215, USA

1. Physics and History of Ideas

The division of 17th century natural philosophy into natural science on the one hand and philosophy on the other – two intellectual activities with different methodologies, different styles, and perhaps different subject matters – has a complex history which I shall not try to recapitulate. I wish only, as an introduction to a discussion of quantum mechanical holism, to remark that in recent decades, there seems to have been a moderate amount of recovery of the old unified discipline of natural philosophy. 20th century scientists have frequently cited classical philosophers not just for the sake of ornament but for the purpose of clarifying revolutionary ideas: e.g., Heisenberg (1962) cited Aristotle's potentiality, d'Espagnat (1995) and others cited Kant's distinction between the phenomenal and the noumenal, Wigner (1960) cited Peirce's evolutionary theory of concept formation, Weyl (1970) cited Pythagoras on intellectual harmony, Stapp (1993) cited James' holistic theory of consciousness, and Primas (1983) frequently refers to classical as well as contemporary philosophical literature.

Several reasons – some stemming from the triumphs of modern physics and some from its anomalies – can be suggested for the renewal of interest in philosophical ideas on the part of serious physicists. The triumphs of physics at the atomic and molecular level have thrown light upon chemical and biological phenomena in an unprecedented way, carrying us a long way toward the integration of knowledge which was part of the dream of 17th century naturalists, and integration is one of the characteristics of the philosophical enterprise. On the other hand, counter-intuitive features of relativity theory and quantum mechanics, and the occurrence of conceptual problems on which there is no consensus in spite of great effort (especially the quantum mechanical measurement problem and quantum mechanical non-locality) have stimulated serious scientists to question assumptions that had become canonical during the period of classical physics; and one has to go back to the great debates of 17th century and Greek natural philosophy in order to find extensive discussions of these assumptions. The occurrence of anomalies in a wonderfully fruitful physics has motivated old and new questions regarding the scope of human knowledge, the nature of scientific explanation, and the character of physical reality.

Although I applaud the renewal of interest in philosophical matters on the part of working scientists, I am far from enthusiastic about the typical manifestations of this interest. Typically one finds an uncritical appropriation of some philosophical terminology or very brief quotations. Almost never is the context of a classical maxim presented and explored, and there is insufficient attention to the shift of meanings of terms and of the aims of inquiry over a period of several centuries or millennia. Attention to the wisdom of the past is unlikely to be enlightening and heuristically valuable unless these defects of scholarship and perspective are repaired. I shall illustrate this thesis by comparing the arguments for holism presented by the purest of all classical rationalist and holist philosophers, Spinoza, with quantum mechanical holism.

2. Some Reflections on Spinoza's Philosophy

Part I (Of God) of Spinoza's *Ethics* (Spinoza 1677) presents a dazzling series of definitions, axioms, and propositions. It is difficult to select passages without great loss, but the following are particularly illuminating for our purposes.

> **Definition III.** "By substance, I understand that which is in itself and is conceived through itself; in other words, that, the conception of which does not need the conception of another thing from which it must be formed."
> **Definition V.** "By mode, I understand the affections of substance, or that which is in another thing, through which also it is conceived."
> **Definition VI.** "By God, I understand Being absolutely infinite, that is to say, substance consisting of infinite attributes, each one of which expresses eternal and infinite essence."
> **Axiom I.** "Everything that is, is either in itself or in another."
> **Axiom III.** "From a given determinate cause an effect necessarily follows; and, on the other hand, if no determinate cause can be given, it is impossible that an effect can follow."
> **Proposition VII.** "It pertains to the nature of substance to exist."
> **Proposition XIII.** "Substance absolutely infinite is indivisible."
> **Proposition XV.** "Whatever is, is in God, and nothing can either be or be conceived without God."

Proposition XVI. "From the necessity of the divine nature infinite number of things in infinite ways (that is to say, all things which can be conceived by the infinite intellect) must follow."

Proposition XX. "The existence of God and His essence are one and the same thing."

Proposition XXIX. "In nature there is nothing contingent, but all things are determined from the necessity of the divine nature to exist and act in a certain manner."

Proposition XXXIII. "Things could have been produced by God in no other manner and in no other order than that in which they have been produced."

In view of Proposition XV, Spinoza is a monist in the sense of admitting the actuality and even the possibility of only one substance. But in view of the infinitely rich nature of the one substance – its infinitely many attributes, each of which is infinite – there is an unequivocal difference between Spinoza's monism and that of Parmenides (1945), according to whom all differentiation and inhomogeneity is mere appearance. Finite entities are acknowledged by Spinoza to exist as modes of substance, even though these do not have the self-explanatory character essential to the only thing which exists without qualification, namely a substance (of which there is only one exemplar). Because substance is indivisible (Proposition XIII), it is – in spite of the plurality of its modes – appropriate to characterize Spinoza's metaphysics as holistic, even though he does not seem to employ that or any similar word.

A thoroughgoing and uncompromising rationalism underlies Spinoza's holism. Axiom III, concerning cause, asserts a rationalist view of causality – it is a necessary relation which, as Spinoza's discussion shows, is not distinguished from logical and mathematical necessity. Proposition XV and others show that for Spinoza the domain of the conceivable is not more extensive than the domain of the existent, whence all that is rationally possible (if all connections are taken into account) necessarily exists. The identification of God's essence with God's existence (Proposition XX) is the epitome of the ontological argument for the existence of God, which is Anselm's (1945) famous rationalistic argument for proving the existence of God without relying upon the order of the observed world or any other empirical data. Spinoza's rationalistic view of the universe is unequivocal in Proposition XXIX, which denies that there is any true contingency in nature, although elsewhere Spinoza admits contingency in the sense of ignorance by a finite mind of some of the causal connections of a thing.

Since my purpose in discussing Spinoza is primarily to point out the profound difference between the ideas underlying his holism and the ideas underlying quantum mechanical holism, I shall not attempt to give a de-

tailed criticism of Spinoza's epistemology and metaphysics. It will suffice for
the purposes of this paper to point out several respects in which his extreme
rationalism is unconvincing. When one looks at the details of mathematical
and of physical knowledge, observing the conceptual autonomy of the former and the inevitable dependence upon sensory observation of the latter,
the conflation of mathematical and physical truth becomes highly implau-
sible. And since mathematics explores domains of possibility far wider than
those which physics selects as actually true or a good approximation to the
truth, the identification of the conceptually possible with the actual is thoroughly unconvincing. Even the most ambitious flights of theoretical physics
to a "theory of everything" rely upon some empirically based framework
(notably that of quantum mechanics) within which the alleged proof of the
inevitable structure of physical law is established. Some of the greatest theoretical physicists, especially Einstein, acknowledge that Spinoza's rational
vision of the world inspired them, but the great gap between inspiration
and substantive reasoning is another reason to be doubtful about the literal
truth of that vision. Finally, Spinoza's endorsement of Anselm's ontological argument, which makes existence a component of the concept of God
– an argument which is the crux of Spinoza's identification of the rational
and the existent – was powerfully challenged by Kant's (1781, pp. 502–602)
criticism of the ontological argument.

What we shall stress in the next section is that quantum mechanics
has led us to recognize holism in nature without postulating Spinoza's
thoroughgoing rationalism, and indeed to formulate a conception of holism
which *depends upon* contingency in physical phenomena. Here is an example
showing that analogues between classical philosophical ideas and the radical ideas of modern physics have to be treated with caution. Before turning
to the quantum mechanical version of holism, some insight, however, will
result from noticing some great differences between Spinoza's world view
and that of classical particle mechanics.

The particle mechanics of Newton and his followers did not pretend to
be a comprehensive philosophy, offering answers to all epistemological and
metaphysical questions that could be raised by reflection upon mechanics.
There are passages in Newton's writings suggesting the mutual relevance
of theology and natural philosophy – e.g., "and thus much concerning God;
to discourse of whom from the appearances of things, does certainly belong
to Natural Philosophy" (Newton 1713; General Scholium to Proposition
XLII of Book III, Cajori translation of 1934, p. 544). Newton did not,
however, leave a treatise, giving the details of his views on the connection of
natural philosophy to theology and to other aspects of what is now classified
as philosophy, comparable to Spinoza's *Ethics* and Leibniz's *Discourse on
Metaphysics*. He does, however, occasionally address metaphysical questions

concerning force, cause, and substance – usually elliptically – but unlike Spinoza and Leibniz he characteristically appeals to experience rather than pure reason, recognizes the fallibility of induction, and confesses ignorance of matters beyond present and possibly even future experience (cf. Stein 1970).

From the standpoint of an autonomous physics a number of crucial propositions are asserted as true on the basis of experimental evidence and without fundamental explanations – without "sufficient reasons". It is left in abeyance whether a more extensive discipline could in principle supply them. Specifically, the law of gravity is affirmed without being derived as an essential property of bodies (Newton 1713; Rule III of the Rules of Reasoning in Philosophy, Cajori translation of 1934, p. 400), and the variety of sizes and shapes of physical particles is attributed to God's choice without any conjecture at a reason for the choice (Newton 1717). The autonomous discipline of Newtonian particle mechanics accepts several classes of propositions without explanation: the physical law and the basic types of particles mentioned in the foregoing citations, and in addition positions and momenta of particles at an initial time. If Spinoza (and also Leibniz) were right, this apparent contingency evaporates when physics is properly extended to a more comprehensive body of knowledge. But an argument going beyond physics would be required to exorcize all contingency, and from the standpoint of Newtonian particle mechanics itself, sheer contingency – brute fact for which no explanation exists – appears to be logically possible.

Newtonian particle mechanics is usually considered to have a pluralistic ontology, and it is anti-holistic because the state of a composite system of many particles derives all its properties from those of the components. But implicit in this classification is a retrenchment from Spinoza's conception of substance. Clearly the dynamics of a many-particle system depends upon interactions, and hence upon the laws of interaction, and cannot be derived from the essential properties of the individual particles in isolation. Therefore no particle is a substance according to Spinoza's definition. But if the law of interaction is a contingency, then neither does the system consisting of the totality of particles satisfy Spinoza's definition of substance, since the dynamics does not follow from the essential properties of the totality. In other words, even the entire Laplacian physical universe is not a substance in the sense required by Spinoza's rationalism.

There is much more to be said about the ontology of Newtonian particle mechanics, especially concerning the status of the space-time framework and the relation of states of particles to space-time (cf. Stein 1967). But enough has been said to indicate a line of cleavage in 17th century scientific thought, in which holism is linked with extreme rationalism and anti-holism with the admission of contingency. One of many ways in which quantum mechanics is a radical revolution of thought is its thorough displacement of this line of cleavage.

3. Quantum Mechanical Holism

In the following discussion I shall draw upon some ontological and episte-mological distinctions which Primas has made in his expositions of quantum theory. I disagree with him concerning the actualization of potentialities, but his careful exposition facilitates the presentation of the disagreement. A fundamental distinction insisted upon by Primas (1994, p. 169) is between "endophysics" and "exophysics":

> "Endophysics refers to a subject-independent reality while exophysics refers to empirical reality. Endophysics aims at metaphysical universal laws, while exophysics aims to give us empirically adequate descriptions. Endo-entities belong to the subject-independent reality, they may be related to hypothetical 'things-in-themselves'... They are hidden from us and certainly not directly observable. [...] In contradistinction, ex-ophysics refers also to tangible objects we can directly see, feel, and touch."

This distinction is reminiscent of Kant's (1781) distinction between the "phenomenal" and the "noumenal", but Primas tempers the Kantian di-chotomy in various ways by linking endophysical ideas to exophysical em-pirical concepts. He recognizes some indirect and intuitive knowledge of the former on the basis of experience, and he allows partial explanation (with qualifications to be noted below) of the latter in terms of the former.

In order to simplify the discussion of quantum mechanical holism, I shall restrict attention to the elementary Hilbert space formulation of quantum mechanics, in spite of the inadequacy of this formulation – strongly em-phasized by Primas – to deal with systems having infinitely many degrees of freedom. Endophysics is concerned with the intrinsic observables of the system, which are represented by the self-adjoint operators on the appro-priate Hilbert space $\mathcal{H}$. If $\mathcal{A}$ is the algebra of the intrinsic observables and $\mathcal{A}^*$ is the dual of $\mathcal{A}$ (i.e., the space of continuous linear mappings of $\mathcal{A}$), an extremal element of $\mathcal{A}^*$ is one which is not a convex combination of any set of distinct elements of $\mathcal{A}^*$.

In more familiar terminology, the individual ontic states are the pure quantum states of the system, represented in a standard way by normalized vectors in $\mathcal{H}$. That is, if u is a vector in $\mathcal{H}$ normalized to unity and $\langle \, , \, \rangle$ is the inner product on $\mathcal{H}$, then the map $\langle u, Au \rangle$ for all $A \in \mathcal{A}$ is an extremal element of $\mathcal{A}^*$. If there is a real number a such that the expectation value $\langle u, (A - aI)u \rangle = 0$, where I is the identity operator, then A has the intrinsic *value a* in the state represented by u. If there is no such number, then A is intrinsically indefinite in the state represented by u. It is a consequence of the structure of a Hilbert space (of dimension greater than unity) that for every individual state there exist observables which are indefinite in that state. Because of the occurrence of objective indefiniteness, Primas (1994, p. 179) characterizes the observables as "potential properties of the universe of discourse".

Without examining in detail the difference in the algebraic structure of the intrinsic observables of endophysics and the contextual observables of exophysics, we note that an observable A that is indefinite in the state represented by u will exhibit a spread of values when replicas of the system of interest, all characterized by u, are subjected to interactions with a suitable measuring apparatus. This spread of values provides indirect evidence for the indefiniteness of A in the endophysics, while the indefiniteness of A in the endophysics provides a kind of explanation for the statistical dispersion of the measurement results. There are some puzzling aspects of this connection which constitute the measurement problem, but there is nevertheless an apparent connection of endo- and exophysics which differs from the Kantian separation of phenomena and noumena. Quantum mechanical potentiality is incompatible with Spinozistic rationalism, because an individual state according to quantum mechanical endophysics implies genuine contingencies – viz., the outcomes of measurements of all those observables that are intrinsically indefinite in that state.

A kind of holism is implicit in what has just been said about the Hilbert space formulation of quantum mechanics. A compound system consisting of two subsystems S and S', associated with Hilbert spaces $\mathcal{H}$ and $\mathcal{H}'$, respectively, is commonly assumed without much argumentation to be associated with the tensor product Hilbert space. A mathematical justification for this assumption, based upon some plausible lattice theoretical propositions, was provided by Aerts and Daubechies (1978).

If $\mathcal{H}$ and $\mathcal{H}'$ are arbitrary Hilbert spaces, then an arbitrary vector η in the tensor product of $\mathcal{H}$ and $\mathcal{H}'$ can be expressed in the form

$$\eta = \sum_i c(i)\, u(i) \otimes v(i), \tag{1}$$

where the $u(i)$ are some orthonormal vectors in $\mathcal{H}$, the $v(i)$ are some orthonormal vectors in $\mathcal{H}'$, and the $c(i)$ are complex numbers. The remarkable fact about this expression is that it involves only a single summation. The expression is called the Schmidt decomposition, and it was proved by Schmidt (1907). A more easily accessible exposition is due to von Neumann (1932, pp. 429–434) or, more recently, Peres (1995).

Although the Schmidt decomposition is not unique, it is easily shown that the number of terms in the sum is invariant under transformation of bases; and hence, in particular, if there is more than one term in one such representation of η, then there is no representation in which η can be expressed as a single term $w \otimes z$, $w \in \mathcal{H}$ and $z \in \mathcal{H}'$. If η has norm unity, then (by the general remarks above) it represents an individual state of the composite system $S + S'$, but if there are more than one term in the Schmidt decomposition, neither system S by itself nor system S' by itself is in an individual state. In other words, even though η determines uniquely a member of the dual $\mathcal{A}^*$ of the algebra of observables of system S, that member is not an extremal element of $\mathcal{A}^*$; and likewise concerning system S'.

Schrödinger (1926) discovered quantum mechanical holism in his pioneering papers on wave mechanics and gave the phenomenon the names "entanglement" and "Verschränkung" in later comments (Schrödinger 1935a, 1935b, 1936). He was so impressed by this conceptual innovation that he remarked, "I would not call that *one* but rather *the* characteristic trait of quantum mechanics" (Schrödinger 1935a). In spite of Schrödinger's immense authority, I contend that he overstated his case. Entanglement depends upon quantum indefiniteness – upon the fact that in an individual state of a system, which is a maximal possible specification of the system permitted by quantum theory, some of the intrinsic observables have indefinite values. Since quantum indefiniteness is exhibited by any quantum system, even an elementary one, while entanglement is exhibited only by composite systems, it follows that quantum indefiniteness is a more general and basic trait than entanglement.

Furthermore, quantum indefiniteness is the key to the peculiar character of the holism exhibited by entangled systems, which is entirely different from the rationalistic holism of Spinoza examined above. If A is an observable of system S with definite value $a(i)$ in the state represented by $u(i)$, and B is an observable of system S' with value $b(i)$ in the state represented by $v(i)$, then the state shown in Eq. (1) constrains the indefiniteness of A and B by exhibiting their correlation. If A is in some way or another actualized to have the value $a(i)$ then in tandem B is actualized to have value $b(i)$, and conversely. If the actualizations of all observables of S and S' were fixed in advance, then the behavior of system S would be fixed without reference to system S' and conversely, and then the behavior of the composite system would be reducible to the behavior of the components. On the other hand, if the contingent actualizations were independent, then again there would be no holism. Thus quantum mechanical holism depends upon both the contingency of actualizations of potentialities and the constraint upon contingency imposed by correlation.

I note, finally, that the extension from the Hilbert space formulation of quantum mechanics to the more general algebraic formulation preserves the possibility of entangled systems, provided that one does not restrict attention to the special case of classical systems, for which all intrinsic observables have definite values.

4. Actualization of Potentialities in Endo- and Exophysics

In the foregoing discussion of quantum mechanical holism the phrase "if A is in some way or another actualized to have the value $a(i)$" was used without comment, as if there is nothing problematic about it. Much was thereby glossed over, particularly since the context of the discussion was endophysics, in which – according to Primas – the physical world is an entangled unity, and the division of that world into an object system and an observing system makes no sense. The typical situation (but not the only one) in which a quantum mechanical potentiality is actualized is a measurement situation, which does presuppose a division into an object system and an observing system, and therefore a transition to exophysics.

But then there is something I find baffling in Primas' philosophy of quantum mechanics. When he speaks of the intrinsic observables of quantum mechanical endophysics, he characterizes them as potentialities. It seems to me, however, that the concept of potentiality is inseparable from the concept of actualization – what is a potentiality *of* other than one of several mutually exclusive outcomes, the choice among which is an actualization? And if the process of actualization requires the transition to exophysics, then the clarity of the concept of potentiality in endophysics is undermined. I shall quote some crucial passages in which Primas speaks of the relation between endo- and exophysics, in order to determine whether he has a satisfactory answer to this problem:

"In contradistinction to quantum endophysics, quantum exophysics is associated with *experimental physics*. Every experiment and every operationally meaningful description requires a division of the endo-world into an object system and an observing system. Yet, the endoworld does not present itself already divided – we have to divide it. Therefore endophysical first principles are not sufficient for an operational description of an experiment, we have to add the particular context which characterizes the cut between the material object and the material observing tools." (Primas 1994, p. 174)

"The fundamental symmetries of the endoworld are not manifest in our everyday experience. It is necessary to break fundamental symmetries [...] All symmetry breakings are contextual [...]." (Primas 1994, p. 175).

"Endophysical first principles are insufficient for a theory of human knowledge. A theory which describes observable phenomena cannot keep the human means of data processing out of consideration. [...] The associated pattern recognition *projects* the holistic, non-Boolean endoworld into an exophysical Boolean registration system. This projection is neither arbitrary nor unique. It is not arbitrary since all possible patterns are preexistent in the endoworld. But these *preexistent* patterns become manifest only in the appropriate exophysical tensor product decomposition." (Primas 1994, p. 182)

Before offering a serious criticism of Primas' division of labor between endo- and exophysics I wish to acknowledge some of its virtues. By assigning to endophysics the function of investigating physical systems as they are apart from human knowledge, Primas' scheme remains realistic, just as in classical physics. By giving to exophysics the task of accounting for the design and the outcomes of experiments, his scheme provides for knowledge of physical phenomena. By the linkage of endophysics and exophysics, in large part by expressing both in the language of algebraic quantum physics, Primas is able to avoid Kant's extreme split of noumena and phenomena. By avoiding symmetry-breaking in endophysics, and hence maintaining a dynamics that is time-reversal invariant, Primas' scheme is conservative with respect to quantum mechanical principles. No nonlinear or stochastic modification of the Schrödinger dynamics is postulated in endophysics; and the breaking of time reversal invariance, as well as the stochasticity of outcomes, in the measurement process are permitted only because the measurement process belongs to exophysics, which is context dependent.

All of these virtues can also be found in the consistent histories and decoherence theories of Griffiths (1984), Omnès (1992), and Gell-Mann and Hartle (1993), but with important differences of conceptual emphasis. A detailed comparison of Primas' theory with these is an important project that has not, to my knowledge, been carried out by anyone, and it certainly is beyond the scope of the present paper. I believe strongly, however, that the central criticism that I shall now present concerning endo- and exophysics is applicable – mutatis mutandis – to the consistent histories and decoherence interpretations.

What I find troublesome in Primas' treatment of endo- and exophysics is the ontological status of the entities of the latter: cuts, contexts, experimental outcomes, appearances, etc. A realist can accept that these entities, which rely upon a distinction between observed object and observing apparatus, do not have the full status of a thing-in-itself in endophysics. A realist treatment of the entities of ordinary life distinguishes between the intrinsic properties of a flower and its appearances to various observers under various conditions of illumination and probing. Each appearance is an effect in a conscious subject who is causally linked with the flower of interest.

A considerable amount of physics, physiology, and psychology is needed in order to understand the entire causal chain establishing this link and thereby to disentangle the contributions from various systems linked by the chain. There is a fairly clear sense in which the flower has a career, representable in an appropriate phase space, apart from any of its appearances, while the appearance has no such separable career. Nevertheless, the appearance has an ontological status of a derivative kind, and an adequate understanding of the composite system consisting of flower, medium, and conscious subject would establish that status accurately.

I have elsewhere proposed a general principle of realist philosophy, called (for lack of a better name) "the phenomenological principle", which asserts that a necessary condition for an adequate ontology is that it accounts for appearances qua appearances, that is, with due acknowledgment of their derivative ontological status (Shimony 1993; Vol. I, p. 36; Vol. II, pp. 278–287). Although this principle is rough-hewn, I maintain that it has some power in philosophical analysis. I deploy it for the purpose of criticizing a reductive physicalist account of psychology and a theory of time that denies the objectivity of temporal transience.

Moreover, the phenomenological principle can be deployed in the interpretation of quantum mechanics by inquiring how the symmetry of the endophysical Eq. (1), in which all the different correlated values $a(i)$ and $b(i)$ have the same status as potentialities, are related to the outcome of an exophysical experiment. If there is no relation, then exophysics becomes an autonomous discipline, as conceived by positivists, and the interpretation can no longer be regarded as a variety of physical realism. If there is a relation, then subsequent to the actualization in the exophysics the symmetry of Eq. (1) is broken in the endophysics. To say this is not tantamount to supplementing Eq. (1) by a hidden variable that determines the outcome of the actualization. There is an abundance of experimental evidence against such hidden variables theories (cf. Redhead 1987). What is being asserted is only that *after* the appearance of a definite experimental outcome in the exophysics the individual state of the pair of systems in the endophysics can no longer have the symmetry of Eq. (1). *Without a broken symmetry in the endophysics there would not be an ontology sufficient to account for the appearance of broken symmetry in the exophysics.*

The conclusion that has been reached is crude but momentous. It does not tell *how* the symmetry is broken in the endophysics – only *that* it is broken. But this crude conclusion suffices to infer that the standard Schrödinger dynamics (or its natural generalization in algebraic quantum theory) cannot be the complete dynamics of endophysics. And this inference provides the motivation for studying non-linear and stochastic modifications of quantum dynamics (e.g., Gisin 1984, Pearle 1985, Károlyházy et al. 1986, Ghirardi et al. 1986, Diósi 1988, Penrose 1989, and Shimony 1991).

It has been known since von Neumann's book (1932) that quantum mechanical holism underlies the measurement problem (equivalently, the problem of the actualization of potentialities), because it is exhibited in the entanglement of object and measuring apparatus. I conclude with the less well known (and speculative) proposition that holism may be an ingredient of a solution to the problem. When a chain of systems are entangled on account of interactions one can obtain a generalization of Eq. (1):

$$\eta = \sum_i c(i) \ u(i) \otimes v(i) \otimes w(i) \otimes \cdots \otimes z(i). \tag{2}$$

It may very well be the case that the earlier systems in this chain – e.g., a microscopic object, a macroscopic measuring apparatus – are systems for which quantum dynamics is so well established experimentally that any effects of non-linear or stochastic modifications of that dynamics are too small to account for the reduction of the superposition in the time of typical measurements. But for later systems in the chain – e.g., the spacetime framework, the human nervous system, the human psyche – there may be little or no direct evidence for the validity of quantum dynamics. Károlyházy et al. (1986) and Penrose (1989), for example, anticipate that the breakdown of the superposition principle, and consequently the breakdown of quantum dynamics, occurs at the level of the spacetime framework.

Because of the entanglement exhibited in Eq. (2), the choice of one term out of the sum due to a spontaneous actualization of a potentiality regarding the nth system in the chain entails the choice of the correlated term for each of the other systems. Quantum mechanical holism could therefore supply the mechanism for transmitting the actualization of potentialities in systems which deviate strongly from the principles of quantum mechanics to simple physical systems, microscopic or macroscopic, which in isolation behave very well in accordance with these principles. The experimental evidence for quantum mechanical holism is so strong that holism is almost certain to survive the next great conceptual revolution in physics. The proposal just stated for the measurement problem is a conjecture about how and with what limitations the physics of the future will retain the conception of quantum mechanical holism.

References

Aerts D. and Daubechies I. (1978): Physical justification for using the tensor product to describe two quantum systems as a joint system. *Helv. Phys. Acta* **51**, 661–675.

Anselm (1945): Proslogium. In *St. Anselm*, ed. and transl. by S.N. Deane (Open Court, La Salle), Chap. III.

Diósi L. (1988): Quantum stochastic processes as models for state vector reduction. *J. Phys. A* **21**, 2885–2898.

d'Espagnat B. (1995): *Veiled Reality* (Addison–Wesley, Reading), p. 7.

Gell-Mann M. and Hartle J.B. (1993): Classical equations for quantum systems. *Phys. Rev. D* **47**, 3345–3382.

Ghirardi G.C., Rimini A., and Weber T. (1986): Unified dynamics of microscopic and macroscopic systems. *Phys. Rev. D* **34**, 470–491.

Gisin N. (1984): Quantum measurements and stochastic processes. *Phys. Rev. Lett.* **52**, 1657–1660.

Griffiths R. (1984): Consistent histories and the interpretation of quantum mechanics. *J. Stat. Phys.* **36**, 219–272.

Heisenberg W. (1962): *Physics and Philosophy* (Harper, New York), p. 53.

Kant I. (1781): *Kritik der reinen Vernunft* (Johann Friedrich Hartknoch, Riga). Transl. by N.K. Smith (Macmillan, London, 1990).

Károlyházy F., Frenkel A., and Lukacs B. (1986): On the possible role of gravity in the reduction of the wave function. In *Quantum Concepts of Space and Time*, ed. by R. Penrose and C. Isham (Clarendon, Oxford), pp. 109–128.

Leibniz G.W. (1947): *Discourse on Metaphysics*, transl. by G.R. Montgomery (Open Court, La Salle).

Newton I. (1713): *Mathematical Principles of Natural Philosophy* (Royal Society, London). Transl. by A. Motte (1729), revised by F. Cajori (University of California Press, Berkeley, 1934).

Newton I. (1717): *Opticks* (London, 2nd ed.). Reprinted 1952 (Dover, New York), Book III, Part I, p. 400.

Neumann J. von (1932): *Mathematische Grundlagen der Quantenmechanik* (Springer, Berlin). Transl. by R.T. Beyer (Princeton University Press, Princeton, 1955).

Omnès R. (1992): Consistent interpretations of quantum mechanics. *Rev. Mod. Phys.* **64**, 339–382.

Parmenides (1945): *Selections from Early Greek Philosophy*, ed. by M.C. Nahm (Crofts, New York, 2nd ed.), pp. 113–121.

Pearle P. (1985): On the time it takes a state vector to reduce. *J. Stat. Phys.* **41**, 719–727.

Penrose R. 1989: *The Emperor's New Mind* (Oxford University Press, Oxford), Chap. 8.

Peres A. (1995): *Quantum Theory: Concepts and Methods* (Kluwer, Dordrecht), pp. 123–126, 147.

Primas H. (1983): *Chemistry, Quantum Mechanics, and Reductionism* (Springer, Berlin, 2nd ed.).

Primas H. (1990): Mathematical and philosophical questions in the theory of open and macroscopic quantum systems. In *Sixty-two Years of Uncertainty: Historical, Philosophical, and Physical Inquiries into the Foundations of Quantum Mechanics*, ed. by A.I. Miller (Plenum, New York), pp. 233–257.

Primas H. (1994): Endo- and exotheories of matter. In *Inside Versus Outside*, ed. by H. Atmanspacher and G.J. Dalenoort (Springer, Berlin), pp. 163–193.

Redhead M.L.G. (1987): *Incompleteness, Nonlocality, and Realism: a Prolegomenon to the Philosophy of Quantum Mechanics* (Clarendon, Oxford), Chap. 4.

Schmidt E. (1907): Zur Theorie der linearen und nichtlinearen Integralgleichungen I, II. *Mathematische Annalen* **63**, 433–476, **64**, 161–174.

Schrödinger E. (1926): Quantisierung als Eigenwertproblem IV. *Annalen der Physik* **81**, 109–139.

Schrödinger E. (1935a): Discussion of probability relations between separated systems. *Proc. Cambridge Phil. Soc.* **31**, 555–562.

Schrödinger E. (1935b): Die gegenwärtige Situation in der Quantenmechanik. *Naturwissenschaften* **23**, 807–812, 823–828, 844–849.

Schrödinger E. (1936): Probability relations between separated systems. *Proc. Cambridge Phil. Soc.* **32**, 446–452.

Shimony A. (1991): Desiderata for a modified quantum dynamics. *PSA 1990* (Philosophy of Science Association, East Lansing), pp. 49–59. Reprinted in Shimony (1993), Vol. II, pp. 55–67.

Shimony A. (1993): *Search for a Naturalistic World View, Vols. I, II* (Cambridge University Press, Cambridge).

Spinoza B. (1677): *Ethics*, transl. by W.H. White and A.H. Stirling (Oxford University Press, Oxford, 1927).

Stapp H. (1993): *Mind, Matter, and Quantum Mechanics* (Springer, Berlin), p. 135.

Stein H. (1967): Newtonian space-time. *Texas Quarterly* **10**, 174–200.

Stein H. (1970): On the notion of field in Newton, Maxwell and beyond. In *Historical and Philosophical Perspectives of Science (Minnesota Studies in the Philosophy of Science, Vol. 5)*, ed. by R. Steuwer (University of Minnesota Press, Minneapolis), pp. 264–287, here: part I.

Weyl H. (1970): *Raum–Zeit–Materie* (Springer, Berlin, 6th ed.), p. 317.

Wigner E. (1960): The unreasonable effectiveness of mathematics in the natural sciences. *Communications in Pure and Applied Mathematics* **13**, 222–237, here: p. 226.

CONTEXTUAL BACKGROUND

Bernard d'Espagnat:
CONCEPTS OF REALITY

One of the amazing features of the early quantum mechanics of the 1920s and 1930s is its epistemological proximity to philosophical Kantianism with its turn towards the perceiving subject, whose a priori characteristics constitute the distinguishing features of the world of appearances. In contrast to a Platonic form of metaphysics or the materialistic metaphysics that was widespread in the 19th century, this "critical" philosophy takes as its starting point the fact that the thing-in-itself, i.e., the very essence of reality, must remain inaccessible to cognition. All cognition is stamped by the categories of the perceiving subject, the so-called synthetic a priori of apperception, so that ultimately this subject recognizes its own modes of cognition even when it directs itself towards an object outside itself. This is particularly true for space, time, and causality which, as Kant demonstrated, are not to be found in the things as they are in themselves but belong to the subject.

Initially this was an inner philosophical position, outlined by Kant at the end of the 18th century in response to the skepticism of Hume, who had himself reacted against conventional metaphysics. At the time it had little or nothing to do with natural science. As the 19th century drew to a close, there was – in belated response to romantic or Hegelian natural philosophy as well as to the materialistic metaphysics of Büchner and others – the academic Neo-Kantianism of Cohen, Natorp, and Cassirer which could also be taken on by natural scientists. As measurement problems began to play a role, it became correspondingly more difficult to draw a clear separation between the investigating subject and his measurement object. "Critical" philosophy began to meet with the approval of natural scientists since it asserted the constitution of objects of cognition by the subject.

The historical coincidence of Neo-Kantianism and quantum mechanics may have been a matter of chance – however, the pioneers of physics at the beginning of this century, especially in Germany, were so well educated that direct influences from philosophy cannot be excluded. If observed objects are seen as depending on the observer, thus questioning the existence of a reality independent of observation, this is a connection which was pre-formulated by Neo-Kantian philosophy. Thus, to a certain extent, a physicist with a philosophical background had at his disposal the categories enabling him to reflect upon his experimental procedures. This was no longer just empirical confirmation or invalidation of physical conjectures, but the creation of a new framework of understanding, within which empirical experience could acquire "meaning" in a new way.

Even if empirically proven facts have paved the way for the end of the classical form of realism (or "standard realism") within the framework of quantum mechanics, we are not necessarily compelled to abandon realism completely. Broadly speaking, once the notion of a mind-independent reality has been clearly distinguished from that of empirically knowable phenomena, d'Espagnat proposes that there are at least two ways in which what is most essential in realism may be salvaged.

One of these ways consists in attempting to retain as conceptual building stones the set of concepts familiar to us from classical physics. This is more or less the guiding idea of the class of ontologically interpretable models of which – following d'Espagnat – David Bohm's model is the paradigm. The other one is to impart such a basic role to the set of principles of quantum mechanics themselves, suitably generalized so as to cover the whole realm of infinite systems but with a few "axioms" removed which happen to directly concern measurements and, at the same time, be only weakly objective. Primas' "nonstandard realism" is discussed in terms of such an approach.

CONCEPTS OF REALITY: PRIMAS' NONSTANDARD REALISM

BERNARD D'ESPAGNAT
Laboratoire de Physique Théorique et Hautes Energies,
Université Paris-Sud, F-91405 Orsay Cedex, France

1. Introduction

In its growth modern science remained on the whole faithful to the general conception of knowledge of its founders, Bacon, Galilei, Descartes, Leibniz, Newton, Huygens and others; and it is not much of an oversimplification to state that according to this conception the purpose of science was to lift the veil of the appearances and reveal "reality as it really is". To be sure, such a judgement calls for some qualification. In particular, it seems that most of these thinkers would have hesitated to claim that science is able to exhaustively disclose the totality of *what is*, including, for example, thought. It is only later that their followers, inspired by the successes of scientific research in a variety of fields, enlarged the scope of the view in question to the extent of declaring its universal validity. But anyhow it may be considered that – taken or not as universal – this conception, because of its simplicity, was, and still is, entertained by a great number of scientists. Let us call it here "standard realism" to be able to refer to it. Within standard realism, no basic distinction is called for between phenomena – "empirical reality" as is also said – and mind-independent reality. This is just because, according to it, appropriate measurements and experiments, when interpreted by means of a safe, well grounded theory, do reveal to us things as they really are, and not merely as they appear.

Very soon, standard realism was questioned, on quite valid grounds, by philosophers. Its hypothetical, speculative, overconfident aspects were emphasized. However, such criticisms did not amount to anything approaching a disproof, particularly since they were balanced by plain but convincing plausibility arguments. Scientific statements are, on the whole, very well grounded and it is a fact that an overwhelming majority of them – including most basic laws and principles – can be expressed in a purely objectivist language; so that the claim that such assertions do *not* describe actual features of mind-independent reality is viewed by most scientists as an unwarranted sophistication. Even those who take the philosophers' approach seriously enough to question our ability at reaching reality per se believe – most of them – that nevertheless science can and must be couched in a universal objectivist language, that is, roughly speaking, a language such that everything takes place as if its statements concerned a mind-independent reality.

This, of course, is related to the fact that most scientists are not quantum physicists. On the whole, quantum physicists are more on their guards, for they are well aware that not all of the basic rules of standard quantum mechanics can be expressed in such a language; and that, in particular, the Born probability rule cannot. But a great number of them consider nevertheless that the only purpose worthy of science – and in particular, physics – is to describe a totally mind-independent reality. Consequently, they have to look either for alternative formulations of quantum mechanics or for a way of making such basic laws as the Born rule compatible with standard realism, at least at the level of those physical systems that we can actually observe (such as instrument pointers).

The first of these two approaches (search for objectively interpretable substitutes to standard quantum mechanics) is here investigated in Sect. 2 and the second one (decoherence and so on) in Sect. 3. The outcome of this quest will be that, for various reasons, neither one is really satisfactory. This means, as explained in Sect. 4, that standard realism has most presumably to be dropped or, in other words, that a genuine distinction must be made between the notions of (mind-)independent reality and empirical reality (or, roughly speaking, between "what is" and the phenomena, in the philosophical sense of the word). It so happens that emphasizing the necessity of such a distinction corresponds to both Primas' standpoint and my own. Between the two there are, however, some differences. They are explained and commented on in both Sect. 5, where Primas' approach is stated with an emphasis on its most illuminative aspects, and Sect. 6, where it is discussed.

Before embarking on all this, however, the notion "objectivist language" should be briefly commented on and that of "separability" should be recalled. By "objectivist language" is meant a language that is basically *descriptive, even about contingencies*, instead of being just *predictive* of them; otherwise said, a language in which *either* physical objects (in a broad sense: particles, fields, etc.) and their structural *and dynamical* properties are assumed to exist per se *or* everything takes place *as if* they so existed, quite independently of us. Note that even the philosophers, such as Kant and his followers, who denied that science can describe anything interpretable as "reality in itself" did consider that the whole of physics (which implies no distinction being made between the microscopic and macroscopic realms) has to be couched in an objectivist language.

But note also that what, in classical science, made the objectivist language operative was a principle of analysis, or of "separability" as it is also called. Since the times of Galilei, Bacon, and Descartes it has more or less been taken for granted that reality – or what may be treated *as if* it were reality – obeys such a principle, which states, roughly speaking, that a complex extended system can – and should! – be thought of as composed of simpler, distinct parts, and that parts that are distant from one another interact on the whole more weakly than those that are near.[1] The trouble, of course, is that neither the wave function nor the Heisenberg operators associated with a complex extended system obey this rule (remember the "EPR correlations"!), and that the same is true concerning the density matrix if we also require, as part of the principle of analysis, that a complete description of the parts and the forces between them should yield – nay, should constitute! – a full description of the whole. This means that standard quantum mechanics is incompatible with a totally meaningful use of the objectivist language.

2. Theories of Reality Alternative to Quantum Mechanics

Not long after the advent of quantum mechanics John von Neumann issued his well-known theorem stating there can be no finer specification of the states of quantum systems than those provided by kets; but it was later shown, particularly by John Bell (Bell 1966) that the proof of this theorem involved an implicit and hardly justifiable assumption. At first sight it could be considered that the latter finding justified the expectations of the physicists who had thought it possible to build up an alternative theory entirely expressed in the objectivist language, consistent with the separability principle and yet reproducing all the experimentally testable predictions of quantum mechanics. But the discovery of Bell's theorem (Bell 1964), which occurred almost simultaneously (and was made by the same man!), ruined such a hope. There are alternative models that obey the first and the last of the three conditions just stated (they are expressed in the objectivist language and reproduce the quantum mechanical predictions) but, in conformity with Bell's theorem, none of them fulfills the above stated separability condition.

---

[1]For a detailed account of the conditions that this principle, if valid, would set on physics, as well as for an explication of its relationship with the twin principle of locality (or "local causality"), a formulation more precise and quantitative than this one is necessary. But of course this formulation calls for developments that cannot be condensed in a few words (see, for instance, d'Espagnat 1995).

These models are nevertheless instructive and interesting. In fact, the longstanding prejudice entertained by the larger part of the physicists' community against their basic idea – the idea that the quantum formalism is incomplete – was mainly due (apart from the influence of the von Neumann theorem proper) to a philosophical mixing up of two acceptations of the word "completeness" (d'Espagnat 1995). In the mind of many – perhaps most – physicists this word means just what was stated above, namely that there can exist no finer specification of a system state – considered as objectively existing – than the one provided by a ket. Let this acceptation be called "strong completeness" in order to distinguish it from the following one, which may be called "weak completeness". This is the assumption that "no theoretical construction can yield experimentally verifiable predictions about atomic phenomena that cannot be extracted from a quantum mechanical description" (Stapp 1972).

The notion of "weak completeness" is, as we see, exclusively focused on the *predictive* aspect of quantum mechanics, whereas "strong completeness" involves the notion of an objective existence of the considered entity (the state) and implies therefore at least the purpose of some descriptive approach. Identifying the two concepts is understandable from the part of those who, viewing *knowledge* as somehow conceptually prior to *existence*, consider that the very notion of "differences between states" having no counterparts in "differences concerning observability" is, in every instance, a void, "metaphysical", inconsistent idea, and that therefore the notion of "state" contains nothing more than just predictivity (by and for human beings). But those physicists who are not prepared to consciously adhere to this philosophical position (which is an extreme one, after all) should be careful not to inadvertently mix up strong and weak completeness. It is true that weak completeness fully agrees with what we know but the physicists in question should realize that within their standpoint this fact is neither a proof nor even an indication that the strong completeness assumption is right.

This means that models in which the state vector is considered as not yielding a description of the "real" (or "objective") state of a system – or as yielding but an incomplete one – may well be physically consistent. Perhaps the best example of such models is the old, Louis de Broglie "pilot wave" model (de Broglie 1927), greatly generalized by David Bohm (Bohm 1952) and nowadays known as the Bohm model. In it (see in particular the very clear account Bell (1987) gave of it) particles are described by their position and velocity, and the wave function has essentially the role of describing a field of "quantum forces" acting on them. This field is real – it is just as real as are the electric and magnetic fields in classical eletromagnetic theory – but it differs from the latter in that it is not a one-point function.

This, of course, is what makes the model nonseparable. On the other hand, it is assumed that we have access only to the particle coordinates so that the wave function cannot be associated to a notion of "state" of individual systems. Other examples are the Ghirardi, Rimini and Weber model (Ghirardi et al. 1986) and the so called "modal interpretation" of quantum mechanics (van Fraassen 1981, Dieks 1990).

By contrast to standard quantum mechanics, which is "weakly objective" only (in the sense that, to repeat, at least one of its axioms, the Born rule, must for consistency be understood as basically referring to human knowledge), all these alternative models are "ontologically interpretable", as Bohm used to say, which means that they can be – and indeed are – formally stated in the objectivist language. In them, nonseparability (which, of course, is what makes them unoperative) takes the form of a violation of Bell's local causality, which means that, in them, some influences are propagated faster than light. This is the deep reason why attempts at making such models relativistic were, up to now, unsuccessful. On the other hand, it would be erroneous to bluntly state that the violation in question amounts to violating special relativity without making precise what conceptually this expression means. For example: while it is true that, due to local causality being violated, influences may be propagated superluminally, it has been pointed out (d'Espagnat 1975) – and later explicitly shown (Eberhard 1978, Ghirardi et al. 1980) – that this does not imply *signals* can be propagated faster than light. Hence if special relativity is understood in a "spirit of positivism" (the one the young Einstein was in when he elaborated it!) it may well be claimed it is unaffected by all this.

To sum up: it is impossible to discard the alternative models by saying that they are disproved. They are not. There are arguments, however, that make them unpalatable. First of all, there are several of them and it is unclear on what rational basis we should select one. Second, for the reason mentioned above the program of making them relativistic seems almost hopeless (whereas standard, weakly objective quantum theory could, at an early stage and rather easily, be generalized in a relativistically satisfactory way, resulting in the quantum theory of fields). Third, it is naturally expected that a model should yield explanations of hitherto unexplained facts and/or novel experimental predictions later verified by experiment. The models here under study proved totally inefficient in this respect. Fourth, these models – or at least Bohm's model – have as their building blocks concepts – forces, localized separable objects endowed with individuality etc. – clearly borrowed from classical macroscopic physics. However, since they lay claim at universality and faithfully reproduce nonseparability (just as faithfully as they do concerning any other testable prediction of quantum physics), they themselves imply that, when all is said and done, the stuff physics – including *macroscopic* physics – deals with cannot be considered as describable "as it really is" by means of the above mentioned concepts (which, to repeat, *include* separability). In other words, if Bohm's model is right the said concepts are *only apparently* fitted to the description of the macroscopic world. But then, as Bitbol pointed out (Bitbol 1996), what is the rationale of having picked them up in the first place? Does it not amount to having been fooled by some fancy appearances?

To this, a supporter of the models in question might well reply that admittedly they all more or less suffer from this defect but that they are strongly objective whereas standard quantum mechanics is merely weakly objective, so that – he would continue – if we demand a strongly objective description of reality we have no choice but to turn to one of them and cannot therefore be too critically minded regarding the nature and "plausibility" of their conceptual building blocks. I think it is at this stage that Primas' ideas may specially catch the interest of the philosopher engaged in such a "conceptual exploration" of the various possibilities: for Primas' quantum endophysics is a strongly objective theory not built up with classical concepts as building blocks, and, in this respect, offers an alternative to the choice in question. But before we come to this, mention should be made of the other tentative approach towards a theory fulfilling the above stated standard realism requests, namely decoherence theory.

3. Decoherence

Questions concerning the notions of reality, existence, etc. are, of course, tightly linked with questions concerning the "truth of statements". This is indeed so much so that some philosophers, such as Dummett (1978), stress the usefulness of characterizing realism with reference to the latter. They then observe that there are various classes of statements. Here we are interested in two of them, the class of statements about general laws – call it *class L* – and that of statements about (contingent) facts – call it *class F*. Dummett defines realism with respect to a given class as being "the belief that statements of [that] class possess an objective truth-value independently of our means of knowing it: they are true or false in virtue of a reality existing independently of us". According to him, anti-realism, on the contrary, is the view that the very *meanings* of the statements of the considered class are tied directly to what counts as evidence for them, in such a way that a statement of that class, "if true at all, can be true only in virtue of something of which we could know and which we could count as an evidence for its truth".

Clearly, realism in Dummett's sense coincides with what was called "standard realism" above. On the other hand, as the very existence of both

Primas' approach and my own clearly shows, in contemporary physics one can be a realist without being a "standard (or Dummett-type) realist", which implies that in this field the term "anti-realism" is not quite suitable for labeling the view Dummett designates with this term. For lack of a better name let us call "anti-standard position" the view Dummett simply calls "anti-realism". Concerning statements of class F, I am – and so, I believe, is Primas – an upholder of the anti-standard position although, since we both consider the mind-independent reality notion as meaningful, we cannot be called anti-realists.

Dummett's definitions are useful but it is obviously necessary to supplement the one concerning realism with some criterion having the effect of discarding right from the outset statements of the "sex of angels" type. More precisely, a criterion is needed selecting the kinds of statements of which it is at all conceivable that they should be endowed with a truth value. Within "standard realism concerning class F" a reasonable choice in this respect is to take up what Dummett proposed as an anti-realist definition of truth but use it not as a definition but just as a *criterion* of meaningfulness and truth.

For example (roundabout but useful in what follows), in quantum mechanics consider a generalized measurement-like process such as

$$\left(\sum_n c_n \Phi_n\right) \Psi_o \rightarrow \left(\sum_n c_n \Phi_n\right) \Psi_n; \quad c_n \neq 0, \tag{1}$$

performed on an ensemble of systems, assume that no interaction with any environment takes place, and consider the statement that, in the final state, the "pointers" A (described by the Ψ's with each Ψ_n corresponding to a definite graduation interval, while the Φ's describe the "measured" systems S) *are not* in definite graduation-scale intervals. According to the just defined criterion this statement has a truth value and is true. This is so in virtue of the fact that assuming the pointers were in definite scale intervals would imply, concerning measurements of some observables pertaining to the composite systems S+A, consequences that are experimentally disprovable (since they are at variance with the quantum mechanical predictions, which are valid by assumption). So the outcomes of such measurements, *which are measurements that we could perform*, would show the considered statement to be true, q.e.d.

Since a "standard realist" must, as we just saw, refer to the same kind of considerations as an upholder of the anti-standard position – even though he uses them as a criterion rather than as a definition – it might at first sight appear that there is not much of a difference between the two. A difference, however, appears in some instances due to the fact that, for the standard realist, the criterion is but a *sufficient* condition (in the sense: for a proposition to be true it *suffices* that we could ascertain its truth), not in the least a necessary one. Since, for him, a statement has an intrinsic truth value independent of our means of ever knowing it, there may quite well be – indeed there *must* be a great number of – statements that do have a definite truth-value although we have no means – not even in principle – of discovering what this value is. For the standard realist a conceptual problem therefore arises that, according to the anti-standard position, would make no sense, namely: how are we to discriminate among all the statements whose truth-values we shall never be able to know, which ones *have* a truth-value – are "meaningful" – and which ones do not?

It is quite unlikely that a general – philosophical – solution to this problem can be obtained. But fortunately there exists a certain class of statements concerning which such a solution is at hand. It is true that this solution rests on accepting a principle, but the validity of the latter can hardly be questioned as we shall see. These are the statements S' the sets of which are just continuations of sets of statements S whose truth-values can unquestionably be ascertained. "Continuation" here means that the statements S' are of the same qualitative nature as the statements S, they involve the same kinds of operations, the same types of concepts etc.; they only bear on systems and concern circumstances that are quantitatively more complex. This increased complexity is, according to the ideas, theories etc. we believe in, the *only* reason why the truth-values of the statements S' could not be known by any means.

The principle just alluded to – let us call it "the continuation principle" – then just is that these S' truth-values, unknowable as they are, still exist: in other words that the S' are meaningful. This is an extremely natural principle for reasons of continuity. Between "meaningful" and "meaningless" there is no continuous transition. A statement that is not meaningful is meaningless and vice versa. Hence negating the principle would imply that, at some level of increasing complexity of systems, a conceptual "sudden jump" would have to be considered. A statement that is meaningful for a system whose complexity is just below this level would be meaningless when applied to a system (perhaps one involving just one more atom!) whose complexity lies above it. Such a hypothesis is too artificial to be considered as being serious.[2]

---

[2]Readers familiar with the problems and theories of complex systems may feel uneasy about the appearance of such a "continuation principle" since in these theories – and quite especially in Primas' approach – increase in complexity is described as implying the emergence of qualitatively new properties, that is as inducing discontinuities. In fact, however, there is no contradiction for in this section we are specifically considering the conception called "standard realism" above, while the just mentioned theories belong to the realm of *non*-standard realism (to be analyzed in the next sections). In Primas' language they are *exophysical* descriptions, requiring "abstraction in the second sense"

This analysis has a bearing on the measurement problem. It is well known that Eq. (1) does not adequately reflect the essence of the measurement process because of the fact that, being macroscopic, the apparatus strongly interacts with its whole environment. If the apparatus-environment interaction is duly taken into account and if no measurement involving the environment (E) together with the system (S) and the apparatus (A) is considered, decoherence holds. Because of this, in an "S+A" ensemble no interference effects can be observed in practice; and this is indeed an argument that many physicists have put forward for claiming that the pointers "are" each lying in one definite graduation interval (thus not restoring, to be sure, realism proper but at least a kind of macro-objectivism). On the other hand, other physicists maintain that this claim is not valid since nothing in principle could prevent us from performing measurements involving observables common to the system, the apparatus, and the environment, and revealing the falsity of the claim in question.[3] A problem then is: which one of these two groups is right? In other words, assuming quantum mechanics is exact, should we or should we not endorse the statement: "it is a mere intersubjective appearance that makes us see the pointers as lying in definite graduation intervals"?

It is at this stage that, according to the foregoing, a standard realist and an upholder of the anti-standard position must differ. The latter will answer: "No, we should not. For the only observables the measurements of which could conceivably reveal the falsity of the assertion are of such an incredible complexity that the instruments for measuring them would have to be constituted of a number of protons larger than the number of protons in the whole universe (Omnès 1994), so that such measurements cannot be done". On the contrary, the standard realist will observe that the continuation principle applies here. Hence, he will have to say that indeed seeing the pointers at definite places is but an intersubjective appearance, not a feature of "reality as it really is".[4]

(see Sect. 6.1), which implies that within the standard realism philosophy they cannot even be formulated consistently. It is only within the latter, in fact presumably too ambitious, conception of our possibility of knowing reality that the continuation principle makes sense, but within it nothing prevents us from considering it natural.

[3]For this argument to be correct it suffices that some such observable should be measurable. For example, in the case of the Zurek model, such an observable can be explicitly produced (see d'Espagnat 1995).

[4]This holds unless the idea of some real collapse à la von Neumann is invoked, either taking place spontaneously or due to the observer's consciousness, far within the macroscopic part of the overall system. The foregoing shows that the differences between physicists (even belonging to one and the same group, see Giulini et al. 1996) concerning the "necessity" of such a collapse are in fact of a philosophical rather than physical nature. They boil down to the difference between standard realists and upholders of the anti-standard position (nonstandard realists or idealists).

Of course, decoherence is quite a general process. Far from applying only to measurement-like interactions it concerns all kinds of physical evolutions involving macroscopic systems of some sort, so that finally, as stressed by many authors (Giulini et al. 1996), it nicely explains "the appearance of a classical world". But this it does basically in the same way as it accounts for the fact that, in a measurement process, pointers appear as localized in definite scale intervals. In other words: for a standard realist, while decoherence does indeed explain such an appearance of realism, it does not make this local realist description true, since it does not remove the objection resulting from the universal existence of Einstein-Podolsky-Rosen (EPR) correlations and the possibilities we would have in principle of "measuring" them. A considerable philosophical difference between classical physics and quantum physics thereby emerges. Admittedly, neither in classical nor in quantum physics could "standard realism concerning class F" be actually proved (and, following Berkeley and Kant, many prominent philosophers of the classical age indeed considered it arbitrary, unlikely, preposterous etc. and finally discarded it). But within classical physics its general validity, as we noted, cannot be disproved. On the contrary, in quantum physics it cannot be generally valid since, in the case of "instrument pointers" it is disproved as we just saw.

For the sake of completeness it should be added that unquestionably decoherence plays, as just noted, a crucial role in explaining the appearance of a classical world but that concerning two questions at least it does not quite lead to results as significant as one might have hoped. One on them is well known. It is that, in a measurement-like process such as schematically described by Eq. (1) (with, now, A standing for "apparatus" plus "environment"), it does not solve the so-called "and-or" riddle. An angel who would know the theory but would never have seen macroscopic objects and, in particular, instruments could not infer from his knowledge that in an ensemble of such processes each pointer is, in the final state, in some definite graduation interval instead of being "spread" over all of them.

The other point concerning which decoherence leaves me unsatisfied concerns its localizing effect (d'Espagnat 1995). As is well known the reduced density matrix relative to an ensemble of, say, dust grains is rapidly brought by decoherence to a quasi-diagonal form in the position basis, which implies that, e.g., if a beam of such dust grains is fired on a Young-type diaphragm, it will behave as a classical one and create no interference fringes. By itself, however, the said density matrix structure does not have all the consequences we should naturally expect. In particular, we should expect that due to decoherence the ensemble becomes a mixture composed of dust grains approximately localized at various places, that is, occupying different regions of space, as is the case concerning beams of classical objects.

Now, while it can be quantitatively shown (d'Espagnat 1995) that such a proper mixture is *compatible* with the density matrix in question, it is only one of those that is compatible. Moreover, it is not the one a theoretically minded physicist would most naturally expect since it is not the one resulting from diagonalizing the said density matrix. Again, the above mentioned angel, informed of the density matrix, could not, from it, infer localization as we understand this concept. Both deficiencies result, of course, from the fact that decoherence theory works with density matrices and that when a density matrix is not a "pure case" (projector), it corresponds not to one but an infinity of distinct proper mixtures.

Remark: It was noted above that, concerning the measurement problem, the anti-realist – contrary to the upholder of standard realism – has no difficulty in asserting that the pointer is in one or another of the graduation scale intervals. That is, his philosophical standpoint allows him to make use in this instance of the objectivist language. It is worth noting, however, that, even for him, the domain of validity of this possibility is limited to the (unfortunately ill-defined) macroscopic domain. Decoherence is of no help for someone who would aim at making this language universal. At first sight, the so-called "consistent histories theories", with their notions of microscopic "events" and the like, may be viewed as attempts to reach the said aim. But, as has been stressed by several authors (d'Espagnat 1995, Omnès 1994), since the propositions concerning these events are – in Omnès' language – not "true" but merely "reliable" in general, it cannot be said that this goal is reached after all.

4. Beyond Standard Realism

What seems to emerge from the content of the foregoing sections is that standard realism is a conception that contemporary physics renders hardly defensible. Many philosophers did not await such scientific developments for declaring standard realism unacceptable. Unfortunately, practically all of them turned to some form of idealism (often giving this basic choice such attractive names as phenomenalism, pragmatism, operationalism, and the like), that is, to a view in which the concept "possibility of knowledge" is so much of a basic concept that it is prior to even that of "existence". Among contemporary philosophers this tendency is still at work under such names as anti-realism and constructivism. But all of these variants of idealism meet two difficulties.

One of them is that these conceptions do not (or at least not adequately) allow for the possibility of errors. The errors referred to here are not those that occur in calculations but those that may be called "rational" errors, that is physical theories that are rationally faultless but do not agree with the data. The very notion of such errors implies a mismatch of some kind. And since, according to the idea under study, there simply is *no* external reality to match our theories with – since, therefore, mind and phenomena are "born together", so to speak – it is not understandable that there should be a place for such a mismatch. In contrast, in the actual world it (unfortunately!) often happens that physicists build up theories that are quite general, logically sound, mathematically beautiful and, nevertheless, turn out to be disproved by experimental data. In other words, we live in a world in which "something says *no*". The magnitude of the intellectual contortions necessary for justifying the idea that *also* this "no" comes from us is definitely too large for the idea to be credible. Hence, there must be something "out there", after all. And this something must somehow influence our observations.

The other one of the two difficulties alluded to above is that maintaining that knowledge is conceptually prior to existence is just impossible. Even the neo-Kantians – who were keen in rejecting Kant's "thing in itself" – did not succeed in doing so. In their philosophical construction, along with the phenomenological self they had to introduce the notions of either a "Transcendental Self" or a set of "Logical Necessities". And it is difficult to grasp in what sense they think these notions do not partake of the general notion of "Being" (alias "Reality").

In view of all this it seems impossible not to consider that the notion of mind-independent reality is, after all, meaningful and necessary: even if this reality is not – or not totally – knowable, and may radically differ from the idea the word "substance" conveys to us. If this standpoint is taken, then departing from standard realism essentially implies that a basic distinction is to be made between mind-independent reality on the one hand and empirical reality on the other hand, the latter expression referring to a set of *phenomena* in the philosophical sense of the word, only some of which may be describable in terms of an objectivist language, and then only conventionally and approximately. The view emphasizing that (i) both notions of reality are to be kept and (ii) they are different is, to repeat it, common to Primas' standpoint and my own and is what distinguishes them from other approaches, be they due to philosophers or physicists. This important parallel notwithstanding, there are some significant differences between the two conceptions, as will become apparent in the course of the following analysis of Primas' one.

5. Primas' Approach

The most outstanding point concerning Primas' approach is that he did not rest content with just emphasizing the – itself often overlooked! – difference between a discourse meant to describe something external to us and another one merely describing *human knowledge*, that is, the difference between ontology and epistemology. Primas also stressed the fact that, in spite of the holistic character of quantum physics, we do speak of molecules and other partial systems as *having* definite shapes and so on, and that it is quite impossible for us not to make use of this objectifying language. Accordingly, he was led to use the word "ontic" in a wide sense, that is, even in connection with aspects of nature of which he himself noted that they are merely weakly objective (i.e., intersubjective: they depend on the abstractions we make but "everybody who makes the same abstractions gets the same observable phenomena" (Primas 1981).

Some of Primas' readers (myself included!) had a hard time trying to understand the use of the word "ontic" – a substitute for the word "ontological" – in this context since "ontological" normally means "totally independent of human approach". I have now grasped that in Primas' mind it referred to some notion akin to Quine's "relative ontologies", that is, to the kinds of "ontologizations of experiential data" that we cannot help building up in our minds. This issue has, I think, been further clarified in recent articles by him (Primas 1993a,b, 1994), in which he introduced the notion of *exophysical* ontologizations and contrasted it not only with the notion of statistical interpretations (bearing on predictions concerning what we shall observe) but also – at the other "conceptual end" – with that of endophysics. Needless to say that what we have to make abstraction of – that is, "do as if" it did not exist – is of the nature of some EPR correlations due to interference terms. This abstraction process is what, for my part, I called the "axiom of empirical reality".

Primas imparts a truth-value to factual propositions lying in the realm of his *exophysical* ontologizations, paradigmatic examples of which are, in my view, propositions bearing on pointer positions subsequent to measurement-like interactions. Consequently he cannot be considered a "standard realist" in the sense of Dummett since, as we saw above, propositions of this type may be called true or false by an anti-realist or, more generally, by an upholder of the anti-standard position, but not by a standard realist. For this reason, I called such propositions "strongly empirically true" (SE-true for short) elsewhere (d'Espagnat 1997). In this expression the use of the qualificative "empirically" (as opposed to "ontologically", taken in its traditional sense) is, of course, related to the impossibility of making the propositions in question refer to a truly mind-independent reality. The qualificative "strongly" (as opposed to "epistemologically", see below), roughly means (for more details see, e.g., d'Espagnat 1997) that the validity of the proposition it applies to is conceived of as not being strictly dependent on – and limited to – the *actually* prevailing conditions.

By contrast, I call "epistemologically empirically true" (EE-true for short) the propositions the validity of which is limited to the mentioned conditions. (After a measurement of the left photon polarization along x in an EPR-Bohm experiment on photon pairs has been made, such is, for example, the proposition "the right hand side photon is polarized along x"). In the latter propositions the verb "to have" has a purely epistemological meaning, that is, a meaning considerably more restricted than the one it has in ordinary language. In fact, it does not express more than just some prediction concerning what we shall actually observe in fully definite circumstances. Clearly, the propositions I call "EE-true" are nothing more than elements of what Primas calls "epistemological formulations" (of physical theories) (Primas 1993a,b, 1994).

In all this, the notion that raises the most interesting questions is, of course, that of SE-true propositions, alias *exophysical ontologizations*. Concerning this notion Primas' approach enables him to introduce a most beautiful, far-reaching idea, not to be found – to my knowledge – in the works of other authors. This is the idea that we may have (in my language) *different* ways of applying the axiom of empirical reality. Otherwise said, it is the idea that, according to the manner we decide to "look at things", one or the other of two (or more) distinct and incompatible exophysical ontologizations becomes valid. More precisely, Primas' view starts from the fact that, in order to get an exophysical ontologization, we must anyhow drop some EPR correlation terms that do exist but he adds that we have some choice as to which ones we shall drop, leading to different exophysical ontologizations.

The cases he points to are most striking. One of them, for example, is the pair of concepts "molecular structure" versus "temperature". From the undivided quantum endophysial reality we get at a description in terms of molecules by dropping some terms and at a thermodynamical description in terms of temperature by dropping other terms instead: so that neither description is more basic than the other. This is a grand view on both physics and natural philosophy and it must be acknowledged that, at least in the case of the example just recalled, it is made possible by Primas' systematic use of algebraic quantum mechanics, a formalism that allows for systems with an infinite number of degrees of freedom to be handled quantitatively.

6. Comments

6.1. ABSTRACTION AND OBJECTS

Concerning the notion of reality there are, as previously noted, strong similarities but also a few differences between Primas' conception and mine.

Some of the differences are, as we know, essentially semantical, and therefore not truly significant. This is the case concerning the words "reality" and "realism", that both Primas and I use for referring to the undivided whole that basic reality is, but that Primas often uses as well – and without qualificative – essentially for referring to the notion I refer to using the term "empirical reality".

A similar difference concerns the use of the expression "mind-independent". Many physicists, including myself, tend to use it in the straightforward sense of "totally independent of any decision we take and more generally of anything having to do with the human mind". But Primas often means it in a different sense. He wrote for instance (Primas 1981): "The notion 'object' is abstraction-dependent but it can be taken as being mind-independent". In this context the expression "mind-independent" is defined by means of the notion of "abstraction", which is one that can itself not be defined without explicitly referring to the human mind: so that Primas' notion "object" is one that, in my language, I would say is weakly objective only.

In fact, at this stage the differences concerning semantics come to merge into problems of meaning. This is the case concerning the term "abstraction". In ordinary language the expression "to make abstraction of" is used in two substantially different contexts. The first context is the one in which we decide, for the sake of brevity, to omit considering certain details that could be mentioned without spoiling our argument but happen to be irrelevant concerning it. The second case is the one in which the details we "make abstraction of" would, if actually stated, prevent the argument we develop or the picture we put forward from being flawless. Trivial, straightforward examples of the first case are readily found. *Models* are examples of the second case. A teacher who (rightly or wrongly) decides, for pedagogical reasons, to limit his teaching concerning atoms to a description of the Bohr model must make abstraction – in the second sense – of a number of well-established facts that simply do not fit with this model (to begin with: the energy level structure of helium).

In the context in which Primas makes use of the term "abstraction", it cannot be taken in the first sense. As he himself stresses, in the cases he considers (for example in the problem of attributing forms to molecules) there are *actually existing* EPR correlation terms, taking account of which would *prevent* us from consistently building up the picture we have in mind (from attributing forms to molecules in the considered example). Hence, clearly, this is an instance of the second case. And within Primas' approach it does not really matter that the existence of these EPR terms cannot be exhibited by means of practically feasible experiments: for his basic postulate is that quantum endophysics correctly describes a basic (unattainable) reality. The Omnès-like question we met above in connection with decoherence theory – the question of deciding whether or not a meaning is attached to measurements that makes fact-like features of the universe unfeasible – does therefore not arise here.

In algebraic quantum mechanics an algebra pertaining to a type of quantum systems may have a center, so that the system may have classical properties. A necessary condition for the system to qualify as an object in the usual sense is thereby met. However, Primas rightly stresses that more is required. The system must also have individuality. According to Primas, this implies that it must not be entangled with other systems. Incidentally, this point is not quite as straightforward as it seems for, on the one hand, wave functions etc. are not "substances", and, on the other hand, the identity that has to be preserved is one of form. I guess Primas' argument must be that, for the identity of forms to be preserved, an obvious necessary condition is that forms should be definable. And to this end it seems unquestionable that, as Primas stresses, some EPR correlation terms should be abstracted from.

Again, however, care should be exerted at this stage not to get abused by words. The combined appearances, in Primas' texts, of the words "abstraction" and "ontic" might induce us to naively believe that from quantum mechanics he was able to derive a world view hardly different from standard realism, for we naturally tend to understand the term "abstraction" in the first sense (defined above) and the term "ontic" as a synonym of "ontological" in the standard usage of the term. Clearly, this would be a faulty understanding. Since the discarded EPR terms actually exist, "abstraction" has to be taken in the second sense and correspondingly "ontic" is to be understood in the "Quinean" sense of "relative ontologies", that is, in a sense that makes it quite definitely relative *to us*.

Consequently, objects are indeed but appearances, although, of course, appearances that hold good for everybody. The well-known comparison with a rainbow is quite suitable concerning them. To repeat, within Primas' writings it seems impossible to impart a signification to the word "ontic" that would be substantially closer than this one to the standard realist views. In other words, stating the existence of such and such an object is – just as stating the existence of such and such properties of this object – but an SE-true (or SE-false) proposition. The thus defined objects are elements of an exophysical ontologization or – in my own language – of empirical reality. The fact that they possess individuality is of course to be set in correspondence with the fact (d'Espagnat 1995) that Bell's nonseparability theorem does not concern empirical reality.

6.2. DO WE EXPLAIN?

The distinction – *within* the realm of empirical reality – between SE-true and EE-true propositions is a most significant one since the robustness of the former makes them useful for practical purposes as if they were ontologically true in the traditional, absolute sense of the word "ontological". For this reason, it is a positive feature of Primas' terminology that it emphasizes very much the distinction in question (through its use of such expressions as "exophysical ontologizations" and "statistical formulations" for referring to the former and the latter, respectively).

However, there are limits to the extent to which the language in question restores the validity of our standard views. This is particularly true concerning the notion of *explaining*. One of the great reasons why we are intuitively inclined to like realism more than operationalism is that we feel operationalism just predicts while realism both predicts *and* explains. In classical physics when, say, a red billiard ball hits a white one we feel we understand why the white one starts moving: it does so because it was hit by a *real* red ball, that is, by a ball that, at any given time, has a definite position, a definite velocity etc. And, of course, our feeling of understanding is not in the least weakened by the fact that, in so describing the properties of the red ball, we abstract from many features such as its precise mass, volume etc. But this is because the said "abstracting" is made in the first (as defined above) sense of the word. In quantum physics, if we are to believe Primas, the picture is quite different. Neither balls exist in any absolute sense since they are totally entangled with the rest of the world. It is we who, through a purely mental operation, separate them from the latter. Under these conditions, also the collision is largely a construction of ours – which makes it difficult to consider it as a genuine explanation.

This remark is not intended to suggest that some other approach would better salvage our intuitive notion of "explaining". It is just meant to point to the fact that, things being what they are, we should not indulge into believing that, because we are offered a means of restoring the *use* of such terms as "ontic" and "reality", we dispose as well of everything that constitutes the basic substance of realism.

6.3. STATUSSES OF QUANTUM ENDOPHYSICS
AND NONSEPARABILITY

To what extent is quantum endophysics credible? This is a big epistemological question and it is one the terms of which, apparently, are new. It seems very doubtful that the epistemologists of the past ever had to cope with a problem similar to this one, so that, unfortunately, there is not much hope of discovering hints at an answer in their works. It is a fact that the implicit answers Primas and I gave to it in our respective texts are very different. As noted above, contrary to many philosophers we both considered that the notion "Independent Reality" has a meaning. We also agreed concerning the idea that, presumably, it is not a "pure X". But from that point on our approaches parted.

Roughly speaking, Primas introduced a definite cut between mind and matter (his "Cartesian cut") and claimed that "matter" is adequately described by quantum endophysics, even though this physics is not operational. I, on the contrary, took up a position of moderate agnosticism: I considered Independent Reality as being conceptually prior to the mind-matter splitting, that is, as more or less identifiable with Jaspers' "das Umgreifende"; and I merely suggested that we may have some – although most uncertain and indirect – glimpses of it. Now, on this issue, who of us is right? It seems most difficult to tell. Personally, I consider Primas' idea referred to above as "most beautiful and far reaching" – the idea that the axiom of empirical reality can be applied in different ways leading to distinct exophysical ontologizations – as being so illuminating that "it should be true". And since it is based in quite an explicit and detailed way on the view that Independent Reality is structured as quantum endophysics says it is, I am inclined to believe that this view is right.

On the other hand, this belief of mine is in no way a "firm belief" (a "conviction"). It is not because, while the argument that an idea is illuminating is a scientifically sound element in a development intended at showing its truth, it is normally not viewed as being a sufficient one. In scientific matters we can be fully satisfied only when we have ascertained that the idea has specific testable consequences of its own and that they are verified by experiment. Concerning the question whether or not quantum endophysics is the faithful description of reality-in-itself it seems difficult that such a condition could ever be satisfied. Consequently, much as I like it, I cannot consider this theory as a firmly established "piece of knowledge". Remember that the evidence for its truth is quite indirect indeed: it is only by *negating* some of its allegedly existing implications that we get at possible validity tests!

One of the consequences of this wary attitude of mine is that, while I do believe, just as Primas does, that Independent Reality is nonseparable (either nonlocal or just simply not composed of parts) I cannot be satisfied, as Primas seems to be, by any alleged "proof" that is merely based on the nonseparability of the wave function. Although proofs along these lines may be refined (I developed such considerations myself in the sixties (d'Espagnat 1965, 1966) they rely much too much on the assumption called "strong completeness" above, which, as already stressed, is an ontological assumption (in the traditional sense of the word) since it consists in *assum-*

ing – without any possibility of proving it – that there is no "subquantum level of reality". This is the reason why I consider Bell's theorem as essential: it proves that nonlocality holds true *even if* there exists a subquantum level.

7. A Speculative Outlook

If the theorems of Bell as well as Kochen and Specker did not exist, Primas' theory of mind-independent reality (alias "quantum endophysics") would, in the mind of realists, have most serious potential competitors. These would be the theories – if not yet existing at least considered as bound to, sooner or later, appear – compatible with "down to earth" ideas. I mean theories grounded on notions, such as space and localized objects, that proved scientifically sound throughout centuries. By showing that the concepts in question *anyhow* cannot serve as ultimate bases for endophysics, the mentioned theorems have to a great extent made Primas' approach rationally more attractive. But, of course, a purely negative piece of information such as this one does not amount to anything approaching a proof. In other words, it seems that in such matters we shall forever have to remain within the realm of the speculative.

If that much is granted, the question that comes to the forefront is: among the – existent or conceivable – speculative pictures of "what exists" fitting with what we know, is there one we have good reasons to like best? Answers to this question are likely to differ from one physicist to the next. Personally, as suggested already above, I tend to believe in an "Ultimate Reality" that is neither mental nor material (or, equivalently, is both), for it is conceptually prior to the mind-matter splitting (the Cartesian cut in Primas' terminology). It is then natural to consider that mind can have – at best! – but glimpses on this "topmost" reality.

Note that here I used the words "material" and "matter" just for convenience (and to follow Primas). In fact it is necessary to realize that what they designate in the present context has hardly anything to do with what they mean in modern usage. In fact, what they do designate must obviously have some, direct or indirect, relationship with empirical reality, and up to now I used not to be very specific concerning the latter concept in my writings. Now, after having (for the purpose of writing this article!) read Primas' works even more thoroughly than I had previously, I come to think that some more preciseness is in order and that, before speaking of empirical reality proper, it might well be appropriate to introduce, as a "topmost level right below the splitting of mind and matter", the reality that Primas' quantum endophysics describes (and that he calls "material"). To repeat: it is essentially as a *conjecture* that I would endorse the idea that the notion of such a level of reality can be meaningfully considered. But it is a conjecture that Primas' beautiful notion of distinct and incompatible exophysical ontologizations (all derivable from endophysics) makes extremely attractive.

Accepting the said conjecture – as I am much inclined to do – then leads us to a conception in which four "levels of reality" may be distinguished. Topmost is Independent Reality proper, prior to the Cartesian cut. "Just below" the said cut we find, on the physical side (the only one discussed here), Independent *Physical* Reality as described by endophysics. Mind, as we saw above, has no direct access to this Reality: it can only guess its structure. But we find that if, following Primas, we assume this structure to be described by algebraic quantum mechanics (with probability rules and the like taken off) then the various (and sometimes incompatible) exophysical ontologizations we use in everyday science can be inferred from it by abstracting from such or such EPR correlation terms, and this makes it credible that the assumption is correct. This process of "abstracting" (in the second sense of the word, see above) gives rise to propositions having definite SE-truth values within these exophysical ontologizations. They are those that define and describe what Primas calls "objects", "ontic properties" of objects and "ontic states".

It must be observed that among those propositions bearing on contingent properties of objects some (such as positions of pointers) have a truth value that depends not only on the abstracting process itself but also on chance. It is at this stage that the "other" quantum mechanical axioms – those involving ensembles and probabilities – come into play. Attempts exist at deriving them from the theory and Primas himself indicated ideas that, in his opinion, constitute valid steps in this direction (d'Espagnat 1976), although he granted at the same time that problems remain. In my view such developments are rather unclear and I consider that at this level – where what is in question essentially has to do with states of our consciousnesses – something must be assumed.

For example, as I suggested many years ago (d'Espagnat 1976, see also d'Espagnat 1995, Sect. 12-2), it is possible to assume that some Hermitean operators correspond to observables and that, when a measurement of such an observable is performed on an individual system, consciousness sets itself in correspondence with *one* eigenket of that observable (no probability considered since we are here at the level of the individual). An argument borrowed from the Everett-Graham theory then shows that in an ensemble of such measurements made by one and the same consciousness the Born probability rule is satisfied. No wave-packet reduction has to be assumed. Admittedly, there is (surprisingly enough!) no "ontologically real" agreement between different observers; but each observer nevertheless has the impression that the other ones agree with him concerning the outcome of any measurement (for if he makes a measurement on the apparatus or brain of his colleague to know "what was registered there" he gets an outcome compatible with the one he got directly).

Finally, it is, in practice, most convenient to extend the "objectivist language" – the one of which such terms as "object", "state" etc. are elements – to a figurative description of the internal development of microscopic processes, of which we have, through measurements, but an overall perception. For this we must use such terms as particles etc., combined within propositions possessing what I called EE-truth values. All this corresponds, roughly, to what Primas calls the "statistical description".

Such a four-level description is avowedly speculative and its opponents will have a point if they compare it with gnosticism. But a criticism of this kind cannot be carried very far. For indeed no science completely keeps to the predictive realm. Every one of them ventures into the fields of the *descriptive*. And, as philosophers know, scientific or not, *any* description is speculative. Moreover it is, after all, quite an essential component of human nature to speculate on what exists, and the real question is therefore merely: "is the proposed description based on sufficiently well established and general facts?" Answering this is a matter of personal appreciation. In the present instance, without committing myself entirely, I am personally inclined to say "yes".

References

Bell J.S. (1964): On the Einstein Podolsky Rosen paradox. *Physics* **1**, 195–200.

Bell J.S. (1966): On the problem of hidden variables in quantum mechanics. *Rev. Mod. Phys.* **38**, 447–452.

Bell J.S. (1987): *Speakable and Unspeakable in Quantum Mechanics* (Cambridge University Press, Cambridge).

Bitbol M. (1996): *Mécanique quantique* (Flammarion, Paris).

Bohm D. (1952): A suggested interpretation of the quantum theory in terms of "hidden" variables I, II. Phys. Rev. **85**, 166-179, 180-193.

de Broglie L. (1927): La mécanique ondulatoire et la structure atomique de la matière et du rayonnement. *J. Phys.* **8**, 225–241.

d'Espagnat B. (1965): *Conceptions de la physique contemporaine* (Hermann, Paris).

d'Espagnat B. (1966): An elementary note about 'mixtures'. In *Preludes in Theoretical Physics: In Honour of V.F. Weisskopf*, ed. by A. De Shalit, H. Feshbach, and L. van Hove (North Holland, Amsterdam), pp. 185-191.

d'Espagnat B. (1975): Use of inequalities for the experimental test of a general conception of the foundation of microphysics. *Phys. Rev.* D **11**, 1424–1435, here footnote 30.

d'Espagnat B. (1976): *Conceptual Foundations of Quantum Mechanics* (Addison-Wesley, Reading), 2nd edition.

d'Espagnat B. (1995): *Veiled Reality* (Addison-Wesley, Reading).

d'Espagnat B. (1997): Aiming at describing empirical reality. In *Potentiality, Entanglement, and Passion-at-a-Distance*, ed. by R.S. Cohen, M. Horne, and J. Stachel (Kluwer, Dordrecht), pp. 71–87.

Dieks D. (1990): Quantum statistics, identical particles and correlations. *Synthese* **82**, 127–155.

Dummett M. (1978): *Truth and Other Enigmas* (Duckworth, London).

Eberhard P.H. (1978): Bell's theorem and the different concepts of locality. *Nuovo Cim.* **46 B**, 392–419.

Fraassen B. van (1981): A modal interpretation of quantum mechanics. In *Current Issues in Quantum Logic*, ed. by E. Beltrametti and B. van Fraassen (Plenum, New York), pp. 229-258.

Ghirardi G.C., Rimini A., and Weber T. (1980): A general argument against superluminal transmission through the quantum mechanical measurement process. *Lett. Nuovo Cim.* **27**, 293–298.

Ghirardi G.C., Rimini A., and Weber T. (1986): Unified dynamics of microscopic and macroscopic systems. *Phys. Rev. D* **34**, 470–491.

Giulini D., Joos E., Kiefer C., Kupsch J., Stamatescu L.O., and Zeh H.D. (1996): *Decoherence and the Appearance of a Classical World in Quantum Theory* (Springer, Berlin).

Omnès R. (1994): *The Interpretation of Quantum Mechanics* (Princeton University Press, Princeton).

Primas H. (1981): *Chemistry, Quantum Mechanics and Reductionism* (Springer, Berlin).

Primas H. (1993a): The Cartesian cut, the Heisenberg cut, and disentangled observers. In *Symposia on the Foundations of Modern Physics 1992*, ed. by K. V. Laurikainen and C. Montonen (World Scientific, Singapore), pp. 245–269.

Primas H. (1993b): Mesoscopic quantum mechanics. In *Symposium on the Foundations of Modern Physics 1993*, ed. by P. Busch, P. Lahti, and P. Mittelstaedt (World Scientific, Singapore), pp. 324–337.

Primas H. (1994): Hierarchical quantum descriptions and their associated ontologics. In *Symposium on the Foundations of Modern Physics 1994*, ed. by K. V. Laurikainen, C. Montonen, and K. Sunnarborg (Editions Frontières, Gif-sur-Yvette), pp. 201–220.

Stapp H.P. (1972): The Copenhagen interpretation. *Am. J. Phys.* **40**, 1098-1116.

CONTEXTUAL BACKGROUND

Harald Atmanspacher and Fred Kronz:
RELATIVE ONTICITY

The multilevel structure of scientific theories of matter has always been a matter of interest for Primas, and he devoted a most original paper to specific aspects of this issue (Primas 1977). In particular he was interested in reductionistic relations between the various levels of quantum mechanical descriptions. Examples are Lorentz-relativistic quantum field theory versus Galilei-relativistic quantum mechanics or traditional atomic physics with phenomenological masses, spins, and charges for electrons and nuclei versus Born-Oppenheimer type quantum chemistry with semiclassical nuclei.

In this regard, there is the interesting question of how philosophical aspects of scientific theories change when the level of description is changed. Is there a sort of "guiding principle" for a correspondence between the philosophical and physical aspects at different levels of description? Or are two different levels of physical description incommensurable in their philosophical aspects? Or can a change of the level of physical description be philosophically exploited in a constructive way, thus providing new possibilities to solve open questions?

The following contribution puts forward the position that the multilevel structure of generalized quantum theories can be used to make progress in understanding the "realism" of physical theories. The starting point is the distinction between ontic and epistemic descriptions of physical systems introduced by Erhard Scheibe (1973) and used extensively by Primas. Broadly speaking, epistemic descriptions refer to the knowledge observers have of systems on the basis of their empirical accessibility – in the sense of engineering science – and ontic descriptions refer to systems as such, independent of external influences such as observation or measurement.

Philosophically, the epistemic realm may be associated with a contextual, empirical realism, the ontic with a context-free holism. The acts of observation and measurement are then crucial for relating these two realms to each other. Though this conceptual scheme is appealing, it is too simplistic if epistemic and ontic elements must be considered at the same level of description. To cover such situations, the concept of "relative onticity" is introduced, meaning that a physical term appearing at two levels of description is interpreted epistemically at one level and ontically at another. This notion is inspired by the concept of ontological relativity as introduced by Quine (1969) and later taken up by Putnam. Both approaches are compared and contrasted with the meaning of relative onticity in multilevel quantum descriptions. It may be speculated, according to the authors, that at some basic level of onticity, the distinction between mental and material domains becomes irrelevant.

References

Primas H. (1977): Theory reduction and non-Boolean theories. *J. Math. Biology* **4**, 281–301.

Quine W.V.O. (1969): Ontological relativity. In *Ontological Relativity and Other Essays* (Columbia University Press, New York), pp. 26–68.

Scheibe E. (1973): *The Logical Analysis of Quantum Mechanics* (Pergamon, Oxford), pp. 82–88. German original: *Die kontingenten Aussagen der Physik* (Athenäum, Frankfurt 1964).

RELATIVE ONTICITY

HARALD ATMANSPACHER
Institut für Grenzgebiete der Psychologie,
Wilhelmstr. 3a, D-79098 Freiburg, Germany
and
Max-Planck-Institut für extraterrestrische Physik,
D-85740 Garching, Germany

AND

FREDERICK KRONZ
Philosophy Department, University of Texas at Austin,
Austin, TX 78712, USA

1. Realists, Realisms, and Realities

A *realist*, in one philosophical sense of this term, is someone who holds that the entities posed by a well-established theory exist (and may be contrasted with an anti-realist who regards such terms merely as convenient fictions). The corresponding notion of *realism* thus characterizes a descriptive concept (a theory), the referents of which can be conceived as elements of *reality*. Every scientist is a realist in a minimal sense insofar as the standard methodology of science requires that models and theories are empirically checked by such elements of reality. This check can confirm or disprove a given hypothesis. It always rests on empirical tools, e.g., measuring instruments, which are presupposed in an unsophisticated, common sense manner. For this reason and in this sense, the concepts of realism and reality are to be understood as *relative to* such tools. In spite of the option to use empirical facts and data for checking models and theories, it is, however, everything else than clear how these two domains are related to each other. There are levels of discussion at which it seems unnecessary to consider any such relationship at all, and there are other levels of discussion which require such a relationship to be explicitly taken into account.

The question of relationships between the material world with its facts or data and its apparently non-material counterpart or complement, the domain of models and theories, respectively, belongs to the oldest, most puzzling, and most controversial questions in the long history of philosophy and the history of science. One of the main reasons for its controversial nature is that the question itself is understood in different ways depending on basic assumptions concerning our conceptions of reality. What makes all approaches toward this question as well as the discussions about those approaches so difficult is the fact that those assumptions are most often implicit rather than explicitly clarified.

For many good reasons, any related inquiry has to take into account the corpus of knowledge we have acquired so far. The contemporary status of the sciences is the result of centuries of history, built upon various lines of empiricist tradition and upon the Cartesian distinction of res cogitans and res extensa. At present, there are quite a number of scientific topics touching this distinction itself. More and more aspects of mind-matter research become timely and sensible research topics, and it may be hoped that the knowledge we have acquired so far provides a sound basis for substantial progress in this field. Of course, this requires detailed work rather than mere verbal assertions or ungrounded speculations.

From the viewpoint of a philosophically informed contemporary physicist (who typically disregards any kind of "mind-over-matter" idealism), there are two general frameworks within which reality can be conceived. (For more details about these topics the reader is referred to the relevant literature, e.g., Chalmers (1996).) One of them is typically denoted as *physicalism* (or *materialism*) and expresses the idea that the basis of reality consists of the material world alone; anything like qualia, consciousness, psyche, mind, or spirit is based on the material elements and fundamental laws of physics. For physicalists, the way in which these apparently non-material higher-level properties can be explained is a follow-up question, again answered differently within different ways of thinking, using conceptual schemes such as, e.g., emergence, supervenience, or reduction. These concepts are tightly related to each other.

In general, it is helpful to keep in mind that emergence is an extremely colorful, often not well-defined concept that has to be discussed together with supervenience and reduction. Some useful references are Silberstein (1998), Scheibe (1997), Chalmers (1996), Crutchfield (1994), Eisenhardt and Kurth (1993), Kim (1984). All these topics have to do with instabilities (of different kinds) and have been addressed in various fields such as morphogenetics, synergetics, complex systems, non-equilibrium thermodynamics, catastrophe theory, and others. It seems to be a good guess that emergence or supervenience is connected with a weak type of reduction insofar as emergent properties must not contradict fundamental laws at a basic level of description, but also neither are uniquely determined nor can be uniquely derived from that level without further (contextual or contingent) conditions. For instance, physical processes in the human brain must not violate any applicable physical laws, but by no means are these laws sufficient to understand any of the higher-level properties and functions the brain has and performs. Nevertheless, the fundamental laws of physics can be assumed to be exhaustive at the basic level, and the existence of higher-level properties does not necessarily require us to add further "fundamental laws".

The other general framework is characterized as *dualism*, ranging from ontological to epistemological and methodological versions. Briefly speaking, ontological dualism maintains that the world *consists* of mind and matter (or other, corresponding concepts) as ultimately separate "substances". Epistemological dualism refers to mind and matter as fundamentally different domains with respect to our modes of gathering and processing knowledge of the world, irrespective of what this world "as such" ("in itself") may or may not be. Methodological dualism reflects an attitude that is neutral to the claims made by the other two variants. It utilizes the mind-matter distinction as a basic, but maybe not the only possible methodological tool to inquire into the structure of the world.

In its weakest (methodological) form, dualism is a prerequisite of any physicalist approach insofar as the latter presupposes a distinction between matter and something that appears to be non-material and – in one way or another – has to be related to, explained by, or even derived from the elements and laws of the material world. Within such a kind of minimal dualism, which is hard to avoid, we may use distinctions such as that of models and data, theories and facts, and so forth (compare Atmanspacher 1994a). In the present article, any dualistic kind of argument is meant at this methodological level.

For a physicalist approach, the concepts of emergence, supervenience, or reduction seem to refer explicitly to the world of material facts; they refer to a reality addressed by a certain type of realism. However, keeping in mind that this reference presupposes the (possibly nonunique) selection of a viewpoint, we may also argue that emergence, supervenience, or reduction primarily refer to our (non-material) descriptions of the material world rather than to elements of that world itself. Depending on the logical structure of those descriptions, they populate the entire spectrum between a naive *realism*, an unreflected belief in an external reality, and a radical *relativism*, hardly found attractive by working scientists who are used to dealing with or relying on the regulative power of events that do "really" happen in the material world.

Emergence, supervenience, and reduction are concepts which have been applied to facts ascribing properties to systems in the material world (i.e., in an assumed material *reality*) or in the sense of our descriptions of those properties (i.e., different kinds of *realism*). In a rough terminology, there are emergent facts and emergent theories. Mixing both of them up, inevitably leads to category mistakes and misunderstandings. The methodological dualism that helps us to avoid this must, however, not at all be understood as a predecision concerning the structure of the world and our knowledge of this world. As mentioned above, it should be understood as a tool to inquire into this structure. It may be a preliminary tool that can, for instance, lead to a precise description of its own limits.

The present contribution deals primarily with realisms (conceptions of reality). It will be argued that neither a naive realism, addressing the material world as a collection of facts that are ready for observation in a theory neutral way, nor a radical relativism with a collection of models posing facts in a theory dependent, more or less arbitrary way, is the right tool to deal with those issues properly. A specific conceptual scheme will be sketched that allows us to combine a certain kind of relativism with the belief that the material world cannot be described arbitrarily. In other words: although facts are certainly model-dependent, they are more than just illusions. There are many (contextually) correct descriptions with well-defined relationships among each other rather than just one (universally) correct description of the world – and there certainly are incorrect descriptions.

2. Ontic and Epistemic Descriptions

Assuming the methodological distinction between a material world with events, facts, or data and a mental world with concepts, models, or theories, it is possible to describe elements of the material world by elements of the mental world. The question then is how to distinguish different elements within the two domains. Modern physics, in particular quantum physics has developed tools to address this question with respect to the material world. A most important distinction in this context is that of systems (objects) and their environment. This distinction is sometimes metaphorically called the *Heisenberg cut* (Heisenberg 1936).

Together with the fact that descriptions of isolated systems are radically different from descriptions of open systems, the Heisenberg cut and the corresponding formal tools play a major role in modern quantum theory. It turns out that a proper understanding of these issues can be achieved using two different descriptions of reality; namely, the ontic and the epistemic, respectively.[1] Primas has developed this distinction in the formal framework

---

[1] These terms are due to Scheibe (1973) and must not be mixed up with the distinction between "ontological" and "epistemological". The distinction between ontic and epistemic descriptions can, for instance, itself be discussed as an ontological or epistemological topic,

of algebraic quantum theory (Primas 1990; see also Atmanspacher 1994b for some indications of possible limitations of this distinction). The basic structure of the ontic/epistemic distinction as it will be used subsequently can be understood according to the following rough characterization (for more details, the reader is referred to Primas 1990,1994a):

Ontic states describe all properties of a physical system exhaustively. ("Exhaustive" in this context means that an ontic state is "precisely the way it is", without any reference to epistemic knowledge or ignorance.) Ontic states are the referents of individual descriptions, the properties of the system are formalized by *intrinsic observables*.[2] Their temporal evolution (dynamics) follows *universal, deterministic laws* given by a Hamiltonian one-parameter group. As a rule, ontic states in this sense are empirically inaccessible. *Epistemic states* describe our (usually inexhaustive) knowledge of the properties of a physical system, i.e. based on a finite partition of the relevant state space. The referents of statistical descriptions are epistemic states, the properties of the system are formalized by *contextual observables*. Their temporal evolution (dynamics) follows *phenomenological, irreversible laws* which can be given by a dynamical one-parameter semigroup if the state space is properly chosen. Epistemic states are empirically accessible by definition.

Although the formalism of algebraic quantum theory is often hard to handle for specific physical applications, it offers significant clarifications concerning the basic structure and the philosophical implications of quantum theory. For instance, the modern achievements of algebraic quantum theory make clear in what sense pioneer quantum mechanics (which von Neumann (1932) implicitly formulated epistemically) as well as classical and statistical mechanics can be considered as limiting cases of a more general theory. Compared to the framework of von Neumann's monograph (1932), important extensions are obtained by giving up the irreducibility of the algebra of observables (not admitting observables which commute with every observable in the same algebra) and the restriction to locally compact state spaces (admitting only finitely many degrees of freedom). As a consequence, modern quantum physics is able to deal with open systems in addition to isolated ones, it can involve infinitely many degrees of freedom such as the modes of a radiation field, it can properly consider interactions with the en-

according to whether its observer-independent *existence* or its observer-dependent status as a *descriptive tool* is addressed. Moreover, Fetzer and Almeder (1993) emphasize that "an ontic answer to an epistemic question (or vice versa) normally commits a category mistake". The literature on mind-matter questions is full of such category mistakes. Numerous examples can also be found in the context of quantum physics.

[2]Note that the term "observable" was historically developed as a technical term for a property of a system. Prima facie it has nothing to do with the actual observability of that property.

vironment of a system, superselection rules, classical observables, and phase transitions can be formulated which would be impossible in an irreducible algebra of observables, there are in general infinitely many representations inequivalent to the Fock representation, and non-automorphic, irreversible (hence non-unitary) dynamical evolutions can be successfully incorporated.

In addition to this remarkable progress, the mathematical rigor of algebraic quantum theory in combination with the ontic/epistemic distinction allows us to address quite a number of unresolved conceptual and interpretational problems of pioneer quantum mechanics from a new perspective. First of all, the distinction between different concepts of states as well as observables provides a much better understanding of many confusing issues in earlier conceptions, including alleged paradoxes such as those of Einstein, Podolsky, and Rosen (EPR, 1935) or Schrödinger's cat (Schrödinger 1935). Second, a clear-cut characterization of these concepts is a necessary precondition to explore new approaches, beyond von Neumann's projection postulate, toward the central problem that pervades all quantum theory since its very beginning: the measurement problem. Third, a number of much discussed interpretations of quantum theory and their variants can be appreciated more properly if they are considered from the perspective of an algebraic formulation.

This applies in particular to the deep (though notoriously vague) deliberations of Bohr, to Einstein's and Schrödinger's contributions, to Bohm's ideas on explicate and implicate orders, to Heisenberg's distinction of actuality and potentiality, or to d'Espagnat's scheme of an empirical, weakly objective reality and an observer-independent, objective (veiled) reality.[3] An important example: the core of the well-known Bohr-Einstein discussions in the 1920s and 1930s (see Jammer 1974) can be traced back to the belief that only one of the mentioned concepts of reality can be (primarily) relevant. While Bohr clearly emphasized an epistemic, contextual realism referring to the results of measurements, Einstein was deeply convinced of an ontically determined realism to which he attached a common-sense type local realism that – as we would say today – applies to an epistemic viewpoint. In the framework of algebraic quantum theory, both kinds of realism play significant roles, and even some of the formal relations between them have been clarified successfully. More details about this issue have been discussed by Howard (1985,1997).

One of the most striking differences between the concepts of ontic and

---

[3]Since detailed discussions of these issues would be far beyond the scope of this contribution, they are omitted here. Some corresponding indications can be found as scattered remarks in recent papers by Primas (1990, 1994a, and others). Among the approaches listed above, d'Espagnat (1995) gives some hints in a non-algebraic terminology but does not substitute a yet-to-be-written systematic algebraic presentation.

epistemic states is their difference concerning operational access, i.e. observability and measurability. At first sight it might appear pointless to keep a level of description which is not related to what can be verified empirically. However, a most appealing feature at this ontic level is the existence of first principles and universal laws that cannot be obtained at the epistemic level. Furthermore, it is possible to rigorously deduce (to "GNS-construct"; cf. Primas 1994a) a proper epistemic description from the ontic description if enough details about the empirically given situation are known. This is particularly important and useful for the treatment of open and macroscopic (quantum) systems.

The distinction of ontic and epistemic states provides an important clue to understand the distinction between holistic and local realisms, i.e., concepts of reality. Ontic states and intrinsic observables refer to a holistic concept of reality and are operationally inaccessible, whereas epistemic states and contextual observables refer to a local concept of reality and are operationally accessible. It is exactly the process of observation, essentially one or another kind of pattern recognition, which represents the bridge between the two. Observation suppresses (or minimizes, respectively) the EPR correlations constituting a holistic reality and provides a level of description to which one can associate a local concept of reality with locally separate (or "approximately" separate, respectively) objects. In this sense it is justified to say that observation generates objects by introducing a Heisenberg cut as a metaphor for the suppression of EPR correlations.

Another way to look at the distinction of ontic and epistemic states and the associated algebras of observables is the following. The ontic holistic realism of quantum theory is related to all sorts of inquiries into a context-, mind-, or observer-independent reality of the outside world. Focusing on an epistemic local realism expresses a change of perspective to the effect that the question "*What is* this independent reality?" is replaced by "*What can we know* about such a reality?" Philosophically the distinction between these two questions is very much in the spirit of Kant's distinction of transcendental idealism and empirical realism, and in this sense one may consider an ontic description as a kind of "idealization" of an epistemic description. As an empirical science, physics addresses only questions of the second kind. But on the other hand, the mathematical formalism that constitutes the formal basis of physics often leads into a way of thinking very much in accordance with the first kind of question. An instructive discussion along these lines, emphasizing those topics as non-standard realism, is due to d'Espagnat (1999).

One of the basic conceptual implications of the distinction of ontic and epistemic descriptions of reality is the fact that it is inadmissible to speak of objects and environments or their observation at the ontic level. Here is the domain of nonlocal, holistic correlations between those properties that are, technically speaking, described by non-commuting operators. Local objects and their environment are generated by a change of perspective from the ontic to the epistemic level, which generally involves the breaking of a symmetry, introduces new contexts (e.g., abstractions that are deliberately made to distinguish between "irrelevant" and "relevant" features), and is intimately related to the distinctions necessary for any kind of observation. This makes it easy to understand why ontic states are non-empirical by definition. Empirical access requires the separation of objects which are not a priori, i.e., ontically, given.

A widespread category mistake resulting from a lack of proper ontic/epistemic distinctions and the associated distinction of holistic/local realism is reflected by the assertion that EPR correlations can be interpreted such that the parts of a holistic system communicate superluminously, i.e. with signal velocities greater than the velocity of light. The state of the system as a whole is an ontic state. If a system as a whole is to be described ontically, then it is in general inadmissible to speak of parts within the same description, and consequently there is no way to talk about communication between such parts.[4] Only if the ontic state of a system is decomposed in order to describe subsystems or parts, the result is a description in terms of epistemic states of those subsystems. They can communicate, but of course not superluminously.

Another consequence of the same category mistake is the misleading interpretation that due to EPR correlations "everything is correlated with everything else". Ontically, there is only "one thing", a system as a whole. Epistemically, where it is admissible to speak of "many things" and consequently of "everything", there are no holistic correlations. Any empirically accessible aspect of those correlations relies on the condition that parts of the environment (e.g., detection instruments) are *not* correlated.[5] All empirical evidence we have for quantum holism is obtained by "destroying" that same holism. Ironically, nonlocality can only indirectly be demonstrated in a local way, conceptually using counterfactual reasoning.

As Primas has discussed extensively (Primas 1998), the transition from an ontic to an epistemic level of description often goes hand in hand with the emergence of properties that are not defined ontically. Almost all known

---

[4]Other terminologies such as "uncontrollable influence" (Bohr 1935) or "passion-at-a-distance" (Shimony 1984) rather than communication or signaling are less suggestive of a direct conflict with the special theory of relativity. They indicate something like an "internal structure" of a system even when it is considered as a whole, an issue that will be taken up in the following section.

[5]Technically speaking, for every quantum system in a given pure state ϕ there is a factorization such that ϕ is a product state. This is to say that there are always (perhaps fictitious) subsystems which are *not* correlated with each other.

classical properties (in the sense of commuting observables) of objects emerge due to contexts that are not given by the intrinsic properties of an ontic description, but have to be selected properly, adapted to the given situation. Some of the examples that are formally well-understood refer to properties such as chirality, temperature, or chemical potential. Another example is the emergence of irreversibility; the time evolution of ontic states is given by a one-parameter group describing a reversible dynamics.[6] The notion of emergence is also used (in a physicalist sense) for much more complicated and fairly little understood properties such as life or consciousness. A common tenet shared by most physicists (not every physicist is a physicalist) is the restriction of the problem of measurement to the material world alone. Consequently, observers are considered as observing apparatuses, and any consciousness of living observers remains disregarded (cf. Primas 1993, Atmanspacher 1997).

3. Relative Onticity

What is a suitable way to address situations which confront us with holistic *and* local features at the same level of description? In such situations, mixtures of ontic and epistemic elements are required at the same level of description, thus forbidding a unique assignment of ontic/epistemic descriptions and holistic/local realisms. (As indicated above, such a mixture is unavoidable from the very beginning since every epistemic description presupposes an ontic description of measuring tools.) This difficulty can be resolved if it is realized that two levels of description are not enough to cover the entire hierarchy leading from fundamental particles in basic physics up to living systems in biology and psychology. It is then suggestive to consider ontic and epistemic descriptions as *relative to* two successive levels in the hierarchy. Concerning material reality, this is particularly relevant to the study of hierarchical complex systems, and some ideas toward a corresponding formal approach have been specified elsewhere (Primas 1994b).

Let us start with an example for such a *relative onticity*[7] in the material domain of reality. From a fundamental viewpoint of quantum theory as sketched above, atoms and molecules are highly contextual objects whose

---

[6] A long-standing misunderstanding in many discussions about the approach to irreversibility as advocated by the Brussels-Austin-group of Prigogine and collaborators can be boiled down to the question whether an ontic or an epistemic description is "primary" – an issue very similar to the Bohr-Einstein-controversies addressed above. The assignment of reversibility and irreversibility to ontic and epistemic levels, respectively, is not controversial: "irreversibility is an emergent property" (Petrosky and Prigogine 1997).

[7] This term has been coined in discussions with Chris Nunn in the context of an attempt to understand archetypes as memes à la Dawkins (1976) and to develop a corresponding hierarchical structure (Nunn 1998, Atmanspacher 1998).

properties can be described by interactions of electrons, nuclei, and their environments. However, from the viewpoint of chemistry one may not be interested in these complicated interactions, but in the shape and other features of molecules, for which it is reasonable to consider the concept of an atomic nucleus ontically as a whole rather than composed of protons, neutrons, or even "more basic" constituents. This leads to the description of a molecule as a contextual object resulting from the interaction of nuclei, electrons, and their environments. The ontic/epistemic distinction can then be shifted from the levels of electrons and nuclei to that of molecules. While molecules are epistemically described within the first realm, they acquire an ontic description within the second. In this manner, the result of a composition of ontic nuclei and electrons (the epistemic molecule) at a certain level can be considered ontically as a basic entity (the ontic molecule) if it is viewed from a successively higher level in the hierarchy.

In a more detailed version of this example one can even address specific relationships between different levels of description. Let us discuss the concept of water as an example. At a rather basic physical level one might think of water in terms of hydrogen and oxygen nuclei and electrons. Leaving the nuclear level of description (protons, neutrons, etc.) involves a change of perspective which, roughly speaking, abstracts from any nuclear forces due to strong interactions, and focuses on electromagnetic (Coulomb) forces. In a general sense, this abstraction leads from a description in terms of ontic states of nuclei and electrons and their properties to the epistemic concept of a water molecule, H_2O.

One of the most important further abstractions in this context leads to the so-called Born-Oppenheimer picture, disregarding the electron mass as compared to the masses of the nuclei. In a corresponding description, the water molecule has properties which H and O nuclei did not have, e.g., the property of a nuclear frame. A special feature of nuclear frames is the chirality (handedness, see Amann 1993) of molecules. Molecular chirality is a property that emerges at an epistemic molecular level of description and is absent at any lower level. However, this is not to say that this property is just a matter of description and has no *real* impact. For instance, thalidomide is a chiral molecule. Today it is well-known that the disastrous consequences of thalidomide-based remedies in the 1960s are caused by only one of the two different chiral species. The remedies were produced as mixtures of both species.

In a thermodynamic description, other properties emerge due to consideration of *many* ($N \to \infty$) entities such as molecules. It is intuitively obvious that one single water molecule H_2O is not wet. The property of liquidity is an emergent property for which the level of a description in terms of individual molecules has to be left and replaced by a statistical or thermodynamical description. The same applies to other properties such as chemical potential or temperature, for which rigorous mathematical derivations are available (Takesaki 1970, Müller-Herold 1980). Again a remark concerning the factual "reality" of a property such as temperature: whoever has burned one's fingers once will have serious doubts that temperature might be nothing else than a descriptive tool that has nothing to do with reality.

For a molecular chemist or biologist, molecules are the building blocks of his mode of description. In this sense, their states and properties are considered ontically. A molecular biologist is not at all concerned with the justification of a molecular (Born-Oppenheimer) picture. He may, however, be interested in the way in which different phosphates (adenine, thymine, guanine, cytosine), so-called nucleotides, can be combined to different DNA sequences. For such a point of view, the phosphate molecules are entities to be described by ontic states and their properties, the different ways they are organized in DNA give rise to an epistemic description with emergent properties (genetic information, e.g., the faculty of self-reproduction) at the level of the DNA. At this conceptual level, there is an analogy between the ontic/epistemic distinction and the distinction of genotypes and phenotypes which deserves further study.

The systems and objects with which the "man on the street" usually deals in everyday life are the trees, tables, bricks, icecubes, and so forth of common sense realism. It would be entirely unreasonable not to include this kind of realism in the framework suggested here. Although for a scientist a tree has to be described as a highly complicated composition of material subsystems with emergent properties of different types (solidity, texture, etc.), common sense realism holds that a tree *is* simply a tree, an object in an ontic state *having* those properties. Many trees together can form a forest, and there certainly are issues for which the forest as a whole is the right object to be addressed rather than many trees. If it is addressed in terms of many trees, the forest is the referent of an epistemic description. If it is addressed in terms of an ecosystem as a whole (see, e.g., Hauhs et al. 1998), the forest becomes relevant as the referent of an ontic description.

The central issue of the general concept of relative onticity is that states and properties of a system, which are treated epistemically at a given level of description, can be considered as ontic from the perspective of a higher level. Objects can be epistemically described to be composed of lower level objects, but alternatively they can be ontically described as wholes, giving rise to "building blocks" of higher level objects. Emergent properties at successively higher levels of description can be *formally* addressed by a change of perspective which is not uniquely given but depends on contexts and conceptual schemes that must be selected properly.

However, this does not imply that *any arbitrary* description is proper. An interesting example for an improper conceptual scheme is given by a supposed "atomic" level of description between nuclear and molecular levels. Since we do not know the interaction between atoms as entities in themselves, molecules must not be conceived as composed of atoms but of nuclei and electrons. Taking atoms seriously in an ontic sense leads to problems and inconsistencies if one wants to use them for the construction of an epistemic molecular picture. Although "atomic physics" doubtlessly was a very important field of research early in the 20th century, a modern point of view suggests that it is more appropriate to consider atoms as a special chapter of molecular physics.

The entire approach discussed so far essentially looks at different levels of description in the sense of increasing diversity. Generally speaking, moving from one level to the next higher level corresponds to a symmetry breaking; in one way or another, a holistic system is considered to be broken up into parts. Such a kind of so-called "bottom-up" approach is usually assumed as a proper way to reflect the evolution of more and more complex systems in the material world. In this framework, it is, however, a natural question whether all conceivable symmetry breakings are to be regarded as feasible, or whether there are some of them which are more feasible than others. For instance, it seems plausible that the symmetry breaking of the ontic state of a *photon pair* in an EPR type situation *before* measurement generically leads to the epistemic states of *two single photons* rather than arbitrary other subsystems *after* measurement.

Teller (1989) has proposed the concept of "relational holism" in very much the same spirit.[8] In Teller's parlance, a local realist tries to interpret EPR-type correlations in terms of nonrelational properties of the relata which underlie any such correlations. On the basis of those (subvenient) properties it should be possible to explain the correlations as supervenient. By contrast, Teller asserts that any EPR-type "correlation – as an objective property of the pair of objects taken together – is simply a fact about the pair. This fact will arise from and give rise to other facts. But it need not itself be decomposable in terms of or supervenient upon some more basic, nonrelational facts" (Teller 1989, p. 222). On the other hand, there are, of course, decompositions of a system as a whole into subsystems, such as the decomposition of a photon pair into two photons with their individual (emergent or supervenient) properties.

Clearly, it would be desirable to have a way of explaining that certain decompositions of a system as a whole are more natural than others. Slight

---

[8]Teller's approach initiated a number of related schemes of interpretation which all highlight the significance of correlations as compared to that of correlata. A short survey can be found in Mermin (1998), Sec. XI.

variations of the context should not in general (but can in exceptional cases) result in different epistemic states. This is a requirement that can typically be taken into account by stability considerations. What we want is that certain decompositions of a holistic system are more stable, more robust than others. A first attempt into this direction has been indicated by Amann and Atmanspacher (1998). This means that any ontic, holistic level of description does already carry some inherent tendencies for more or less stable decompositions. A forest is more likely to be decomposed into individual trees rather than into strange mixtures of them. In this sense, holistic systems are *not* totally void of internal distinctions. It is an unresolved problem how such "preformed tendencies" for the stability of certain decompositions can be taken into account formally. It may be speculated that elements of "top-down" thinking could play a role in this regard, thus closing a self-referential loop between any pair of ontic and epistemic frameworks at any level of description. Such a scheme would imply that ontic and epistemic elements of a description mutually depend on each other, thus rendering any ultimate "primacy" of one over the other as ill-posed.

4. Ontological Relativity

The formal concept of "relative onticity" resembles to some extent the (less formal) discussion of "ontological relativity" as introduced by Quine (1969). In this essay, Quine argues that if there is one ontology that fulfills a given theory, then there is more than one. This claim is the crux of his doctrine of ontological relativity, claiming that it makes no sense to say what the objects of a theory are, beyond saying how to interpret or reinterpret that theory in another. Moreover (Quine 1969, p. 53):

> "Ontological relativity is not to be clarified by any distinction between kinds of universal predication – unfactual and factual, external and internal. It is not a question of universal predication. When questions regarding the ontology of a theory are meaningless absolutely, and become meaningful relative to a background theory, this is not in general because the background theory has a wider universe. One is tempted ... to suppose that it is; but one is then wrong. What makes ontological questions meaningless when taken absolutely is not universality but circularity. A question of the form 'What is an F?' can be answered only by recourse to a further term: 'An F is a G.' The answer makes only relative sense: sense relative to the uncritical acceptance of 'G'."[9]

---

[9]In this spirit, the concept of quantum holism does only make sense relative to the uncritical acceptance of measuring tools that are *not* EPR-correlated.

For Quine, any question as to the "quiddity" (the "whatness") of a thing is meaningless unless a conceptual scheme is specified relative to which it is discussed. It is not the uniqueness of such a scheme, e.g., any "theory of everything" with universally given referents, but the faculty of reinterpretation of one scheme in another which belongs to the important features of scientific work. Nevertheless, Quine encourages "ontological commitment" in the sense that a most proper conceptual frame should be preferred for the interpretation of a theory. The circularity which he mentions as the crucial point of ontological relativity expresses itself in an inscrutability of reference. This stresses his conviction that the issue of reference causes the problems necessitating ontological relativity, not the unique assignment of referents as objects in the external world of a realist (cf. Gibson 1995).

After his farewell to functionalism (cf. Chalmers 1996), Putnam (1981, 1987) has developed a related kind of ontological relativity within an approach rejecting both naive (spectator) realism and relativism. His approach rather attempts to reconcile the two and was first called "internal realism", later sometimes modified to "pragmatic realism". Ontological (sometimes conceptual) relativity is a central feature of Putnam's internal realism, but it differs from Quine's usage of the term in an important detail. In an interview with Burri (1994),[10] Putnam characterized Quine's ontological relativity as due to the impossibility of a uniquely fixed relationship of our concepts to the totality of objects which those concepts refer to. Putnam's own position is more radical insofar as he questions that we know what we mean when we speak of a totality of objects (Burri 1994, p. 185):

"If we start with the notion of a totality of objects it becomes entirely untransparent how our terms – maybe except those referring to sense data – can refer in a fixed way. But from this I have not concluded that no term other than sense data terms refer in a fixed way; rather I have concluded that the premises leading to such a conclusion must be wrong. In this context I basically think of the assumption that we know what we mean when we speak of a totality of all objects."

Considering the perspective of quantum holism, this position is highly sensible. If an object can only be reasonably defined within the framework of some preselected conceptual scheme, as Putnam's internal realism holds, then it is evident that any definition of an object is only relevant within a given context, i.e., objects are "ontologically relative" entities. But Putnam's point of departure is *not* quantum holism; it is our common sense realism, referring to "a usage of the word 'object' which we cannot change without loss of its meaning. The notion of an object roots in speaking of

[10]To our knowledge, this interview of January 14, 1994 (given at Cambridge, Massachusetts) is the most recent source directly addressing the issues of interest for the present discussion.

tables, chairs, and bricks. Tables, chairs, and bricks are objects in a fundamental sense of the word." (Burri 1994, p. 182) "The actual problem is to
work out the difference between our common sense realism on one side and
a transcendental or physical, respectively, realism on the other. Currently
I try to criticize physical realism from the viewpoint of common sense realism." (Burri 1994, p. 177) Concerning the main features of this position,
Putnam admits (Burri 1994, p. 177f) that his

> "ideas keep being subject to change. At present I see the crucial points
> other than immediately after the turn. The publications which I wrote
> by the end of 1976 ... finished a period of my thinking in which I began
> to see more and more clearly that the semantics underlying classical
> realism are hopelessly metaphysical. In particular, I became convinced
> that numerous concepts of metaphysical realism are untenable, for ex-
> ample the idea that one can reasonably talk about 'all entities' – as if
> the terms 'entity' or 'object' had a unique, fixed meaning – as well as
> the illusion that there is an answer to the question of which objects
> the world consists. Later I called this conviction 'internal realism' or
> 'conceptual relativity'. It rests upon the idea that there is a real world,
> but it does not dictate its own descriptions to us. Internal realism does
> not imply 'anything goes' but rather accounts for the fact that there are
> many descriptions of the world, depending on our interests and ques-
> tions, and on what we intend to do with the answers to those questions.
> The assumption that certain descriptions cover the world as it is in itself
> seems to be pointless to me."

In his version of ontological relativity, Putnam wants to maintain a
meaningful concept of reference and gives up the concept of a totality of
uniquely defined objects as a precondition for any attempt to fix references
once and for ever. Objects cannot be uniquely defined, but they can be
defined with respect to conceptual frames. Within a preselected frame, it is
then possible to establish reference without inscrutability. Putnam's discussion of water on a twin earth is illuminating in this context. If inhabitants
of a "twin earth" use the term water to refer to a chemical substance other
than H_2O (say XYZ), then Putnam holds that due to the best of our knowledge ("expert knowledge") the *proper* referent of "water" – at the level of
a molecular description – is H_2O in the external world; H_2O is the extensional referent of "water". Putnam's example can only roughly be sketched
here (for more details see Putnam 1975, Putnam 1981, Sect. II). It seems
to be related to what Quine calls ontological commitment, but other than
Quine's, Putnam's ontological commitment explicitly takes extensions into
account. Quine's ontological relativity relates to inscrutability of reference,
Putnam's relates to non-uniqueness of objects. (For other viewpoints in this
discussion, see Searle 1983, Sec. 8, and Chalmers 1996, Sec. I.2.4.)

In Putnam's thinking, conceptual schemes serve a purpose very similar to contextual representations in the framework of quantum theory. In this regard, relative onticity and ontological relativity are tightly related to each other. Both Putnam's internal realism and the realism of quantum theory agree with regard to a basic assertion according to which there is a "real world as such". The starting point of Putnam differs, however, from the starting point of a quantum theoretical perspective. For Putnam, objects in a fundamental sense are common sense objects such as "tables, bricks, icecubes". From the viewpoint of quantum theory, a universe of discourse in a fundamental sense does simply not consist of objects (although every quantum theoretical statement presupposes such objects, e.g., observational tools). Objects at each level of description are generated by symmetry breakings within a holistic universe of discourse, addressed by a holistic realism. The concept of relative onticity entails a recursive application of formal transformation principles (some classes of which are well-known, see Primas 1998), translating between successive levels of description. Such principles or even rules of transformation are called for (at least by Quine), but not given in the framework of Putnam's or Quine's ontological relativity. In the scheme of relative onticity, common sense objects are objects just in a very special contextual framework, high up in the hierarchy of descriptions. But nevertheless they are considered as "real" objects in an external reality.

In his more recent writings, Putnam often refers to Kant and his distinction between (empirical) realism and (transcendental) idealism. It seems that the philosophy of Kant had an important impact on Putnam's way of thinking:

"Only after I rejected metaphysical realism I began to understand what is correct in Kant's philosophy. Nevertheless, I am not a Kantian idealist. But he was the first philosopher who saw that we do not simply represent the world. To describe the world does not mean to represent it. It seems to me that this is an important insight." (Burri 1994, p. 178)

"[Kant] does not doubt that there is some mind-independent reality; for him this is virtually a postulate of reason. He refers to the elements of this mind-independent reality in various terms: thing-in-itself (Ding an sich); the noumenal objects or noumena; collectively, the noumenal world. But we can form no real conception of these noumenal things; even the notion of a noumenal world is a kind of limit of thought (Grenzbegriff) rather than a clear concept. Today the notion of a noumenal world is perceived to be an unnecessary metaphysical element in Kant's thought. (But perhaps Kant is right: perhaps we can't help thinking that there is somehow a mind-independent ground for our experience even if attempts to talk about it lead at once to nonsense.) At the same time, talk of ordinary empirical objects is not talk of things-in-themselves but only talk of things-for-us." (Putnam 1981, p. 61)

"Internal realism says that the notion of a 'thing in itself' makes no sense; and *not* because 'we cannot know the things in themselves'. This was Kant's reason, but Kant, although admitting that the notion of a thing in itself *might* be 'empty', still allowed it to possess a formal kind of sense. Internal realism says that we don't know what we are talking about when we talk about 'things in themselves'. And that means that the dichotomy between 'intrinsic' properties and properties which are not intrinsic also collapses – collapses because the 'intrinsic' properties were supposed to be just the properties things have in 'themselves'. The thing in itself and the property the thing has 'in itself' belong to the same circle of ideas, and it is time to admit that what the circle encloses is worthless territory." (Putnam 1987, p. 36)

According to these selected quotations, it is Putnam's view that we can only reasonably talk about the empirically accessible world of ontologically relative objects. Their relativity is due to different conceptual schemes with extensional referents in a real world. The concept of things-in-themselves has to be rejected not only as empirically empty but primarily because they do not make sense. It has to be added that for Putnam "making sense" means more precisely: making sense in the sense of common sense. In other words, it is the absurdity of things-in-themselves that causes Putnam to reject them – although he admits that the concept of a (noumenal) world independent of empirical access may be an unavoidable idea.

The perspective of modern quantum theory offers an interesting alternative to Putnam's viewpoint. Putnam's (and Kant's) empirical realism is the local realism of any working scientist. Objects as the referents of local realism are always contextual, they are relevant with respect to a conceptual scheme corresponding to a preselected level of description. The states of those objects are epistemic at this level. On the other hand, it is also possible that there are ontic states at the same level, e.g., molecules described as wholes rather than described as consisting of nuclei and electrons. These ontic states refer to a holistic realism which – from the perspective of an empirical (local) realism – seems as "absurd" as Putnam claims. However, it does so for other reasons and with other implications.

Quantum holism invalidates the concept of objects at any level to which it is applied in terms of an ontic description. In this regard, things-in-themselves are not relevant as empirically accessible entities and it does indeed not even make sense to address them as separable entities. (As discussed in the preceding section, one might nevertheless think of some kind of tendency that objects with emergent properties can be obtained by moving to another, higher level of description.) However, quantum holism indicates that this has to be understood as an encouragement to question an unrestricted application of common sense realism beyond its significance as a necessary precondition for gaining empirical access to quantum holism by classical, uncorrelated measuring tools.

This is a decisive difference from Putnam's viewpoint. He interprets absurdity in the sense of common sense as an argument for rejecting conceptual schemes that are absurd in this sense. Quantum holism interprets such absurdity as an argument for questioning common sense if it is applied beyond its proper domain.[11] Of course, arguments of this latter type have to be investigated extremely carefully before they can be accepted. The present state of discussions in the foundations of quantum theory with its necessarily indirect, but overwhelming empirical evidence for holism provides strong evidence that its apparent "absurdity" must be taken seriously. Quantum holism might give us the right hint to understand Kant's transcendental idealism more properly than in terms of things-in-themselves.

5. Speculative Remarks on Mind and Matter

There is a considerable non-mainstream tradition of physicists who have suggested that quantum measurement has to do with consciousness. One of the pioneers of this conception is Wigner, among its more recent advocates are – with different arguments – Penrose and Stapp. Quite a number of publications addressing the relationship between the philosophy of quantum theory and the philosophy of mind over the last decade (cf. the overview by Butterfield 1995) show that there is a steadily growing interest in this idea. Already in the mid 1940s, and presumably as an offspring from his extensive discussions with Pauli, Jung discussed a distinction similar in spirit to that of epistemic and ontic states with respect to conscious and unconscious levels in the mental world.[12] In an afterword to his essay "On the nature of the psyche" (Jung 1971), Jung quotes Pauli with the statement that "the epistemological situation with regard to the concepts 'conscious' and 'unconscious' seems to offer a pretty close analogy to the ... situation in physics. ... From the standpoint of the psychologist, the 'observed system' would consist not of physical objects only, but would also

---

[11]In the early days of quantum theory, in 1922, Heisenberg once asked Bohr: "If the interior structure of the atoms is so inaccessible to any illustrative description as you say, if we actually don't have any language to talk about this structure, will we ever be able then to understand the atoms?" Bohr hesitated for a moment, then he said: "We will. But at the same time we will understand the proper meaning of 'understanding'." (Heisenberg 1969, p. 64)

[12]Pauli's position in this regard was ambivalent: though he always stressed the fact that quantum theory refers to the material world alone, there are letters by Pauli in which he expressed his uneasiness with that state of affairs (see Atmanspacher and Primas 1996).

include the unconscious, while consciousness would be assigned the role of 'observing medium'." In other words: mental objects and their mental environments are conceived to be generated by the transformation of elements of the unconscious into consciously and empirically accessible categories.

Analogous to the material world, it might be appropriate to consider the possibility of different levels of descriptions, regarded as elements of a mental world, providing a whole spectrum between the most fundamental and the most contextual ones. One end of this spectrum would refer to a "most ontic" level of description, serving as a limiting case, meaning that it has no broken symmetry at all. At the other end, we would find a "most epistemic" level of elements of a cultural environment, manifesting themselves in individual human psyches. A nice example is given by national or regional versions of cultural key ideas by contrast with more general versions. A certain element of a cultural environment may have ontic meaning with respect to a local environment whereas it is regarded as epistemic in a larger scope. Cartesian dualism is epistemic from the viewpoint of a worldwide cultural perspective, from which it can be regarded as a regional version of the more general principle of duality. However, it represents a concept that has implicitly acquired almost ontic (collective and unconscious) features within the narrower scope of traditional Western science and technology. An additional appealing feature of such a multilayered scheme is the fact that there are many ways to draw distinctions (break symmetries) at every level. Each distinction is contextual relative to the preceding level and generates its own specific features.

Applying the idea of relative onticity, it is conceivable that under suitable conditions epistemic elements at a certain level of description can be transformed into ontic elements when considered from the perspective of the next higher level. In other words: explicit elements of the sociocultural environment at a certain epoch can become implicitly ontic elements in a later epoch, thus leading to additional (archetypal) features in the collective unconscious in Jung's parlance. In addition to the Cartesian distinction, other basic concepts of traditional science such as determinism, causality, and locality may serve to provide further examples. What was once explicitly "invented", has later to be "discovered" as an implicit assumption underlying a new epistemic level. Such processes can be expected whenever a new epistemic level in the hierarchy of descriptions *emerges*, rendering the preceding one as its own ontic basis. More details of this picture, particularly with respect to the concept of archetypes, have been addressed by Nunn (1998) and Atmanspacher (1998).

Jung and Pauli (and others) have speculated that at a level which is "ontic enough" the symmetry breaking according to the Cartesian distinction of matter and mind dissolves, providing an "unus mundus" in which fundamental physics and depth psychology refer to the same unbroken reality (see Atmanspacher and Primas 1996). Such a scenario points toward an interesting alternative to the idea that consciousness (mind) emerges as a higher level property of the brain (matter) just as, roughly speaking, liquidity emerges as a higher level property of water (see, e.g., Searle 1984). The Pauli-Jung approach considers the mind-matter distinction as a fundamental symmetry breaking at a very primordial level of description. In this scheme, *both* mind and matter are emergent domains of description (not only mind emerges from matter), used to describe the world in terms of the corresponding distinction. The holistic features of modern quantum theory might induce and even support speculations of this kind. At present, however, the available knowledge about these extremely difficult issues is far from sufficient to flesh out the corresponding ideas. It remains mandatory to distinguish sound results from wishful thinking.

References

Amann A. (1993): The Gestalt problem in quantum theory – generation of molecular shape by the environment. *Synthese* **97**, 125–156.

Amann A. and Atmanspacher H. (1998): Fluctuations in the dynamics of single quantum systems. *Stud. Hist. Phil. Mod. Phys.* **29**, 151–182.

Atmanspacher H. (1994a): Objectification as an endo-exo transition. In *Inside Versus Outside*, ed. by H. Atmanspacher and G.J. Dalenoort (Springer, Berlin), pp. 15–32.

Atmanspacher H. (1994b): Is the ontic/epistemic distinction sufficient to describe quantum systems exhaustively? In *Symposium on the Foundations of Modern Physics 1994*, ed. by K.V. Laurikainen, C. Montonen, and K. Sunnarborg (Editions Frontières, Gif-sur-Yvette), pp. 15–32.

Atmanspacher H. (1997): Cartesian cut, Heisenberg cut, and the concept of complexity. *World Futures* **49**, 333–355.

Atmanspacher H. (1998): Commentary on Chris Nunn's 'Archetypes and Memes'. *Journal of Consciousness Studies* **5**, 355–361.

Atmanspacher H. and Primas H. (1996): The hidden side of Wolfgang Pauli. *Journal of Consciousness Studies* **3**, 112–126. Reprinted in *Journal of Scientific Exploration* **11**, 369–386 (1997).

Bohr N. (1935): Can quantum-mechanical description of physical reality be considered complete? *Phys. Rev.* **48**, 696–702.

Burri A. (1994): Interview with Hilary Putnam, in *Hilary Putnam* (Campus, Frankfurt), pp. 170–189. Translation into English by Atmanspacher/Kronz.

Butterfield J. (1995): Worlds, Minds, and Quanta. *Proceedings of the Aristotelian Society* **69**, 113–158.

Chalmers D. (1996): *The Conscious Mind* (Oxford University Press, Oxford), particularly parts I and II.

Crutchfield J.P. (1994): Is anything ever new? Considering emergence. In *Complexity – Metaphors, Models, and Reality*, ed. by G.A. Cowan, D. Pines, and D. Meltzner (Addison Wesley, Reading), pp. 515–537.

Dawkins R. (1976): *The Selfish Gene* (Oxford University Press, Oxford).

Einstein A., Podolsky B., and Rosen N. (1935): Can quantum-mechanical description of physical reality be considered complete? *Phys. Rev.* **47**, 777–780.

Eisenhardt P. and Kurth D. (1993): *Emergenz und Dynamik* (Junghans, Cuxhaven).

d'Espagnat B. (1995): *Veiled Reality* (Addison-Wesley, Reading).

d'Espagnat B. (1999): Concepts of reality: Primas' nonstandard realism. This volume.

Fetzer J.H. and Almeder R.F. (1993): *Glossary of Epistemology/Philosophy of Science* (Paragon House, New York), p. 100f.

Gibson, R.F. (1995): Quine, Willard Van Orman. In *A Companion to Metaphysics*, ed. by J. Kim and E. Sosa (Blackwell, Oxford), pp. 426–428.

Hauhs M., Dörwald W., Kastner-Maresch A., and Lange H. (1998): The role of visualization in forest growth modeling. In *Proc. of a Conference on "Empirical and Process-Based Models for Forest Tree and Stand Growth Simulation"*. In press.

Heisenberg W. (1936): Prinzipielle Fragen der modernen Physik. In *Neuere Fortschritte in den exakten Wissenschaften. Fünf Wiener Vorträge, fünfter Zyklus* (Franz Deuticke, Leipzig), pp. 91–102.

Heisenberg W. (1969): *Der Teil und das Ganze* (Piper, München). Translation into English by Atmanspacher/Kronz.

Howard D. (1985): Einstein on locality and separability. *Stud. Hist. Phil. Sci.* **16**, 171–201.

Howard D. (1997): Space-time and separability: problems of identity and individuation in fundamental physics. In *Potentiality, Entanglement, and Passion-at-a-Distance*, ed. by R.S. Cohen, M. Horne, and J. Stachel (Kluwer, Dordrecht), pp. 113–141.

Jammer M. (1974): *The Philosophy of Quantum Mechanics* (Wiley, New York), Chaps. 5 and 6.

Jung C.G. (1971): Theoretische Überlegungen zum Wesen des Psychischen. In *Gesammelte Werke, Band 8* (Walter, Olten), footnote 129, pp. 261f. English translation: On the nature of the psyche. In *Collected Works, Vol. 8* (Princeton University Press, Princeton 1969), footnote 130, pp. 229f.

Kim J. (1984): Concepts of supervenience. *Philosophy and Phenomenological Research* **45**, 153–176.

Mermin N.D. (1998): What is quantum mechanics trying to tell us? *Am. J. Phys.* **66**, 753–767.

Müller-Herold U. (1980): Disjointness of β-KMS states with different chemical potential. *Lett. Math. Phys.* **4**, 45–48.

Neumann, J. von (1932): *Mathematische Grundlagen der Quantenmechanik* (Springer, Berlin). English translation: *Mathematical Foundations of Quantum Mechanics* (Princeton University Press, Princeton 1955).

Nunn C. (1998): Archetypes and memes. *Journal of Consciousness Studies* **5**, 344–354.

Petrosky T. and Prigogine I. (1997): The Liouville space extension of quantum mechanics. *Adv. Chem. Phys.* **XCIX**, 1–120, here p. 71.

Primas H. (1990): Mathematical and philosophical questions in the theory of open and macroscopic quantum systems. In *Sixty-Two Years of Uncertainty*, ed. by A.I. Miller (Plenum, New York), pp. 233–257.

Primas H. (1993): The Cartesian cut, the Heisenberg cut, and disentangled observers. In *Symposia on the Foundations of Modern Physics*, ed. by K.V. Laurikainen and C. Montonen (World Scientific, Singapore), pp. 245–269.

Primas H. (1994a): Endo- and exotheories of matter. In *Inside Versus Outside*, ed. by H. Atmanspacher and G.J. Dalenoort (Springer, Berlin), pp. 163–193.

Primas H. (1994b): Hierarchic quantum descriptions and their associated ontologies. In *Symposium on the Foundations of Modern Physics 1994*, ed. by K.V. Laurikainen, C. Montonen, and K. Sunnarborg (Editions Frontières, Gif-sur-Yvette), pp. 201–220.

Primas H. (1998): Emergence in exact natural sciences. *Acta Polytechnica Scandinavica* **Ma 91**, 83–98. See also Primas (1983), *Chemistry, Quantum Mechanics, and Reductionism* (Springer, Berlin), Chap. 6.

Putnam H. (1975): The meaning of 'meaning'. In *Philosophical Papers Vol. I: Mind, Language, and Reality* (Cambridge University Press, Cambridge), Chap. 12.

Putnam H. (1981): *Reason, Truth, and History* (Cambridge University Press, Cambridge).

Putnam H. (1987): *The Many Faces of Realism* (Open Court, LaSalle, Ill.).

Quine W.V.O. (1969): Ontological relativity. In *Ontological Relativity and Other Essays* (Columbia University Press, New York), pp. 26–68.

Scheibe E. (1973): *The Logical Analysis of Quantum Mechanics* (Pergamon, Oxford), pp. 82–88. German original: *Die kontingenten Aussagen der Physik* (Athenäum, Frankfurt 1964).

Scheibe E. (1997): *Die Reduktion physikalischer Theorien. Teil I: Grundlagen und elementare Theorie* (Springer, Berlin).

Schrödinger E. (1935): Die gegenwärtige Situation in der Quantenmechanik. *Naturwiss.* **23**, 807–812, 823–828, 844–849.

Searle J.R. (1983): *Intentionality* (Cambridge University Press, Cambridge).

Searle J.R. (1984): *Minds, Brains, and Science* (Harvard University Press, Cambridge).

Shimony A. (1984): Controllable and uncontrollable non-locality. In *Proceedings of the International Symposium "Foundations of Quantum Mechanics in the Light of New Technology"*, ed. by S. Kamefuchi et al. (Physical Society of Japan, Tokyo), pp. 225–230, here p. 227.

Silberstein M. (1998): Emergence and the mind-body problem. *Journal of Consciousness Studies* 5, 464–482.

Takesaki M. (1970): Disjointness of the KMS states of different temperatures. *Commun. Math. Phys.* **17**, 33–41.

Teller P. (1989): Relativity, relational holism, and the Bell inequalities. In *Philosophical Consequences of Quantum Theory*, ed. by J.T. Cushing and E. McMullin (University of Notre Dame Press, Notre Dame), pp. 208–223.

CONTEXTUAL BACKGROUND

Vittorio Hösle:
RATIONALISM, DETERMINISM, FREEDOM

In all areas related to mathematical practice – electrical engineering, quantum mechanics, mathematics itself – Hans Primas is known for taking great pains in his scientific work to meet the professional standards of the area in question. Yet his relationship to philosophy is ambiguous and cannot be judged adequately by purely professional criteria. Although philosophy was a concern of his from very early on, and although he built up a respectable collection of philosophical works for his private library, his attitude toward philosophy was and is that of a stonemason to a quarry: he extracted what seemed useful to him and paid little attention to the rest.

In particular he remained non-receptive to the distinctive features of traditional philosophy – its care and attention in dealing with texts and contexts, with the origins of the connections in philosophical thinking and the secular shifts of meaning in terminology – in short, non-receptive to those very things that make up the philological components of philosophy. The same applies to his relationship to philosophical systematics. What he did instead was to foster the enthusiasm of the philosophical layman for interesting individual thoughts and ideas, not unlike the musical dilettante and his delight in beautiful passages. Thus it comes as no surprise that there are no entrenched "Primas positions" in any philosophical questions. To put it in non-philosophical terms: he simply thinks against the grain – rather than to come up with ready-made philosophical propositions.

To a certain extent, this attitude reflects biographical developments. His early confrontation with philosophy was autodidactic, via the medium of books. The opportunity for regular, lively and direct philosophical discussion came about for Primas after he had turned fifty, with the appointment of Paul Feyerabend to the newly created chair for Philosophy of Science at the ETH in Zürich in 1979. Feyerabend's profoundly liberal and critical attitude was very much in tune with Primas' own convictions. Feyerabend had gained Primas' particular respect with his early work on Niels Bohr, a respect which was to grow and develop as a result of Feyerabend's independent thinking and the painstaking way he dealt with his sources. His whole personality – his humor, his intellectual stature, not to forget his human warmth – were a veritable gift not just for Primas but for the ETH Zürich as a whole.

For his book "Chemistry, Quantum Mechanics and Reductionism" (Primas 1981), Primas asked Feyerabend for critical comments, particularly with respect to the epistemological and ontological problems dealt with in the book. Feyerabend agreed, but confined himself to correcting the English grammar, style and punctuation rather than commenting on the actual subjects specifically requested. Feyerabend's reserve was certainly due to the fact that he was basically skeptical about strongly formalized theories such as algebraic quantum mechanics or formal philosophy of science. In his lectures he tended to present them as not very fruitful. At any rate he assigned them not to the creative phase of a discipline but to its academic phase of consolidation. It was only when Feyerabend brought Abner Shimony over from Boston as a guest professor for the winter semester of 1983/84 that Primas found a philosophically professional discussion partner in basic questions of quantum mechanics. This resulted in his first joint seminar with a philosopher – facilitated by the fact that Shimony's background is physics as well.

Feyerabend retired in 1991. Among all the potential candidates for the position, Primas was particularly enthusiastic about Vittorio Hösle, a brilliant philosopher from Tübingen and then a guest professor at the Department of Environmental Sciences at the ETH. Hösle was well-known for reviving elements of Hegel's philosophy, for his interest in philosophical aspects of ecology and ethics, as well as for his commitment to Apel's theory of "Letztbegründung". Hösle marks Primas' first encounter with a philosopher who had made his way completely outside the field of natural sciences. Hösle was the very embodiment of exotic intellect which had always fascinated Primas. Here for the first time was the opportunity for daily talks with a highly trained protagonist of the *philosophia perennis*. Primas was not only influenced by how well-versed Hösle was in the philosophical tradition but above all.by his incisive analytical powers.

After 1994, once the plans to bring Hösle to Zürich had finally failed, Primas embarked on his most recent and comprehensive experimentation with great philosophy when Michael Esfeld, a young philosopher from Münster, came to Zürich. After completing his doctorate on Hobbes, Esfeld had decided to undertake a critical examination of philosophical holism as part of his habilitation thesis and opted to work with Primas' research group. In the years that followed, the intense daily discussions between Primas and Esfeld became permanent fixtures in the physical chemistry institute. Even *after Primas' retirement, joint seminars by himself together with Esfeld provided young students with the opportunity to delve into the fundamental questions of quantum mechanics in the context of great philosophy.*

In these questions, the issue of determinism plays a crucial role. Despite claims to the contrary in popular literature in particular, the formalism of quantum mechanics does not exclude determinism. There are, of course, aspects that must be understood in a non-deterministic manner. Nevertheless, the pronounced trend in the 1920s to celebrate quantum mechanics as the savior of free will on the grounds that it disproves a deterministic world view (compare von Meyenn 1994) is of little attraction today.

Primas' distinction between endophysics and exophysics (Primas 1994) is based on a distinction between the realm of the ontic and the epistemic. Formally, the time evolution of (the ontic) states in endophysics is deterministic and is characterized by a unitary group. The time evolution of (the epistemic) states in exophysics exhibits non-deterministic aspects and presupposes a breaking of time-reversal symmetry. Hösle's contribution discusses philosophical arguments that can be related to these distinctions. More precisely, he addresses a number of classical philosophical arguments in favor of determinism, followed by the most important objections to these arguments, and offers strategies for how these objections can be overcome. For a fair account of all these issues, distinctions have to be made between different kinds of determinism. Hösle draws particular attention to one of them: a non-materialistic determinism that is very close to a position advocated by Leibniz.

References

Meyenn K. von, ed. (1994): *Quantenmechanik und Weimarer Republik* (Vieweg, Braunschweig).

Primas H. (1981): *Chemistry, Quantum Mechanics, and Reductionism* (Springer, Berlin).

Primas H. (1994): Endo- and exo-theories of matter. In *Inside Versus Outside*, ed. by H. Atmanspacher and G.J. Dalenoort (Springer, Berlin), pp. 163–193.

RATIONALISM, DETERMINISM, FREEDOM

VITTORIO HÖSLE
Forschungsinstitut für Philosophie,
Gerberstr. 26, D-30169 Hannover, Germany

An excellent book about determinism ends with the following advice (Earman 1986, p. 250): "As a practical 'solution' I recommend the ostrich tactic: don't think too closely or too long on the issues raised here, and in daily life continue with the presumption that the 'I' that chooses and the self to which we attach value judgments are autonomous. Let those who want to call themselves philosophers bear the risk to their mental health that comes from thinking too much about free will."

Belonging unfortunately to the high risk group called "philosophers", I want to dedicate at least some time and energy to reflecting upon the relation of freedom and determinism. Since the allegedly non-deterministic character of quantum mechanics was one of the causes of the interest which the theory sparked also outside of physics, such a reflection may deserve the attention of Hans Primas, with whom I had the pleasure to engage in several conversations which provided me with philosophical stimulation no less profound than that gained by the discussions I was able to conduct with some of the most challenging philosophers of our time. However, the conceptions I propose will necessarily leave Professor Primas unsatisfied. First, not being a natural scientist nor even a philosopher of natural science in the narrower sense of the word, I am not qualified to contribute to the question of whether quantum mechanics or the other relevant physical theories are really deterministic. As far as I understand, quantum mechanics as "endophysics" (to use Primas' term) is in Primas' interpretation a deterministic theory, but I am not able to judge how profoundly this result is altered by the non-deterministic character of what Primas calls "exophysics". The unsolved problem of finding a satisfying interpretation of quantum mechanics regards obviously not only the question of determinism but also the question of realism or phenomenalism and a variety of concrete ontological questions, such as the relation of parts and wholes, the violation of the locality principle, etc. The issues are closely interconnected which renders their analysis even more difficult: for one can hardly decide about the deterministic or non-deterministic character of a theory without certain ontological assumptions about which properties exist and can, or cannot, be determined.

But my incompetence in this field is not the main reason why I am afraid to disappoint Hans Primas. For, secondly, the whole tendency of my contribution is a defense of determinism – a philosophical position not regarded with sympathy by Primas. I do not claim in this paper its truth, but I want to make as strong as possible a case for it. At least I want to reject some gross misrepresentations existing about determinism which have led to exaggerated expectations with regard to quantum mechanics – as if only this theory could free us from a horrible and ultimately immoral vision of the word. There is no doubt about the fact that quantum theory signifies a profound challenge for ontology; but its philosophical importance would remain great enough, if, e.g., it enforced a revolution in mereology (the doctrine of the relation between parts and the whole) without, at the same time, undermining determinism or even realism. In any case, quantum theory is not the only way to overcome determinism, and perhaps there is no urgent need to overcome determinism.

Planck's and Einstein's seemingly stubborn refusal to embrace a non-deterministic interpretation of quantum mechanics is explained by psychological categories of reductionistic kind only by those persons who are not acquainted with the arguments in favor of determinism. Such arguments played a role already in ancient philosophy; it is, however, not difficult to see why in late antiquity and in the middle ages determinism became a more concrete position, attractive for philosophers of all three monotheistic religions. The doctrines of divine prescience and, even more, of divine omnipotence are, to say the least, more easily compatible with a deterministic universe than with a non-deterministic one, even if much effort was dedicated to various attempts to show that free will was not excluded by those two doctrines. In early modernity, finally, a determinist view of the world was largely accepted, partly on theological grounds – albeit the changes in the concept of God which took place in this time were profound –, partly also by agnostic or even atheistic positions. It is significant that one of the classics of determinism – perhaps the best known – antedates Newton's *Philosophiae Naturalis Principia Mathematica* of 1686: Spinoza's *Ethica* appeared (posthumously) in 1677. This shows that the triumph of determinism did not presuppose the emergence of classical mechanics, even if it would be misleading to deny that it was favored by it.

And yet there are philosophical arguments for determinism which do not depend on the state of the art of physics; and as determinism did not need Newtonian physics in order to be articulated as a philosophical position, it cannot be confuted either by the replacement of Newtonian physics by other paradigms. I do presuppose here, for the argument's sake, that Newtonian mechanics is a deterministic theory. This is, however, a position not shared by everybody. Earman denies it explicitly – even if by introducing solutions of the relevant mathematical equations which could be considered as not genuinely possible physically (Earman 1986, p. 33ff.). He regards special relativistic physics as more friendly toward determinism. But I shall not discuss these issues in this paper. (I have to ignore the problem of statistical laws to an even greater degree.)

I must restrict myself in this context to a very rough concept of determinism. By "determinism" I understand an ontological, not an epistemological position; predictability is therefore not a necessary moment in the concept here presupposed, even if it played an enormous role in its history.[1] The universe should be called deterministic if whatever will happen is already implicit in what has happened earlier and in the natural laws, if – to be more precise – the present is compatible with only one future development. In order to understand this definition it is not necessary to analyze in detail the difficult concept of causation. There are, however, obvious links between determinism and the principle of sufficient reason, even if it is wrong to regard the statement "every event has a cause" as the equivalent of determinism. This principle is only implied by determinism, since an earlier state of the world can be regarded as the cause of the later one, but the mere statement in itself does not yet imply determinism. This is at least true as long as we do not add something like: the same causes always have the same effects. One could imagine a world in which every event had a cause, but the same causes always produced different effects, and it would be absurd to call such a universe "deterministic".

Yet one could counter that this addition was already implicit in our proposition. In fact, this proposition, well understood, and even more determinism, is not as much an assertion about "causes" as an assertion about the universal character of certain relations. It presupposes a metaphysics of natural laws, another point which unfortunately has to be ignored in this paper. Here I want only to draw attention to the fact that the idea that the same causes must have the same effects is formally similar to the basic principle of ethics and law that equal cases have to be treated equally. It suggests some more general principle in the architectonics of our reason, a principle prior to the split between theoretical and practical reason. I am not presupposing in my approximate definition of determinism that natural laws are coextensive with physical laws – in fact, such an assumption would obviously be false. It can rightly be doubted whether the laws of chemistry

---

[1]Cf. Laplace's famous first chapter in the *Essai philosophique sur les probabilités* with the allusion to an "intelligence" able to predict everything.

can be reduced to those of physics (see Primas 1981); and it is manifest that the laws of psychology will never be reduced to the laws of physics, which do not contain concepts about the life of the mind.

The problem of determinism and freedom is not only, and perhaps not even mainly, a problem of the philosophy of physics. It is linked to various fields of philosophy – one could even risk the statement that there are few philosophical issues so tightly connected with so many other philosophical disciplines. As we shall see, epistemological options influence strongly the rationality or irrationality of deterministic assumptions; and the ancient, Diodorean form of determinism shows that questions of logic (particularly of modal logic) are also at stake.[2] Whether the world is deterministic or not is an important metaphysical issue. There are few features which characterize the structure of the world as profoundly as this one – the very concepts of being, substance, and time change if we accept determinism. But the general interest in determinism is not limited to nature in the narrower sense of the word; the existentially relevant question is whether it applies also to human actions. Their relation to nature and consciousness, along with the whole body-mind problem, is at the core of the determinism-freedom controversy. This controversy has important consequences for ethics, particularly for the doctrine of sanctions, and since decisive fundaments of our conceptions of law and state consist in questions of criminal law and of punishment, it is also of grave concern for legal and political philosophy. The links with philosophical theology concern in part eschatology (a topic related to the last one, but ignored in this paper), in part the relation of freedom and necessity in God.

In the following pages I shall first develop some classical arguments in favor of determinism (Section 1); secondly, I shall name the most important objections against it and the main strategies used to avoid it, and suggest why these strategies remain problematic (Section 2); finally, I shall explain why, after all, certain concerns of the critics of determinism can be dealt with in a subtler form of determinism (Section 3). In fact, one of the purposes of this essay is to distinguish different forms of determinism and to show that, while some forms are morally repugnant and even self-contradictory, others are more interesting and challenging. The conception proposed here of a non-materialistic determinism is very close to Leibniz's philosophy. It is something the truth of which I do not yet wish to assert but simply to discuss with Hans Primas and others who perhaps are too prone to deduce from the unacceptability of some forms of determinism the impossibility of all forms. In the course of my essay, I shall briefly sketch the positions of several philosophers of the past, for I have never been able

---

[2]On the logic behind ancient determinism – which I have to ignore in this essay – cf. Schuhl (1960) and Vuillemin (1984).

to convince myself that the later positions in the history of philosophy are always the better ones.

1.

One of the determining features of early modern philosophy is its rationalism. By "rationalism" we mean a strong trust in reason as the ultimate intellectual capacity. In a broader sense of the word, rationalism can be ascribed to those early modern authors who insist on the importance of experience. Therefore one may regard even the empiricists, Locke, Berkeley, and Hume, as belonging to the larger family of rationalists; for they regard it as rational to ground knowledge on experience. The central idea common to both rationalists (in the narrower sense) and empiricists is their opposition against authority and tradition as last justifications of validity claims. But the Latin word "ratio" from which "rationalism" stems means not only "reason"; it also means "cause" and "ground". Therefore rationalism usually is committed to the acceptance of some form of the principle of sufficient reason. The application of this multifaceted principle to events leads (with the above-mentioned addition) to determinism, and therefore determinism can be regarded as implied by rationalism in the broader sense of the word. The two forms of rationalism, however, are merely connected, not logically equivalent. Descartes is a rationalist in the epistemological sense of the word, but denies a determination of human actions, as he also rejects necessity in God. Hume is not a rationalist in the narrower sense of the epistemological term, but he defends some form of methodological determinism with regard to events (actions included). And yet, despite the logical independence of the two forms of rationalism, a certain connection between the two is obvious, and this renders it reasonable to begin with some general reflections in favor of epistemological rationalism.

The main reason for early modern rationalism was the profound desire for freedom. The powerful traditions of the middle ages were felt as limitations of intellectual and political freedom, whose vindication is the main purpose of a work as seminal as Spinoza's *Tractatus Theologico-Politicus*. Traditional beliefs ought to be justified, and their reasons should be clarified; this is one aspect of modern rationalism. At the same time the project emerged to liberate humankind from the seemingly perennial problems as hunger, plagues, and wars, which clearly limit the freedom of human actions. It was soon understood that only an unbiased analysis of nature and society could help to bring about this aim, and therefore an appeal to reason or to experience had to replace the traditional philosophies of nature, the human, and the state. One had to explain, to find the causes of such problems, in order to gain the chance to overcome them. But why are the projects of justification and explanation connected with "reason"? In a very rude approximation one may say that reason is the human capacity which asks the original question "why?". This capacity is already present in children, and the rejection of the corresponding question by the educators, although sometimes important and even inevitable for the stability of a society, is often detrimental to a philosophical development of the individual concerned. The question "why?" is indeed the link between rationalism in the epistemological sense and determinism, and the intermediate ring in the chain is the principle of sufficient reason.

The idea that every process presupposes a cause is found already in Plato,[3] and in Boethius we find the explicit argument that for something to happen without a cause would contradict the (Eleatic) principle that nothing comes from nothing.[4] In Spinoza's *Ethica* something like the principle of sufficient reason is stated as the third axiom of the first book, and even if Spinoza does not distinguish terminologically between causes and reasons, it is clear that in his vision of the ontological structure of the world a clear difference is made between things or events on the one hand and natural laws (the laws of the divine nature) on the other. Both are "caused", but in different ways: single events only by other events on the basis of general laws, general laws by other, more general laws culminating in the *causa sui* (which can be best understood in terms of the ontological proof). One could speak of horizontal and of vertical levels of "causality". We would say today that only events could have causes; laws, on the other hand, have reasons – if such reasons are conceivable at all. In fact, Spinoza's attempt to ground the general laws is utterly dissatisfying; it is not even clear whether he would like to defend the position of panlogicism, according to which the propositions about natural laws are analytical.

More elaborated are his assertions about the horizontal level of "causality". It is on this level that one can speak of determinism, even if the principle of sufficient reason encompasses both the horizontal and the vertical levels. Spinoza explicitly applies determinism to the two intelligible attributes of substance, to thought as well as to extension; every human action, every thought, is caused and predetermined. As a consequence of this determinism, Spinoza denies the character of substantiality to all but God – God is the only substance. This signifies an unutterably profound

---

[3]See his Philebos 26 e; Timaios 28 a,c.

[4]Boethius, *De consolatione philosophiae* V 1: "Nam nihil ex nihilo exsistere vera sententia est, cui nemo umquam veterum refragatus est, quamquam id illi non de operante principio, sed de materiali subiecto hoc omnium de natura rationum quasi quoddam iecerint fundamentum. At si nullis ex causis aliquid oriatur, id de nihilo ortum esse videbitur; quodsi hoc fieri nequit, ne casum quidem huius modi esse possibile est, qualem paulo ante definivimus." About a case so defined it is said just before: "Quis enim cohercente in ordinem cuncta deo locus esse ullus temeritati reliquus potest?"

break with Aristotelian ontology which has as its starting point the assumption of different, sensible substances (the Spinozian conception, however, has certain traits in common with Platonic metaphysics). For Aristotle as well as for his ancient and medieval followers, this plant, this cat, this man are entities within their own rights. For Spinoza they are only modi, local functions of a general extension which itself is but one attribute of a more universal structure which alone can cause itself: the divine substance. As caused, the single modi manifest only natural laws, but they do not subsist on their own and must therefore not be called "substances",[5] even if there are pragmatic reasons for the observer to isolate single "slices" of the *res extensa*.

Spinoza's physics is the strongest challenge to atomistic thinking one can conceive – it is a form of "field ontology" (Bennett 1984). Time does not create anything new; to understand the world as necessary means to understand it "sub specie aeternitatis" (II p. 44 cor. II). While God can be called "free" insofar as he exists based only on the necessity of his own nature (p. 17), a freedom of the will is impossible (I p. 32, II p. 48). Teleological arguments, which had been so important for ancient and medieval philosophy and science, are rejected in the appendix to the first book. With regard to the four traditional causes, Spinoza is interested mainly in the *causa efficiens*. Assertions about the teleological behavior of organisms or human beings have to be translated into an efficiently causalistic language. If something appears as accidental, i.e., as undetermined, this is due only to our own ignorance (I p. 33); and in fact one has to concede to Spinoza that it is difficult, if not impossible, to exclude the possibility of hidden parameters determining a process.

The profound influence which Spinoza exerted on Leibniz is manifest. Despite major differences between the personalities, the careers, the methods, and the styles of the two thinkers, one cannot deny that they share a similar program of rational theology and that, for both, rationalism in the epistemological sense implies determinism. At least the two following aspects distinguish, however, the contents of their philosophies. First, even if Leibniz shares Spinoza's rejection of atomism, he insists on the substantial character of the monads, the subjective centers acknowledged also by Spinoza, but immersed in the one attribute of thought.[6] There are not atoms, since matter can always be divided, but there are individual unities which serve as basis of the stream of consciousness and which are ontologically different from each other. Second, Leibniz is much more interested in the "vertical" series than is Spinoza, and he understands that the panlog-

---

[5]Cf. already the remark in Descartes, *Principia Philosophiae*, I 51.

[6]"Mihi nondum certum videtur, corpora esse substantias. Secus de mentibus", Leibniz writes in his notes on the *Ethica* (Leibniz 1890, I 145).

ical program cannot be fulfilled. The world as a whole is contingent, not logically necessary; therefore not logical, but only moral reasons which were rejected by Spinoza can explain why the world is as it is.

But all this does not deny the principle of sufficient reason – on the contrary it presupposes it. Leibniz is the first to bestow on it an importance equal to that of the principle of contradiction: "Nos raisonnements sont fondés sur deux grands Principes, celuy de la Contradiction ... Et celuy de la Raison suffisante, en vertu duquel nous considerons qu'aucun fait ne sauroit se trouver vray ou existant, aucune Enontiation veritable, sans qu'il y ait une raison suffisante, pourquoy il en soit ainsi et non pas autrement, quoyque ces raisons le plus souvent ne puissent point nous être connues."[7] As arguments in favor of this principle Leibniz gives the following: "Sans ce grand principe, nous ne pourrions jamais prouver l'existence de Dieu, et nous perdrions une infinité de raisonnements tres justes et tres utiles, dont il est le fondement: et il ne souffre aucune exception, autrement sa force seroit affoiblie. Aussi n'est il rien de si foible que ces systemes, où tout est chancelant et plein d'exceptions."[8] Leibniz fears that even a single exception to this principle would endanger the work of reason – for if we grant that there are facts without causes or reasons, then we can never exclude that the search for causes or reasons in a given case is meaningless and that those are right who are satisfied with simple facticity. In particular, he is afraid that his arguments for the existence of God would fail, if the principle lost its absolute validity.

The persuasive force of determinism must have been powerful indeed if not only the majority of the great philosophers of the seventeenth and eighteenth centuries were convinced of it, but if even the thinker to whom we owe the greatest revolution in our concept of causality remains committed to some form of determinism, at least as a form of thought of the human mind, without any ontological commitment. I must ignore here the difficult question of whether the epistemological projects of *A Treatise of Human Nature* and of *An Enquiry Concerning Human Understanding* are similar or at least compatible with each other; but it can be stated safely that David Hume regards in both works the idea of a free act of the will not determined by anything as an empty idea. Our whole social intercourse presupposes regularities in the behavior of our fellow human beings which are not significantly different from the regularities of natural bodies. "There is no philosopher, whose judgment is so riveted to this fantastical system of liberty, as not to acknowledge the force of moral evidence, and both in speculation and practice proceed upon it as upon a reasonable foundation. Now, moral evidence is nothing but a conclusion concerning the actions of

---

[7] *Monadology*, par. 31f. (Leibniz 1890, VI 612).
[8] *Théodicée*, par. 44 (Leibniz 1890, VI 127).

men, derived from the consideration of their motives, temper, and situation" (*Treatise*, II.III.I). Hume acknowledges liberty as "a power of acting or not acting according to the determinations of the will" (*Enquiry*, par. VIII, Part I). But this liberty shared by everybody who is not a prisoner and in chains is compatible with factors determining the will, being themselves functions of character and the situation. As John Locke put it, freedom implies the existence of the will which itself therefore cannot be called "free" (Locke, *An Essay Concerning Human Understanding*, II 21, particularly 16).

Hume is not the only philosopher who simplifies Spinoza's and Leibniz's subtle and complex conception of determinism by eliminating the ontological and the cosmological proof originally connected with the program of rationalism and by taking interest merely in the horizontal series of events. The atheistic and materialist philosophers of the eighteenth and the nineteenth centuries pursue a similar project, although on the basis of a dogmatic epistemology. Even if Schopenhauer can be regarded only in a very superficial way as a materialist, it makes sense to focus on his position as paradigmatic for this alternative type of determinism, since he has dedicated more explicit reflections to the principle of sufficient reason than all other materialist philosophers I know of. Furthermore, Schopenhauer has elaborated determinism in its application to human actions in a way which, even if it is not really original, is more concrete than all the earlier applications. I do not claim in the least to render justice to Schopenhauer's philosophy as a whole – I must ignore both its core, the metaphysics of the will, and the strange mixture of transcendental idealism and realism –, but I will try to name the main features of his peculiar type of determinism. In fact, his dissertation *Über die vierfache Wurzel des Satzes vom zureichenden Grunde* as well as his work *Über die Freiheit des menschlichen Willens* signify important steps in the history of our problem. When in popular discussions the ghost of determinism appears, one often associates with it fragments of argumentations developed by Schopenhauer.[9] He was influential also because he developed, as before him Hobbes and Spinoza, an immanentistic ethics which is built on merely descriptive sentences. The project of a justification of ethics was integrated into the "horizontal" deterministic world view.

The main intent of Schopenhauer's dissertation was the distinction of four classes of objects to which the principle of sufficient reason is applied, assuming different forms. Schopenhauer recognized that these forms have

[9]Similar arguments were developed later by Nietzsche whose awareness of the methodological problems linked with determinism is, however, very limited – Nietzsche was not, whatever his other merits may have been, talented in the "harder" branches of philosophy; his reflections on epistemology are dilettante and even self-contradictory.

a common feature; all four forms guarantee a unity of our conceptions, "vermöge welcher nichts für sich Bestehendes und Unabhängiges, auch nichts Einzelnes und Abgerissenes, Objekt für uns werden kann" (Schopenhauer 1977, V 41, par. 16). The first class is formed by empirical representations; for this class the principle of sufficient reason appears as the law of causality, as the *principium rationis sufficientis fiendi*. It states that every change is caused by another; there can be no first cause, but only an infinite series of events. Corollaries of this principle are the principle of inertia and the law of the conservation of substance. The changes presuppose something stable, namely matter, but also the forces of nature. The laws determining the actions of these forces are eternal and cannot be explained; the cosmological proof is rejected as strongly as the ontological one. Causality manifests itself in three forms: in the inorganic world as a cause in the narrower sense of the word, in plants as a stimulus and in animals (including humans) as a motive. Motives presuppose cognition and therefore a process of mediation which in humans is more complex than in other animals, but this does not change the deterministic character of the world. Given the character and the motive, the actions of a person follow with the same necessity as the fall of a body in a gravitational field. Only the principle of causality can transform the amorphous mass of sensations into a structured whole, into an objective world – through the assumption that the sensations have an external cause. With this reflection Schopenhauer wants to ground the apriori nature of the principle of causality, while he rejects Kant's demonstration which had insisted on the causal relation as the only way to guarantee an objective time order (Kant, *Kritik der reinen Vernunft*, A189ff/B232ff). Schopenhauer, it should be noted, does not aim at grounding the validity of the principle of sufficient reason. Such an attempt he regards even as absurd, since it would presuppose the principle it tried to prove.[10]

The second class of applications for the principle of sufficient reason dealt with by Schopenhauer consists of concepts; in this realm the principle becomes the *principium rationis sufficientis cognoscendi*. Judgments have to be justified, and Schopenhauer acknowledges four types of reasons for the truth of propositions, according to whether they are logical, empirical, transcendental, or metalogical truths. It is, however, clear that the justification soon comes to an end – either with an empirical fact or with a transcendental principle such as the principle of causality (Schopenhauer 1977, V 172, par. 50). – The third class consists of the apriori forms of intu-

---

[10]Schopenhauer 1977, V 38, par. 14: "Wer nun einen Beweis, d.i. die Darlegung eines Grundes, für ihn fordert, setzt ihn eben hiedurch schon als wahr voraus, ja, stützt seine Forderung eben auf diese Voraussetzung. Er geräth also in diesen Cirkel, daß er einen Beweis der Berechtigung, einen Beweis zu fordern, fordert."

itions; here Schopenhauer treats the demonstration of mathematical truths *(principium rationis sufficientis essendi)*. Of special interest is his idea that a cogent demonstration may nevertheless fail to grasp the ontological reason for a mathematical theorem. – The fourth class finally consists of one's own subjectivity, i.e., of the subject of one's own will. As the *principium rationis sufficientis agendi*, the principle for this class becomes the law of motivation already discussed within the first class, but now based on introspection. For Schopenhauer the will is the essential feature of a person; intellect and reason are only its tools. In *Über die Freiheit des menschlichen Willens*, Schopenhauer explains with the peculiar character of introspection the illusion of the free will, an illusion favored furthermore by the theological desire to find an exoneration of God in the free will, as far as it can be made responsible for evils.[11] It remains remarkable, however, that Schopenhauer's work ends, surprisingly, with an invocation of Kant's transcendental freedom, which he regards as necessary in order to allow for the possibility of moral imputation.

2.

This is indeed the first, if not the main objection against determinism: that it seems impossible to regard persons as "responsible" for their actions if whatever they have done and will do is predetermined by an earlier state of the world. Punishment and even weaker forms of social sanctions seem to presuppose that the person could have acted otherwise and therefore are not applied when the person was, e.g., forced to commit the reproachable deed; but in a certain sense of the expression this cannot have been the case if the universe is a deterministic system.[12] It is always safe to state that the person would have acted otherwise if he or she had taken another decision; but the problem is that he or she did not take the decision and could not have done so, given the laws of nature and an earlier state of the universe (Inwagen 1975). One can readily grant to Leibniz that the necessity at stake is not a logical one, but it is still a necessity, given the factual world in which we happen to live.

Now "compatibilism" has always taught that our system of sanctions can survive even if we accept the truth of determinism – we should only interpret it in a different way, related to the future and not to the past. The

[11] Schopenhauer 1977, VI 107, section IV. A similar explanation of the antideterministic convictions is given by Hume (*Treatise*, II.III.II).

[12] I do not distinguish in this essay between the proposition P: "Everything can be explained in a deterministic way" and the proposition Q: "Our world is a deterministic system". P and Q are not strictly equivalent (cf. Kutschera 1982, p. 279ff.), but one needs only to introduce all the independent laws presupposed by P as axioms of a theory in order to get Q.

so-called hard determinists, on the other hand, recognize an incompatibility between determinism and our practice of social sanctions; but, while they deduce from it that our practices are not appropriate, the indeterminists, on the contrary, see in our practices a proof of the absurdity of determinism. It is not only the love for habits inherited from times immemorial which prevents the indeterminists from reforming our practice of sanctions. They argue that the wrongdoer must deserve the punishment (which is not the case if the main justification for punishment is its deterrent effect) and that he is honored by being regarded as responsible. Even if Strawson belongs to the large family of compatibilists, one could try to find material against determinism in his brilliant essay *Freedom and Resentment* (Strawson 1962). In fact, the change from the reactive to the objective attitude which takes place when we come to the conviction that a certain person is a psychotic individual and which has the consequence that we no longer resent his behavior but regard it as some calamity which simply has to be brought under control is not necessarily in the interest of the wrongdoer. Certainly a world would be poorer emotionally in which individuals would know only objective attitudes toward each other, because they believed that reactive attitudes made no sense in a determinist universe.

But not only does our practice of sanctions seem to contradict deterministic beliefs – our self-understanding does so as well. We regard ourselves as free and react angrily against those who pretend to anticipate our decisions. The conviction of our own freedom is perhaps even a presupposition of our acting. Therefore, some critics of determinism argue that this position must lead to fatalism, namely to the refusal to act because whatever will happen will happen also without our own contribution. Particularly a physiological determinism which denies the causal power of mental states could invite quietism: people should simply attend to how their neurons will behave. In any case one cannot deny that the belief in our own freedom is one of the strongest intuitions we have. If we accept an intuitionistic epistemology, we should take such an intuition very seriously.

In general, an intuitionistic epistemology has to reject central tenets of rationalism in the epistemological sense of the word. On its basis, it is utterly impossible to ask, as Leibniz did, for a justification for every assertion, because this would lead to an infinite regress. There are final certitudes which cannot and need not be grounded. Even Schopenhauer defends such a position with regard to reasons, while at the same time he thinks that every change has a cause. Can the principle of causality itself be grounded? Schopenhauer denies this question; he tries to justify only its apriori status, not its validity. But if the principle of causality is not justified in a cogent way, why should we accept it? Nobody will deny that it is important and useful – yet this does not imply that we should sacrifice to it one of our most cherished intuitions, namely the intuition of our freedom. This is all the more convincing, as determinism is nothing more than a general program, not realized completely even for the domain of physics. We are very far from understanding all the factors determining human behavior – even if one can hardly deny that, e.g., criminology has shown us several causes of criminality, individual as well as social ones. It is, however, always possible to regard these causes as merely rendering a certain behavior more probable, not as sufficient causes, since nobody will, at least in the near future, be able to name all the factors which together could be a sufficient condition for a certain action.

However, this way of thinking shows only that we need not accept determinism – not that we must reject it. Furthermore, it shares the general weaknesses of intuitionism: first of all, that my intuitions are not necessarily also the intuitions of other persons (a point which endangers their claim of truth, since truth is necessarily intersubjective, and even elicits the suspicion that certainties may be functions of social factors, as power in its various forms, including education[13]); secondly, that even in my own set of intuitions there may be some which contradict each other – in this case, which one should I prefer? Intuitionism usually does not include a criterion for solving conflicts between contradicting intuitions. If we are frank, we must confess that most of us accept both some version of the principle of sufficient reason and the belief in our own freedom (perhaps also in the freedom of other human beings); and therefore the insistence on the second intuition can easily be countered by pointing to the first one.

It is in this context that one type of solution has been proposed which could be called "perspectivistic". According to this conception which exists in various variants, both determinism and indeterminism are necessary perspectives of our mind, but they are valid on different levels. The most famous version comes from Kant, whose subtle and even astute argument has the following form: the principle of causality is necessary for science, for physics as well as for psychology. But this necessity cannot be grounded on experience which already presupposes the principle; therefore it stems from reason. Since Kant conceives reason fundamentally as a subjective faculty, the causal determination concerns only the *phenomena*, i.e., the world as it appears to us, not the *noumena*, the world in itself. Therefore it is possible to assume that in the real world entities exist which are not determined by the past but still have the capacity to begin anew a causal series. This assumption – which in the domain of theoretical reason is only a possibility – becomes a necessity in the domain of practical reason. We must believe for moral reasons in the transcendental freedom of moral agents. Kant's

---

[13]This suspicion is omnipresent in Wittgenstein's *Über Gewißheit*.

subjective-idealist limitation of the claims of experience and science has been very influential, even in our century, and it is sufficient to liberate us from the threat of determinism. Those philosophers of quantum mechanics who interpret it in a non-realistic and in a non-deterministic way may have good reasons to do so, but they go too far if their only aim is to overcome determinism. A non-deterministic and realistic interpretation of the theory is sufficient for this purpose as well as a deterministic and phenomenalistic interpretation.

Kant's solution has a great merit lacking in many other "perspectivist" solutions (which after the "linguistic turn" and the late Wittgenstein now prefer to speak of different "language games"). Kant offers a clear hierarchy of the two positions. He presupposes – perhaps with a certain naiveté – the superiority of the point of view of practical reason, because he does not regard the moral law as something merely subjective, while he does so with natural science. If one does not share this presupposition, one would be at a loss; for we would have no criterion to decide which perspective is, in the last instance, the right one. In fact nothing is achieved by granting that there are different legitimate perspectives. As long as they cannot be simultaneously true, one has to choose between them.

In his famous lecture *Vom Wesen der Willensfreiheit* (Planck 1979), Max Planck proposed a solution of our dilemma which insisted on the deterministic character of the laws of physics (also of quantum mechanics) and of nature in general while granting at the same time the irreducible freedom of the subjective will from the point of view of introspection. "Von außen, objektiv betrachtet, ist der Wille kausal gebunden; von innen, subjektiv betrachtet ist der Wille frei" (Planck 1979, p. 310). This is supposed not to lead to a contradiction; Planck even appeals to relativity theory to explain why from different systems different statements can be made with exactly the same right. But the comparison is grossly misleading – for movement, not, however, determination, is a relative category. Planck himself seems to recognize this when he states that the acting individual only feels free (Planck 1979, p. 312) – which, of course, is compatible with his or her being determined. One cannot be at the same time determined and not determined – one can only say that there are two different positions with regard to this issue. But then the question unavoidably arises: which position is the right one? Kant tries to answer this question, Planck does not.

While compatibilism aims at showing that determinism does not endanger our common intuitions and while perspectivism wants to demonstrate that determinism and the belief in free will are positions appropriate to different levels of our thought, there are also attempts to confute determinism, to show that it is wrong and perhaps even self-contradictory. It is tne achievement of Ulrich Pothast to have categorized the great number of such arguments, i.e., to have reduced them to some few elementary types. It is further to his credit that he has shown that none of these arguments is really cogent. His book, *Die Unzulänglichkeit der Freiheitsbeweise* (Pothast 1980),[14] deserves particular praise because of the remarkable capacity of treating with equal competence both analytical and "continental" arguments; he succeeds in showing a common logical structure wrapped in very different languages. What are the elementary types?

One group of authors insists on the impossibility to predict the future with absolute certainty. Such an impossibility is, by the way, an immediate consequence of one physical theory of recent decades, namely chaos theory: infinitesimal deviations from a given value can lead to a very different behavior of the relevant physical system. Since there are limits to our approximation of physical values in measurement, humans will never be able to anticipate which course the system will take. But it is very easy to object against this argument that no reasonable person has ever defended epistemological determinism and that epistemological indeterminism does not entail ontological indeterminism. Quantum mechanics might lead to a destruction of ontological determinism; chaos theory, on its own, certainly does not.

More interesting are those arguments which do not make use of concrete physical theories, but are more general. So Popper claims to dispose of an argument valid also within classical mechanics against determinism, an argument based on the impossibility of a complete description of the world by a system which always must leave itself out, at least in some aspect (Popper 1950). Furthermore one encounters the argument that one cannot know today how one will decide in ten days – otherwise either the decision would be taken today and not in ten days, or the categorical difference between prognosis and decision would be undermined.[15] The argument does make questionable presuppositions; but even if we grant them, it will show as little as Popper's that my decision is not predetermined, but only that I cannot know it before I take it. Nobody denies that the I-perspective which for so many aspects is indeed a unique privilege implies also certain restraints: one cannot objectify oneself as other persons. But it remains unintelligible why this should prove freedom in an ontological sense of the word.

The second group works with the distinction between reasons and causes (see, e.g., Melden 1967 and Kenny 1975). To understand an action, it is

[14]Pothast edited also important texts on our problem in the volume *Seminar: Freies Handeln und Determinismus* (Pothast 1978). I owe much to these two books.
[15]Besides Planck in the above mentioned essay also Hampshire has defended the argument, compare, for example, Hampshire (1965), chap. 3.

argued, we must recognize the intentional character of psychic acts; but the logic of intentionality is completely different from the logic of causes. The merit of these authors is that they clearly recognize that indeterminism is in the best of cases a necessary, but never a sufficient presupposition of freedom. To act in an utterly unpredictable way is not yet to act freely; an action can be regarded as my action only if it is willed and caused by me. Some philosophers have tried to deduce from this fact that free will even involves determination (cf. Hobart 1934). But even if this claim is too far-reaching, it is clear that a model of self-determination is needed if we want to have more than hazardous behavior. If this self-determination is to transcend determinism, it must have certain further qualifications which, however, are hard to specify and to conceive because we have given up the categorical thread of causality.[16] We have to express ourselves in the following way: The action is caused by the person, and the person itself is utterly free in causing the action (a quality several philosophers and theologians would ascribe only to God); a causal explanation of the relevant act is therefore inconceivable.

I will not discuss such attempts (for an important example, see Chisholm 1964), but return to the clear distinction between causes and reasons. Our authors analyze the peculiar nature of responsible decisions in which arguments *pro* and *con* for a possible action always play a role. They are completely right to reject a naturalistic ontology which knows only causes and ignores reasons. Not only could such an ontology never be justified, because justifications presuppose reasons; such an ontology would deny the difference between humans and the other animals – for humans are animals able to grasp reasons. But the difference between reasons and causes, as important as it is, does not imply indeterministic consequences. Reasons as such cannot cause anything; but it is the understanding of reasons, i.e., a mental act (or its physical pendant), which, together with a series of other factors, may cause human behavior. The free person, according to a profound concept of freedom, is not the person whose actions cannot be accounted for. The free person is the person who follows the strongest reasons. The capacity to follow reasons, however, may well be caused by different factors, such as education, features of personal character, intelligence, etc. In any case the essential distinction between reasons and causes can be easily integrated into a deterministic system (see Hösle 1997a, p. 234ff.).

----

[16]Some philosophers of causality argue that it is our own intervention in the physical realm which generates the idea of causality which could never be deduced from the relation between bodies and that therefore our self-determination is a clearer concept than that of normal causality. Now this may be genetically true, but does not yet solve the validity question; furthermore, the allegedly more evident idea has to do with the cause-effect connection between our body and an external body, not with the internal self-determination of the self.

In order to confute a position one may try to show that it contradicts some assumptions regarded as true by the opponent, who, however, might be willing to give up these assumptions if this is the price he has to pay for sticking to that position. It is therefore better if the critic can show that a position is immediately self-contradictory. The contradiction may subsist on the propositional level, or it may be a contradiction between the position itself and the presuppositions necessary for its performance. Arguments showing a contradiction of the last type have been called "transcendental", and they are a powerful tool for grounding fundamental principles of epistemology and of metaphysics. It is therefore no surprise that such a transcendental argument for freedom has been elaborated in the discussion about determinism, and – what is particularly interesting – both by pre-analytical and by analytical philosophers (Rickert 1921, p. 302f.; Boyle et al. 1976).

The argument has the following structure: In order to claim the truth of determinism, one has to appeal to a norm of rationality. Therefore one must in principle be able to judge according to this norm. But this presupposes freedom: if our mental acts and our behavior were only functions of a blind causal process, we could not determine ourselves according to truth and rationality. The argument is indeed sufficient to reject a naturalistic determinism which can never arrive at justifying truth claims. It is impossible to say: I am determined by a blind series of causes, and I state this as true – for the statement can be taken seriously only if my statement is more than the function of a causal process. But does anything exclude that it is both the function of a causal process and something else? Could it not be that I am determined to argue well and to accept the best argument which I can find? And would not such a determination be a better reason to trust me than a *liberum arbitrium indifferentiae* which would still grant me the possibility to contradict the insights of reason?

I am even willing to go a step further and concede that the fact that I am able to argue and that there are in general persons able to follow reasons cannot be a contingent truth. This has important consequences with regard to the understanding of modalities which cannot be analyzed here; the task to bridge the gap between the formal and the transcendental concept of necessity has yet to be well understood. But if I accept the idea that the world must contain persons able to follow reasons and to discuss rationally truth claims with regard to questions such as determinism, then nothing prevents us from believing that the existence of such persons is brought about by a causal process. Indeterminism is not the only possible consequence which can be drawn from our argument; a conceivable solution is also a non-materialistic determinism which accepts teleological restraints to the whole system of the world, without, however, violating the causal order by the concrete interference of ends. The ends determine the laws of nature and the initial conditions – but with this their task is fulfilled.

3.

This corresponds quite thoroughly to the complex deterministic conception of Leibniz to which I shall return after my further discussion of the indeterministic strategies will have shown that their arguments are not at all cogent. Not even the Kantian strategy is really convincing. On the one hand, the price Kant has to pay for his transcendental idealism is high; he has to assume a separate world of unintelligible entities, the things in themselves, about which he nevertheless has to make statements. On the other hand, it remains dubious whether indeterminism is really necessary for the general traits of our practice of sanctions (and it cannot be accepted as a valid argument for indeterminism that without it our more cruel ways of punishing could no longer be justified[17]). It may be true that the criminal could not act otherwise, given the person he or she is; but it still remains true that this is the person he or she is. Therefore criminals cannot complain about their punishment, for they could do so only in the name of a metaphysics which denies the substantiality of their own selves, and then it would no longer be they who complained about it, and scarcely could they claim to have any rights. In a statement like "I believe that my actions are the product of causes existing already long before my birth", there is a complete abstraction from the own actions; but as even an author such as Pothast writes (Pothast 1980, p. 392): "Mit ebenso viel oder wenig Recht, wie sich jemand mittels dieser Beschreibung von seinen Entscheidungen distanziert, könnte er sich überhaupt von sich selbst distanzieren."

I would go even further: they could not even distance themselves from themselves, for such an act of distancing presupposes, on the performative level, a subjective act of the I. The I cannot be eluded; in this unavoidability there is a hint of the absolute which justifies the peculiar rights of the I and which may claim substantiality even if the concept in this use has a meaning very different from that in Aristotelian metaphysics. A society would collapse not only empirically, but in its own claim to be taken seriously under a moral perspective if it would accept a distancing from one's own actions such as the one suggested. Nevertheless, Strawson is right when he writes that the decision for an objective behavior rather than a reactive one has nothing to do with the problem of determinism. There is something in the essence of a person which determines whether we resent his or her actions or begin to objectify his or her behavior, and it may well be that this

[17]Theodor W. Adorno – whose contribution to our problem is generally garrulous and confuse – rightly reproaches Kant for a repressive desire to punish (Adorno 1973, p. 257).

something is determined in both cases. Therefore we can at the same time in which we resent the behavior of wrongdoers have a certain compassion with them; and this compassion should prevent us from applying sanctions which destroy them. I believe that the last aim of punishment is the future prevention of evils, even if this aim can be achieved only if we do so as if the criminal had been free in the past. True freedom is moral rationality, but we have a chance to raise wrongdoers to it only if we presume that they are free also in evil acts (Hösle 1997b, p. 833ff.). As compassion toward the wrongdoer, also a certain modesty with regard to oneself and even a sense of gratitude toward the creator will be peculiar to the moral person who is convinced of this kind of determinism – qualities which should, I think, recommend the more considered forms of determinism.

Furthermore it is clear that epistemological indeterminism is indeed a presupposition of our actions – but this does not yet prove ontological determinism. Precisely because we do not know what will happen, we have a duty to do our utmost to realize the good. Fatalism is not at all implied by ontological determinism, but only by epistemological determinism. Leibniz dedicated much of the energy of his subtle and noble mind to confute "la raison paresseuse" which sees in determinism an invitation to laziness. Yet determinism does not teach that something will happen as such, but only that something will happen if something else happens, e.g., if our action takes place; and ontological determinism does not claim to anticipate our actions. Leibniz rightly asserts that the argument – or, better, sophism – proves too much: Not even the defenders of fatalism will drink a poison, saying "If I have to die, I'll die, if not, I won't". "C'est qu'il est faux que l'évenement arrive quoyqu'on fasse; il arrivera, parce qu'on fait ce qui y mene; et si l'évenement est écrit, la cause qui le fera arriver, est écrite aussi. Ainsi la liaison des effects et des causes, bien loin d'établir la doctrine d'une necessité prejudiciable à la practique, sert à la détruire."[18]

I am willing to concede that the belief not to be determined in one's own actions and decisions is a necessary and even a healthy illusion. When I have to make a decision, a reflection on the causes determining me is utterly useless, because I have to concentrate on the relevant reasons. But this does not signify that the causes no longer exist. As psychoanalysis teaches us, even unconscious causes may continue to operate. Therefore I should try to become conscious of the unconscious motives of my behavior; this may even lead to an alteration of my motivational structure. In fact, those determinists err who teach that we can only act as we will but that we cannot will what we will – one can try to change one's desires, even if this is a long and tortuous process. Yet even if such a capacity exists and can

---

[18] *Théodicée*, Preface (Leibniz 1890, VI 33).

be called a freedom of the second degree, I do not want to contradict those determinists who insist on the fact that there are causes for the existence of this capacity in certain human beings and its non-existence in others.

The arguments against determinism may all seem weak; but if there are not positive arguments for it, why should we take it so seriously? One cannot agree with Schopenhauer that the principle of sufficient reason simply needs not be proved – it would be awkward it the principle asked for something which it itself would not satisfy. Schopenhauer sees something important when he states that a proof for it would be circular, but he does not grasp that the impossibility to ground a principle otherwise than in a circular way may be a mark of its being a first principle. However, this circle has to be distinguished from the vicious circle, also because this trait has to be accompanied by the further one that the principle cannot be denied without being simultaneously presupposed (Apel 1976, Hösle 1997a, pp. 163ff). This applies, e.g., to the principle of contradiction. However, one has to concede that the principle of sufficient reason does not enjoy the same logical status, for it is not contradictory to deny it, even if every attempt to deny it with the help of an argument will presuppose some form of it.[19] Leibniz in any case does not dispose of satisfying reflections on the foundation of principles, and certainly one of the greatest lacks in his metaphysics is his complete inability to ground (and even name in a satisfying way) the moral criteria which allow God to choose between the possible worlds. The idea of a transcendental foundation of ethics is completely alien to Leibniz, even if ethics or, better, axiology in the framework of Leibniz's metaphysics acquires the status of First Philosophy.

Nevertheless, Leibniz has some important arguments which explain why he sticks so stubbornly to this principle, also and particularly in the context of his philosophical theology. Perhaps even greater than in the *ordo cognoscendi* is the function of this principle in the *ordo essendi* of Leibniz's rational theology: God himself has to apply it in the creation of the world. Leibniz rejects the idea that God could have created another world as well as the real one, that there was no sufficient reason to prefer the existing one to possible alternatives. He concedes, as I have already said, that there are no logical reasons for the necessity of the actual world, but he insists on the existence of moral reasons determining the choice of the real world. God must create the best possible world,[20] although this necessity is obviously not anything external to God, but God's own self-determination. He regards as particularly repellent the voluntaristic conception according to

---

[19]See already Sextus Empiricus, *Adversus Mathematicos*, IX 204.
[20]The concept of the best possible world presupposes quite a lot – e.g., that there is only one world with a maximal axiological value. This presupposition is very strong, but I shall not discuss possible alternatives here.

which not only the structure of the world but also the moral duties depend on an arbitrary act of will of God. Such a position – as it was defended by Descartes and Hobbes – would render God indistinguishable from an almighty tyrant, for no moral criterion beyond the divine power would exist in order to evaluate it.[21] Leibniz regards as even more repugnant the conception that there are objective criteria of good and evil, but that the capacity to violate the moral norms is the true expression of freedom and something higher than the obedience toward them – be it God or the human who owns this "freedom". This conception is familiar to the friends of German literature from Eberward Schleppfuß's lectures, described in the thirteenth (!) chapter of Thomas Mann's *Doktor Faustus*. Schleppfuß is Privatdozent of theology at Halle, but his lectures even more than his name and appearance suggest that he is one of the manifestations of the devil in Mann's sublime novel.

According to Leibniz, freedom and moral necessity coincide not only in God, but also in moral human beings. Leibniz rejects passionately the idea that the irrational and immoral person could claim to have more freedom than the person dedicated to reason. In the *Nouveaux Essais*, Philalèthe states "que Dieu luy même ne sauroit choisir ce qui n'est pas bon et que la liberté de cet Estre tout puissant ne l'empeche pas d'estre determiné par ce qui est le meilleur". And he adds: "Estre determiné par la raison au meilleur, c'est estre le plus libre. Quelqu'un voudroit-il estre imbecille, par cette raison, qu'un imbecille est moins determiné par de sages reflexions, qu'un homme de bon sens? Si la liberté consiste à secouer le joug de la raison, les foux et les insensés seront les seuls libres, mais je ne crois pourtant pas que pour l'amour d'une telle liberté personne voulût estre fou, hormis celuy qui l'est déja" (II 21, par. 49f.; Leibniz 1890, V 184). Like Spinoza, Leibniz cannot take the idea of the *liberum arbitrium indifferentiae* seriously: The determinant factors are unknown to us, but this does not mean that they do not exist. Even when one commits an act contrary to one's own interest in order to show one's freedom, one is in fact determined by the will to demonstrate one's freedom.[22] The good person cannot act otherwise than morally, as even Schelling will acknowledge in *Über das Wesen der menschlichen Freiheit*, which in many aspects breaks with the earlier tradition of rational theology. But Schelling agrees with regard to the individual (Schelling 1974, p. 111): "Schon der Wortbedeutung nach läßt Religiosität keine Wahl zwischen Entgegengesetzten zu, kein *aequilibrium arbitrii* (die Pest aller Moral), sondern nur die höchste Entschiedenheit für das Rechte, ohne alle Wahl." At the same time, Schelling's work constitutes an im-

---

[21] *Théodicée*, Discours préliminaire (Leibniz 1890, VI 71, par. 37).

[22] Par. 25; Leibniz 1890, V 168. The same argument is found in Schopenhauer's *Über die Freiheit des menschlichen Willens* (Schopenhauer 1977, VI 82, section III).

portant step forward in the phenomenology of evil – an advance which in principle can be integrated into a deterministic system.

The challenge of Leibniz's determinism is all the greater, as he rejects – again with Spinoza – any form of "interactionism". Since Descartes' monumental discovery that mental states and physical states cannot be reduced to each other but have to be characterized by two different, mutually exclusive classes of predicates, the body-mind-problem has been vexing philosophers. If we accept such a dualism (which I regard as unavoidable, even if I readily grant that it, alas!, creates many problems happily alien to ancient and medieval philosophy), there are four combinatorial possibilities to determine the relation between mental and physical states. Either there are causal interactions between the two domains (this position will be called here "interactionism"; sometimes this specific possibility is meant when one speaks of "dualism" in a narrower sense of the word), or the mental states are functions of the physical states (epiphenomenalism), or the physical states are projections of the mental acts (subjective idealism), or there exist causal connections within the two domains, but not directly from one to the other.

While interactionism seems the most natural position and has been reworked in the last decades by several philosophers (as, e.g., Popper and Eccles 1977, Jonas 1981), the objections against it, as they were understood already in the seventeenth century, are powerful. First of all, it is not clear how a causal relation between two domains so different could be conceived without undermining the ontological difference between them.[23] And secondly, the assumption that a physical movement could be caused by something immaterial endangers the physical laws of conservation (and even opens the door to magical beliefs[24]). Even if Descartes is one of the earliest philosophers who sees in the conservation laws of physics an expression of God's immutability and even if he adduces as an example the conservation of momentum,[25] he still regards momentum as a scalar magnitude, not as a vector. Therefore he can believe that the *res cogitans* may influence the mere direction of the *spiritus animales* without altering the quantity of momentum. But already in the seventeenth century the vectorial nature of momentum was discovered, and therefore the Cartesian belief of a possible change of direction without a violation of the corresponding conservation law had to be abandoned.[26] Occasionalism was one of the at-

----

[23]See already Spinoza's criticism of Descartes in the preface to the fifth book of the *Ethica*.

[24]Schopenhauer interprets with a certain plausibility parapsychological phenomena as an extension of the power of the mind upon the own body to other bodies in *Über den Willen in der Natur* (Schopenhauer 1977, V 307).

[25]*Principia Philosophiae* II 36.

[26]One could try to argue that the conservation laws are only idealizations, that the

tempts to cope with the new situation, and there can hardly be a doubt that, compared with it, Leibniz's doctrine of the preestablished harmony represents a considerable progress.[27]

According to Leibniz, no physical event – no action either – is caused by anything mental, but only by antecedent physical states. However, the world is structured in such a way that simultaneously with certain mental events the corresponding physical events take place, and vice versa. When I want to lift my arm, it is not my will which lifts it, for my will cannot cause anything physical, but only other mental states, as, e.g., satisfaction or frustration; the cause of the lifting is a physical state (e.g., the state of my brain). But it is not by chance that I usually lift my arm when I want to lift it – God guarantees such a correspondence between the mental and the physical states. The development of the mental states follows a special logic grounded in the peculiar nature of the single monad: This distinguishes Leibniz's position distinctly from the epiphenomenalist one. While epiphenomenalism must deny even the ontological continuity of mental life (every mental state is caused by a physical one and unable to produce another mental state) and must deny even more forcefully its power to act in the physical world, Leibniz strongly defends the first quality – even to the point that all propositions about mental acts of a monad are ultimately analytical propositions. And even if with regard to the monad's power to act Leibniz denies a direct impact on physical states, he certainly recognizes that the creation of the physical world by God was done with the intention to guarantee a correspondence with the mental states of the different monads. This means that the essence of the created monads – not, however, acts of the existing monads – does determine physical events, via the choice of the best possible world by God. It would go beyond the task of this article to analyze Leibniz's conception in greater detail – its main problems are that it is not easy to justify on its basis the assumption of an external world and even more difficult to imagine sufficient reasons for the correspondence of physical and mental states. But I think that parallelism deserves a revival in our time, for which the body-mind problem has regained an importance comparable to the relevance it had already attained in the seventeenth century.

I repeat that it was not the purpose of my essay to argue for the truth of determinism. I regard a theory which assumes a plurality of entities capable of beginning a causal series on their own as a very serious philosophical

changes caused by the *res cogitans* are minimal or even, as Hans Jonas, that there is a deviation from the laws in both directions – both in perception and in action – so that the two forms cancel each other. But all these solutions are utterly unsatisfying.

[27]Cf. Leibniz's criticism of occasionalism: *Systeme nouveau de la nature et de la communication des substances* (Leibniz 1890, IV 483): "pour resoudre des problemes, il n'est pas assez d'employer la cause generale, et de faire venir ce qu'on appelle Deum ex machina."

alternative, and I am often tempted by it (even if I do not believe that it really solves the theodicy problem with any greater ease: the pain inflicted by nature on sentient organisms is still huge). However, I think that it then becomes unavoidable to deny God's omnipotence and perhaps also his omniscience, as Jonas, one of the great philosophical theologians of our time, had the courage to do (Jonas 1987). The price for this step is high, e.g. in terms of the philosophy of history. But it is certainly worth trying to develop the strongest possible arguments for such a theory. What I wanted to show in this essay is that there are different types of determinism, and that a determinism of the Leibnizian type copes remarkably well with some of the questions other determinisms are unable to answer. It will not be surprising if I finish with the wish that this type of philosophy, too, might be made as cogent as is possible and needed in our time, so that the competition between the two systems might be both fair and interesting.

Acknowledgments: I want to thank my dear colleague Prof. Dr. Richard Schenk for various discussions on the subject.

References

Adorno T.W. (1973): *Negative Dialektik* (Suhrkamp, Frankfurt).
Apel K.-O. (1976): Das Problem der philosophischen Letztbegründung im Lichte einer transzendentalen Sprachpragmatik. In *Sprache und Erkenntnis, Festschrift für G. Frey*, hg. von B. Kanitscheider (Amoe, Innsbruck), pp. 55–82.
Bennett J. (1984): *A Study of Spinoza's "Ethics"* (Hackett, Indianapolis).
Boyle J., Grisez G., and Tollefsen O. (1976): *Free Choice* (University of Notre Dame Press, Notre Dame).
Chisholm R.M. (1964): *Human Freedom and the Self* (Kansas City).
Earman J. (1986): *A Primer on Determinism* (Reidel, Dordrecht).
Hampshire S. (1965): *Freedom of the Individual* (Chatto &Windus, London).
Hobart R.E. (1934): Free will as involving determination and inconceivable without it. *Mind* **43**, 1–27.
Hösle V. (1997a): *Die Krise der Gegenwart und die Verantwortung der Philosophie* (Beck, München), 3rd edition.
Hösle V. (1997b): *Moral und Politik* (Beck, München).
Inwagen P. van (1975): The incompatibility of free will and determinism. *Philosophical Studies* **27**, 185–199.
Jonas H. (1981): *Macht oder Ohnmacht der Subjektivität?* (Insel, Frankfurt).
Jonas H. (1987): *Der Gottesbegriff nach Auschwitz* (Suhrkamp, Frankfurt).
Kenny A. (1975): *Will, Freedom, and Power* (Blackwell, Oxford).
Kutschera F. von (1982): *Grundfragen der Erkenntnistheorie* (deGruyter, Berlin).
Leibniz G.W. (1875–1890): *Die philosophischen Schriften*, ed. by C.J. Gerhardt (Berlin). Reprinted 1961 by Olms, Hildesheim.
Melden A.I. (1967): *Free Action* (Routledge, London).
Planck M. (1979): Vom Wesen der Willensfreiheit. In *Vorträge und Erinnerungen* (Wissenschaftliche Buchgesellschaft, Darmstadt), pp. 301–317.
Popper K.R. (1950): Indeterminism in quantum physics and in classical physics. *British Journal for the Philosophy of Science* **1**, 117–133; 173–195.
Popper K.R. and Eccles J. (1977): *The Self and Its Brain* (Springer, Berlin).
Pothast U. (1978): *Seminar: Freies Handeln und Determinismus* (Suhrkamp, Frankfurt).

Pothast U. (1980): *Die Unzulänglichkeit der Freiheitsbeweise* (Suhrkamp, Frankfurt).

Primas H. (1981): *Chemistry, Quantum Mechanics, and Reductionism* (Springer, Berlin).

Rickert H. (1921): *System der Philosophie. Erster Teil: Allgemeine Grundlegung der Philosophie* (Mohr, Tübingen).

Schelling F.W.J. (1974): *Über das Wesen der menschlichen Freiheit* (Reclam, Stuttgart).

Schopenhauer A. (1977): *Werke in zehn Bänden* (Diogenes, Zürich).

Schuhl P.M. (1960): *Le Dominateur et les Possibles* (Presses Universitaires de France, Paris).

Strawson P.F. (1962): Freedom and resentment. *Proceedings of the British Academy* 48, 187–211.

Vuillemin J. (1984): *Nécessité ou Contingence: L'aporie de Diodore et les Systèmes Philosophiques* (Edition de Minuit, Paris).

CONTEXTUAL BACKGROUND

Wilhelm Just:
CUTS AND SYMBOLS

For quite a time it has been well-known among his colleagues and students that Hans Primas has a long-standing interest in the school of psychology developed by Carl Gustav Jung, also known as analytical psychology. More precisely, it is the interaction between Jung and Wolfgang Pauli to which Primas has devoted and still devotes a considerable amount of time and attention. The way in which this issue comes to the fore in discussions with him as well as in his publications, demonstrates impressively a unique blend of strong criticism of dogmatic reasoning and an enormous openness toward ideas which he finds attractive.

This interest dates back to the days when he was attending the Vocational School in Zurich. When it turned out that Primas had a knowledge of chemistry far above average for a laboratory technician, he was allowed to read books during the lessons and the practical courses (what a good school!). This led to occasional differences of opinion. Primas himself reports (Primas 1991): "One of my teachers, Prof. Emil J. Walter, spent hours discussing with me, trying to convince me that Emanuel Kant (Vom ewigen Frieden) and Carl Gustav Jung (Psychologie und Alchemie) were not suitable reading matter for an eighteen-year-old; instead he recommended Karl Marx and Sigmund Freud – albeit not very successfully. Only many years after I had become acquainted with Pauli in his lectures on theoretical physics did I learn that the most impressive dreams in 'Psychology and Alchemy' were actually Pauli's."

Following Wolfgang Pauli's strong response to the ideas of Jung (compare the relevant chapters in Pauli 1994 and the overview by Atmanspacher and Primas 1996), Primas uses, transfers, and applies certain Jungian concepts – not all – in respects he thinks are important for both science and society. In these contexts, he is mainly concerned with the structural, non-personal aspects of Jungian psychology rather than those clinical domains dealing with therapy and analysis. The first publication by Primas in which he explicitly addresses elements of Jung's psychology is his book *Chemistry, Quantum Mechanics and Reductionism* (Primas 1981). Chapter 2 of this book contains section titles such as "There is no insight without inner pictures" or "Are theories dangerous?" – subjects clearly indicating the decisive role that can be played in scientific work by what is referred to in Jung's terminology as the collective unconscious.

Primas serves as a board member of the Jung Institute in Küsnacht and he serves as a member of the Pauli Committee at CERN in Geneva. He was one of the main initiators of the conference on Pauli and Jung that took place at Monte Verita in 1993 (see Atmanspacher et al. 1995). The dialog which Pauli and Jung pursued from 1930 to 1958 touched on a number of extremely interesting and important subjects (see Meier 1992). Primas belongs to the small group of respected scientists who are indefatigable in their endeavours to make it commonly known that certain elements of this dialog should not be dismissed as a historical curiosity but at least be taken as a rich source of inspiration for future developments in the sciences.

His interest in the Pauli-Jung dialog reveals more clearly than most of his scientific papers the idea of a "background reality" which is not directly accessible by the empirical methods of engineering science, but which is nevertheless real. Jung's collective unconscious and its archetypal contents refer to domains that have been addressed by such wide-ranging philosophical concepts as Plato's ideas, Kant's noumena, Hegel's world spirit, Schopenhauer's will, or Bergson's élan vital. Of course, professional philosophers will be uneasy with a list combining these highly sophisticated but very different approaches. However, since a critical evaluation of Jung's ideas in a systematic philosophical framework is lacking so far, it is hard to offer any more detailed comparison.

It is rather obvious that a concept such as a non-empirical, "transcendental" background reality cannot be easy to accept in scientific communities for which sober operationalism was of formative influence. In the earlier phases of the development of quantum theory, there were heated debates about the kind of reality to which the theory refers; the discussions between Bohr and Einstein are the best known example. By contrast to their controversies as to which is the right and which the wrong position concerning physical reality, it is a revolutionary turn to seek the right balance between an observer-independent reality and the engineering world of the empirical sciences. For decades Primas has investigated the status of these domains with respect to each other and their mutual relationships.

The non-clinical part of the Pauli-Jung dialog, on which Primas' attention is focused, goes beyond the description of the material world of physics. In addition to a wholeness within the physical world, without so-called Heisenberg cuts separating objects from their environments, the realm of

*archetypal contents is conceived in terms of an even more holistic whole-
ness without any Cartesian-like distinction between mind and matter. In a
corresponding unus mundus (as Jung used to say), the archetypes act as
entities arranging events that look like meaningful, "synchronistic" coincidences between mental and material events once the wholeness is broken.
In a recent paper on chance and synchronicity (Primas 1996), Primas has
suggested some speculative arguments in this regard, characteristic of his
boldness in exploring novel ideas in the context of existing knowledge.*

*The unprecedented success of the sciences for centuries notwithstanding,
Primas continues to express his uneasiness with the repression of the shadow
sides of this success. "Good scientists need to be fascinated by their subject, but fascination is dangerous" is his conclusion in a recent manuscript
on fascination and inflation in the sciences (Primas 1997). When scientists
become fascinated and, in Jungian parlance, inflated by archetypal contents, their sense of responsibility can become severely distorted; there are
many examples confirming this. Of course, there is no immediate recipe nor
remedy against such tendencies. In particular, the dilemma is not solvable
by straightforward scientific ways of thinking alone, since the fundaments
of this thinking are at the very core of the dilemma itself. But there is the
option to understand more explicitly what is going on – in other words,
to confront the problems consciously rather than being their unsuspecting
victim.*

*The contribution by Wilhelm Just deals with some of these issues in the
light of Jungian analytical psychology. As a trained physicist, the author
worked in the solid state physics laboratories at Grenoble for quite a time
before he started to study Jung's psychology in the late 1970s. Since then
he has been working as a psychologist at Linz (Austria). The starting point
of his contribution is a parallel between the process of becoming conscious
and the quantum physical process of observation which Pauli pointed out
long ago (see Jung 1990). Just discusses the step from an undivided whole
to the distinction between physical objects and their physical environment
as a projection of an inner psychic process, leading from the unconscious
to conscious concepts and categories. He develops this idea further by considering the relationship of both processes to the origin of concrete reality
as recounted in creation myths.*

*The archetype of origin is encountered in different symbolic expressions.
In creation, the process originates in the divine realm and eventually cul-
minates in the making of man – the conscious, reflecting being which is
distinguished from the rest of creation by its special role. This distinction,
or briefly, this cut, is experienced as suffering and is understood as guilt
in religions. The deepest need in man, so Just, is to overcome the cut.
This archetypal urge shows up in the different sciences as the attempt to*

328   EDITORS

design a complete and unified picture of reality. The Babylonian creation epic *Enuma Elish* is used as an example to demonstrate typical patterns of creation in myths. The original oneness is slain, and the world is made out of its dead body. After the creation of man, it is up to him to fulfill the gods' duties and suffer death.

The price for regaining the previous oneness is the sacrifice of the pretense to dominance of a one-sided rational consciousness, an act which is symmetrical to the former (self-) sacrifice of the divine origin. In an interpretation going beyond that of Freud, the myth of Oedipus circles around this psychic dynamics. In his one-sided reliance on human rationality, Oedipus is an archetypal image of modern man.

References

Atmanspacher H. and Primas H. (1996): The hidden side of Wolfgang Pauli. *J. Consciousness Studies* 3, 112–126 (1996). Reprinted in Journal of Scientific Exploration 11, 369–386 (1997).

Atmanspacher H., Primas H., and Wertenschlag E., eds. (1995): *Der Pauli-Jung-Dialog und seine Bedeutung für die moderne Wissenschaft* (Springer, Berlin).

Jung C.G. (1990): Theoretische Überlegungen zum Wesen des Psychischen. In *Grundwerk Band 2* (Walter, Olten), Fußnote 133 zum Nachwort. English: On the nature of the psyche. In *Collected Works 8*, Princeton University Press, Princeton 1969.

Meier C.A., ed. (1992): *Wolfgang Pauli und C.G. Jung. Ein Briefwechsel 1932-1958* (Springer, Berlin).

Pauli W. (1994): *Writings on Physics and Philosophy*, ed. by C.P. Enz and K. von Meyenn (Springer, Berlin).

Primas H. (1981): *Chemistry, Quantum Mechanics, and Reductionism* (Springer, Berlin).

Primas H. (1991): Symbol und Irrationalität. Paulis Ideen zu einer neuen Naturwissenschaft. In *Programmheft Cortona Woche 1991*, ed. by P.L. Luisi (ETH, Zürich).

Primas H. (1996): Synchronizität und Zufall. *Zeitschr. Grenzgeb. Psych.* 38, 61–91.

Primas H. (1997): Faszination und Inflation in den Naturwissenschaften. Unpubliziertes Manuskript eines interdisziplinären Gespräches des C.G. Jung-Institutes und der Schweizerischen Gesellschaft für Analytische Psychologie.

CUTS AND SYMBOLS

WILHELM JUST

Donaulände 12, A-4100 Ottensheim, Austria

1.

La nature est un temple où de vivants piliers
Laissent parfois sortir de confuses paroles:
L'homme y passe à travers des forêts de symboles
Qui l'observent avec des regards familiers.

Comme de longs échos qui de loin se confondent
Dans une ténébreuse et profonde unité
Vaste comme la nuit et comme la clarté,
Les parfums, les couleurs et les sons se répondent.

(Charles Baudelaire 1857)

Every scientific concept, in general every content of consciousness, possesses a symbolic aspect. Without symbolic content, no development, no further differentiation would be conceivable. Every content of consciousness, however isolated and rigorously defined, is suspended and embedded in a net of connections, conscious and unconscious, by which its meaning is constituted. For the purpose of this exposition, we will understand the term "symbol" in the sense of analytical psychology. Jung defines a "symbol" in the following way (Jung 1976b, pars. 814–818):

"The concept of a *symbol* should in my view be strictly distinguished from that of a *sign*. Symbolic and *semiotic* meanings are entirely different things. [...] A symbol [...] presupposes that the chosen expression is the best possible description or formulation of a relatively unknown fact, which is none the less known to exist or is postulated as existing. [...] A view which interprets the symbolic expression as the best possible formulation of a relatively *unknown* thing, which for that reason cannot be more clearly or characteristically represented, is *symbolic*. [...] So long as a symbol is a living thing, it is an expression for something that cannot be characterized in any other or better way. [...] Every psychic product, if it is the best possible expression at the moment for a fact as yet unknown or only relatively known, may be regarded as a symbol, provided that we accept the expression as standing for something that is only divined and not yet clearly conscious. Since every scientific theory contains a hypothesis, and is therefore an anticipatory description of something still unknown, it is a symbol. [...] Whether a thing is a symbol or not depends chiefly on the attitude of the observing consciousness; for instance, on whether it regards a given fact not merely as such but also as an expression for something unknown. Hence it is quite possible for a man to establish a fact which does not appear in the least symbolic to himself, but is profoundly so to another consciousness."

With an understanding of symbols in this psychological sense, we will try to investigate the psychological content and background of the Heisenberg cut, used as a metaphor for the distinction between a quantum system and its environment in the quantum theory of measurement (see Primas 1992a, 1993). A psychological and symbolic way of looking at a scientific topic is certainly justified, since in doing so the actual aspirations of science are connected with those needs and questions which mankind has been dealing with ever since and which have always worried man, but also have given him satisfaction and meaning.

The very nature of consciousness is to fix and tie down, to isolate. In this way it removes itself from the original state of being undifferentiated and interwoven with everything. This original pre-conscious existence is experienced in retrospect, as a paradise lost and also as a threat and a danger of getting dissolved again. The emancipation of the collective consciousness from being intertwined with the unconscious (and nature, respectively) made a giant leap since the Renaissance and seems to have advanced with unprecedented speed since then. Sounding like a marching order for the coming centuries was René Descartes' assertion of a fundamental split of the realm of human experience into *res extensa* and *res cogitans* – i.e., the outer world taken as objectively given at one side, and the inner world of the human subject at the other. Descartes assumed the outer, corporeal world to be created in space and time, following the strictly determined, unchanging laws of nature. Those laws are to be comprehended by human consciousness and can be formulated mathematically. Accordingly, natural phenomena are to be described exhaustively in mathematical equations. Psychologically speaking, this is an attempt to represent the outer world by conscious contents exclusively. Finally, nothing dark ought to remain; what seems dark to us now is just taken to be *not yet* known.

This depotentiation and reduction of everything unknown in philosophy and the sciences in modern age are consequences of the centuries-old tradition of driving the pagan demons out of forests, trees, ponds, and waters, thus exorcizing nature. Clearing nature from uncanny influences played a decisive role in the victory of Christianity over "primitive" pagan beliefs. Exorcizing demons and declaring them non-existent, however, also meant robbing nature, and with it eventually the human body, of its soul. In the blooming natural sciences of modern age, this brought forth the assumption that the cosmos and nature, and also the living body, function like a clockwork which was originally set in motion by the Creator and is rolling on, following some given laws of nature which can be comprehended by human reason. Eventually outer reality and its representation in rational conscious terms was more and more taken as the only real and objectively given root of existence.

In sharp contrast to the heroic efforts to explore and comprehend the external material world and the triumphant progress of the sciences appears the disinterest and helplessness toward the inner world as a reality *sui generis*. As strong as modern man has proven to be in dealing with outer reality, he appears lost and uneasy when confronting the soul as a reality. That there is a world of inner pictures independent of external stimuli and not reducible to such, was and is simply ignored or denied (see, e.g., Primas 1992b, 1995). Even psychic experience was to be understood purely mechanistically in terms of *res extensa*, thus to be derivable from the facts and laws of the outer world.

The attempt to find such a foundation also for the world of inner experience was not recognized as being circular for a long time. Soul and consciousness were assumed to be identical on the one side, consciousness and rational knowledge on the other. The stubborn attempts to dissolve the subjective reality as such and the unquestioned assumption that it could eventually be understood rationally, however, have finally pushed the importance of the subject and its consciousness back into the center of interest. The circularity of the enterprise became evident at the beginning of this century, even though the old habit of one-sided reductionistic approaches is still alive today.

That the circularity of reductionism showed up especially in the exact sciences is an impressive example of the psychic fact of "enantiodromia", the sudden spontaneous emergence of the repressed opposite (Jung 1969, par. 425). On the one hand, the efforts of establishing a final and absolute foundation of mathematics on the grounds of formal logic were frustrated and proven to be in vain; on the other hand, in confronting the problem of measurement in quantum physics, a level of comprehension was touched which resists being dissolved in mechanistic terms alone, although physics had been so successful with this method until then.

Measurement in the quantum realm could not further be neglected any more – as in classical physics – but had to be acknowledged as a central issue in the interpretation of quantum theory. Ever since the early approaches to express quantum measurement in a coherent mathematical formalism and to interpret this in the concepts of our daily world experience physicists struggle to find a commonly agreed-upon understanding. The many different interpretations and the fascination which the question of measurement exerts for physicists and philosophers of science are indications that this element of quantum physics carries symbolic meaning, and each individual physicist is experiencing his or her own (mostly unconscious) contents in it. This, of course, is always the case whenever fascination is present; in the case of measurement, however, it seems to be particularly difficult to disentangle personal projections from the outwardly given facts and their scientific description. Despite considerable progress in understanding measurement over the past decades, it remains enigmatic.

Basically, the question is where and how a quantum physical process under observation results in a macroscopically registered fact which is perceptible and perceived by our senses. The problem touched by such a transition from the state of a system before measurement to the macrophysical event was exposed to ridicule by Schrödinger in the well-known burlesque case called "Schrödinger's cat" (Schrödinger 1935). It cannot be possible, Schrödinger argued, for an objective physical event to take place just in or by the act of a subjective observation by the experimenter, as quantum theory seems to suggest. Von Neumann attempted a first rigorous mathematical analysis of the process of measurement using Markov chains. He pointed out that, from a purely formal point of view, there is no logical reason to terminate the infinite chain somewhere – rather, it has to be cut off deliberately. Only the conscious perception of a registered result could bring about such a cut of the chain which otherwise would remain infinite. Therefore, von Neumann argued, one may speculate whether the cut is caused by consciousness.

In Bohr's answer to the thought experiment by Einstein, Podolsky, and Rosen, in Wigner's approach to the problem of measurement, in Everett's "many worlds theory" – to name just a few of the attempts to deal with the riddle of quantum measurement[1] – a perceiving consciousness or a conscious observer is assumed as an essential part of the process of observation, or enclosed as an inseparable part of the process as a whole. This role of consciousness must not be understood as some obscure psychic influence on the process of measurement by the observer; it rather has to do with the fact that a complete measurement is always connected with and based

---

[1]For quite a comprehensive collection of different approaches, see Wheeler and Zurek (1983).

on the precondition that there is someone who has prepared the quantum system for measurement by isolating it from its surroundings, performing the measurement in a well-defined way, registering the result via his senses, and integrating it into his state of knowledge. Human consciousness – understood as inspired by *spirit* by way of a spontaneous impulse, of curiosity, of unrest, or whatever else might provide the incentive to put a question to nature – cannot be eliminated from the dynamics seen as a whole. It is rather the very basis on which science is carried out and in which it is always contained.

The presence of such a spirit cannot be put into a formalism and can never be explored completely. It can manifest and incarnate itself in the human soul, and it always needs consciousness to be recognized. The Cartesian cut was an essential step in freeing the human being from an overwhelming spiritual world on the one side, and in banning spirit from what was then taken to be the only objective world on the other. Eventually it also enforced a more differentiated understanding of the activity of the spirit which cannot be abolished by human dictum. It may die and become ineffective in one form, but it is its very nature to transform and reappear resurrected in another. The Heisenberg cut is an attempt to grasp reality via reason alone, analogous to the Cartesian cut. It represents a necessity in physics and is an auxiliary construction by which the measurement problem is reduced to a purely formal procedure, for example as the split of a system as a whole into parts.

By means of the Heisenberg cut as well as the Cartesian cut (cf. Primas 1993), man has created tools and prerequisites for grasping aspects of reality and comprehending them rationally. In this very act of creation he separates the incomprehensible part from that which can be delimited rationally. What was previously a numinous whole splits into the incomprehensible, irrational background and that which can be inspected via reason. However, the decisive preparatory events basically happen in the unconscious, as we shall see in the next section. Rational man could be tempted to ascribe these preparatory processes in the unconscious entirely to his own arts and skills. Cartesian cut and Heisenberg cut are rational descriptions of what has already taken place, to a great extent, in the unconscious.

From the point of view of depth psychology, one could say that in the concept of such an operation the philosopher or physicist, respectively, pins down what mythical man had tried to come close to and to formulate with awe and marvel in his creation myths. Again and again man was and is confronted with the question of how the reality he perceives arises or rather is already arisen and with the urge of searching for what is lasting, fundamental, and incomprehensible behind this world of phenomena which are constantly changing. In depth psychology this transition is called "becoming conscious", a shimmering concept, difficult to pin down precisely. The spectrum of what is understood as consciousness ranges from the physiologically measured reaction of the senses to external stimuli to the enlightened consciousness of a mystic or the integral consciousness, as Gebser (1977) called it.

Creation, becoming conscious, Cartesian cut, Heisenberg cut, measurement, and so forth appear like symbolic expressions of one and the same mysterious issue: the emergence of reality. In the end it cannot be reached by reason, but is implanted into the human soul as possibility and task. What mythical man ascribes to the act of a deity or divine demiurg in creation myths is governed by an element of uncertainty not further reducible in the scientific approaches of physics.·

The connections among creation myth, quantum physics of measurement, and the emergence of consciousness in psychology can be seen in some central motifs common to all three. From whatever position man approaches his existence, he ends up circling around the event from which everything had started. In creation myths there is a primordial oneness before anything comes into existence, in the form of a divine undivided being, a divine primordial unity, often described as the all-containing chaos. In the empirical study of consciousness, the *participation mystique*[2] or archaic identity exists before any conscious activity, "an a priori oneness of subject and object" (Jung 1976b, par. 781) and unconsciousness. In an individual life, this primal state is often projected into the "good old times", the "happy childhood". In quantum physics, it recurs as the superposition of all possible states of a quantum system before measurement, a state containing all potentialities before one state becomes actual and real.

From the all-containing oneness emerges a first split into fundamental dichotomies which in myth belong to the divine realm though they do already exhibit tender patterns of more concrete properties. Each creation myth exhibits such opposites, from which a world can emerge, or within which a world can unfold and grow (see Fig. 1), whether in the form of light and darkness, upper and lower world, heaven and earth, etc. Often these opposites represent differentiations which are very important and familiar in the concrete daily life in the culture concerned, like salt and fresh water oceans in the Babylonian creation myth, the *Enuma Elish*, which we will examine in more detail in Section 2. As a particularly interesting analogy in the problem of quantum measurement there is the dichotomy of ontic and epistemic levels of descriptions (Primas 1993, see also Atmanspacher 1996),

---

[2]Jung (1976b, par. 123) took this term from Lévy-Bruhl (1923, 1926) who used it to characterize the collective mentality of "primitive man". The term means an unconscious identity, a oneness of the inner and the outer, a being dissolved in the collective: "Instead of individuality, we find only collective relationship" (Jung 1976b, par. 12).

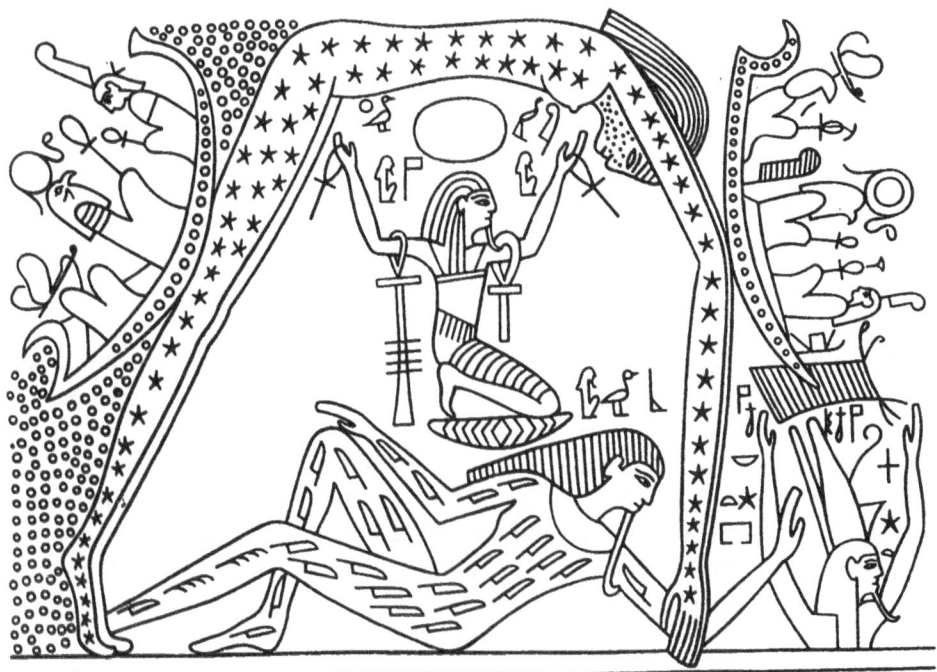

Figure 1. According to the mythical tradition of the old Egyptians, Geb (earth) and Nut (heaven) are separated, so that between them and from them concrete reality can emerge. They are pushed apart by Shu, the wind god. (Reproduced from Hornung 1983, p. 59.)

which can be related to additional pairs of concepts such as reversible and irreversible dynamics, waves and particles, and many others.

The fundamental dichotomy in psycholgy is that of conscious and unconscious, a dichotomy which arises only in the course of development of consciousness and through it, as the ego and the other. With the growth of consciousness, reality becomes more and more differentiated. In the unconscious itself there is no differentiation; everything is tied up with everything else, as Freud (1953a) showed in his great work *The Interpretation of Dreams*. Space and time are not stretched out into above and below, before and after; or rather they are, and then again they are not. In creation myths this is symbolized as the primordial chaos, emptiness, or as a primordial divine being containing everything in nuce.

At the end of a creation myth, man, as created by a divine being, faces the world and is released into it by the gods. A more careful study shows that creation myths deal with the origin of the conscious world in the human being rather than the world outside (Franz 1995). The blurring of inner and outer worlds is an expression of the aspect that mythical man does not distinguish between them as clearly as we who hold them distinctly apart in our experience. Mythical man still lives close to the state of *participation mystique*. The unconscious with its imagery is what is immediately given, not the man made objects. Subject and object are not separated as distinctly as Descartes postulated. The wonderment at existence is still in the foreground.

2.

Ich ging an einem Frühlingsmorgen draußen; die Saaten grünten, die Vögel sangen, der Tau blitzte, der Rauch stieg, hie und da ein Mensch; ein verklärendes Licht lag auf allem; es war nur ein kleines Stücklein Erde; es war nur ein kleiner Moment ihres Daseins; und doch, wie ich das mit immer mehr sich weitendem Blicke auffaßte, schien es mir nicht nur so schön, sondern so wahr und klar, daß es ein Engel ist, der so reich, frisch und blühend und dabei so fest und in sich einig in den Himmel geht, sein lebendiges Antlitz ganz dem Himmel zuwendend und mich selbst in den Himmel tragend, daß ich mich fragte, wie sich die Ansichten der Menschen je so verpuppen konnten, in der Erde einen trockenen Klumpen und die Engel darüber oder daneben in den Leeren des Himmels zu suchen, um sie nirgends zu finden. Doch diese Anschauung heißt Phantasterei. Die Erde ist ein Globus, und was sie sonst noch ist, ist in den Naturalienkabinetten zu finden.

(Gustav Theodor Fechner 1861)

Let us now turn to creation myths more closely and look for similarities to the emergence of reality in physics and psychology. In comparing creation myths, those of primitive tribes and cultures seem harder, stranger in their imagery, less definite and more unintentional in their unfolding. Creation myths of more developed cultures deal with a definite creation of a finished world and human beings in just a few steps and in a rather straight course. Such myths reflect a level of development where consciousness has already reached a rather autonomous and independent position. Purposeful powerful action is then predominant. Looking from within the subjective stage as it is immediately experienced, the growing of consciousness from the depths of the unconscious is not so straightforward, but happens in breaks, detours, and seemingly accidental branchings. It might even end in an impasse or an abortion of what has been achieved already. Pictures as exhibited in primitive creation myths are certainly closer to the actual process of becoming conscious. This can be verified readily by one's own experience of the process of becoming conscious, especially when the unconscious side is predominant and we can hardly disentangle the two from each other. Then we are the scene of a primitive creation myth. The process of becoming conscious can only be reconstructed in hindsight. It is only at the end of the creation myth that man, the reflecting being, is created. Before that, he is still dissolved and contained in the whole in *identité archaique*.

Not unlike the creation myths of more developed cultures, modern cosmologies and evolutionary theories, and also, of course, the theory of measurement, do not at all deal with the subjective, mostly arduous and painful part in the process of becoming conscious. They do faithfully reflect Descartes' guidelines to distinguish radically between *res extensa* and *res cogitans*, and thereafter to sweep the latter under the carpet and label it as subjective and therefore devoid of reality. Primitive creation myths are in sharp contrast to such a style; they deal extensively with the subjective moods and states of the creator and stick closer to the very process of creation.

Myths do not fall short at the event of creation as a sober (and outer) fact, but also look for its inner necessity as well as its meaning for mankind – a meaning which is felt as binding or satisfying and in which man is centered. In this aspect myths naturally go beyond the modern scientific approach to the world and to reality. Creation myths can be used to make conscious what mostly remains unconscious for the scientist of today and nevertheless meets deep human needs. It is the essence of creation myths to convey the meaning of man's existence and to draw a coherent picture of how the divine and heaven and man and earth are related.

This aspect remains unconscious in the projections of modern science – if we dare to call their stories of the origin as such – and is even tabooed in the Cartesian tradition. Everything which pushes toward consciousness and could be made conscious but is not dealt with due to some preconception or inadequate one-sidedness will exert its effects out of the unconscious – then, however, mostly in a negative, destructive manner. The unconscious has its own ways: when one of its contents emerges which it wants to incarnate in human reality, then there is no way for man to cheat. If such a content wants to live with us, so to say, it has to be respected and related to as a living opposite; and it is up to us to find a proper "ritual" for integrating it to our benefit. In creation myths, the evil of this world is often attributed to forces which emerged or were created as attempts of new forms of existence early in the creation process – in Greek mythology the Titans, for example, or the devil and his horde in the Judaeo-Christian creation story – but for some reason these forces were not suitable for survival in this world. They turned out to be unsuccessful attempts of creation which were often not allowed to further exist in this world, i.e., to become conscious. As a consequence, they exert a destructive, demonic influence out of their underground, repressed (unconscious) existence and hamper and endanger the progress of life. In this context it seems appropriate to draw attention to the concept of *projection* in analytical psychology. According to Jung (1976b, par. 783),

"projection means the expulsion of a subjective content into an object; it is the opposite of *introjection.* Accordingly, it is a process of *dissimulation*, by which a subjective content becomes alienated from the subject and is, so to speak, embodied in the object. The subject gets rid of painful, incompatible contents by projecting them, as also of positive values which, for one reason or another – self-depreciation, for instance – are inaccessible to him. Projection is based on the archaic *identity* of subject and object, but is properly so called only when the need to dissolve this identity has already arisen. This happens when the identity becomes a disturbing factor, i.e., when the absence of the projected content is a hindrance to adaptation and its withdrawal into the subject has become desirable. The term projection therefore signifies a state of identity that has become noticeable, an object of criticism, whether it be the self-criticism of the subject or the objective of another."

Dissolving and withdrawing the projection into the subject in the case of quantum measurement means to separate off the aspect of consciousness, i.e., the projected content, from the sequence of physical events. This is precisely done by introducing the Heisenberg cut and the Cartesian cut, if they are understood in terms of sweeping the projected inner process under the carpet of reason. If this happens, it will inevitably reappear somewhere else and provoke negative effects there. Such inner processes have to do with an autonomous creative spirit, something which belongs intimately to man and nature.

Taking the *Enuma Elish*, the Babylonian creation epic, as an example, we will see how mythical man understood the evolution of consciousness from the unconscious. This epic was found on tablets in the library of Ashurbanipal (668–626 BC) in Niniveh, but it can be traced back to much earlier epochs and has its roots in Sumeria in the third millennium BC. "Enuma Elish" are the words with which the epic begins: "When the heavens above were not named ...".[3] As is characteristic for myths in general, the story of creation was celebrated and performed ritually on special occasions, as for example at the beginning of the circle of the year, on the assumption of power by the king, or periodically to renew his potency and power. In this way a situation in outer reality was connected with its origin. It was re-

---

[3]The English text is my translation, following the German translation by Garelli and Leibovici in Eliade (1980), p. 134ff. See also Sproul (1979), p. 91.

newed by re-enacting the primordial drama of creation. A modern scientist unconsciously does the same when he tries to explain a physical problem starting from "first principles" and also, of course, when searching for primordial reasons in a "grand unified theory". It is also common practice in psychoanalytical work with patients – often without acknowledging the mythical aspect and power of the primordial reason as the symbol which is effective – to go back to early childhood or even further, in order to come to an understanding of the actual situation and to dissolve it by comprehending its origin.

"When the heavens above were not named, when below the earth did not have a name, when even Apsu, the primordial, the begetter of the gods, Mummu Tiâmat, who has given birth to all of them, mingled their water into one..." These are the first lines of the myth. The formless original state is described as the mingling of the two primordial reasons, Apsu and Tiâmat. Apsu is a male deity; he will later be the fresh water ocean, but is to be understood more as a principle or potentiality, an idea in the Platonic sense rather than in the sense of concrete fresh water. Out of it will come the subterranean ocean upon which the earth will be floating, according to Babylonian belief. Tiâmat is the later abyss of the salt water ocean. Salt water and fresh water were the two elements on which life in ancient Mesopotamia was based. They are introduced as deities, primordial elements, archetypal principles, as molds in the unconscious. After their differentiation they will appear personified. How Mummu should be understood is not quite clear. It could be an epithet of Tiâmat, meaning something like "mother", or it could be some nebulous third, the mist hanging over the mingled primordial waters (Eliade 1980, p. 112).

The state before creation is addressed by way of negation, as being not (yet) named, a state of one-ness of what will appear as distinguished and opposite in conscious reality – an attempt to express the inexpressible, the primordial, the original chaos. We also refer to it when we call that which cannot be grasped in a clear, distinct way the unconscious. By using the negation in the myth, the realm where there is no definite form and image yet is addressed, a realm of which we can dimly be aware when we open ourselves as a totality and let ourselves get taken over by the here and now of our psychic background, when the control of consciousness is suspended. It is especially present directly in creative experience; it cannot be grasped and dissected rationally and with conceptual language. Once it is named and pinned down, it is dead.

As in various other creation myths, naming things in relation to their existence is touched on in the opening lines of the *Enuma Elish*. From the point of view of psychology, we can only talk of something of which we have an (inner) picture, a name so to speak (Primas 1992b). "Is the moon there if nobody looks?", Einstein asked in order to expound a crucial question posed by quantum theory. Does the world exist outside and independent of our being conscious of it, as the Cartesian dogma asserts? Is there a choice only between a lifeless and life-killing reductionism and a sterile solipsism? The question of reality and its objectivity cannot be treated without taking into account the psychic fact of projection, i.e., without the discoveries of psychology of the unconscious (not as "psychology without psyche", as Jung used to refer to a psychology which negates the existence of a psychic reality *per se*). Without taking the psyche and its dynamics into account, the essential which is searched for so desperately will appear again and again in further projections while remaining unconscious itself. Instead of repressing the unconscious by projection, it certainly would be salutary to acknowledge its effectiveness and reality.

Similar descriptions of a primordial state can be found in many other creation myths. For example, the ancient Indian *Rig Veda*, reflecting a fairly advanced state of consciousness, is even more radical in conveying the essence of what was before consciousness; the negation of everything which can consciously be conceived: "In that time there was neither being nor not-being, there was no space in the air, nor were there heavens, ..., death did not exist, nor did life ..." (Deussen 1963).

The original oneness from which the opposites and, in further differentiation and evolution, the concrete reality emerges is addressed in the *Enuma Elish* as mingling of Apsu and Tiâmat, fresh and salt water. In the above mentioned passage of the *Rig Veda*, this oneness is described: "Only the One listened windless through itself and out of itself there was no other". How this oneness and namelessness, which contains everything in nuce, brings forth reality eludes every attempt to be definitely pinned down in rational analysis. A reflection of creation in myth is the question in psychology of how a conscious content emerges from the unconscious, in physics of how a concrete and registered result is produced from an initial superposition of all possible states of a quantum system, and in general of how the new originates. In the *Tractatus Logico-Philosophicus*, Wittgenstein confronts the approach of the exact sciences with the attempts of the myth (Wittgenstein 1961, 6.371, 6.372):

"The whole modern conception of the world is founded on the illusion that the so-called laws of nature are the explanations of natural phenomena. Thus people today stop at the laws of nature, treating them as something inviolable, just as God and Fate were treated in past ages. And in fact both are right and both wrong: though the view of the ancients is clearer insofar as they have a clear and acknowledged terminus, while the modern system tries to make it look as if everything were explained."

The *Enuma Elish* continues: "When no god yet had appeared, not named by name, fate was not appointed to them, then out of the womb of Apsu and Tiâmat the gods were born." At the beginning of what will become reality, there is first a differentiation of the one as the creation of the gods; "... [in] the waters still precipitated ..." is the symbolic expression for a process in the unconscious. Water is to be understood as the opposite of concrete earth which indicates the stability and firmness of consciousness. We should not overlook the point, expressively shown here, that even the gods are only created and given fates now; i.e., they assume and carry characteristic traits according to which man experiences them as distinguished from each other. From a psychological point of view, the gods represent different archetypes, i.e., patterns of psychic experience which man finds himself exposed to or afflicted by (Pauli 1954, 1984). It is they who structure and form the human psychic experience.

Here we are faced with the central mystery of how the definite and differentiated can arise from the one, nameless, formless, and unspecific. Both principles are already present, according to the myth, "long before" man is created, i.e., "long before" reflecting consciousness exists. The myth is dealing with a dynamic flow of events in the unconscious. If the origin is all-too quickly approached and pinned down, its numinosity vanishes and becomes blurred, the original oneness splits up – as many creation myths show. Even a myth can suffer this fate if its interpretation is taken too absolutistically. This can be observed by looking at attempts to equate the psychic dynamics to one myth alone, as is frequently done with the myths of Oedipus or Narcissus. What was supposed to be primordial behavior has the tendency to split up in a manifold of aspects and becomes a complicated theory if taken too rationalistically.

The opening lines of the *Enuma Elish*, stating that the gods were created in the waters, reminds us of the *ignotus per ignotius*,[4] the well known motto of the alchemists. It runs completely counter to the conscious claims of science. On one side, there is the attempt to bring light into the dark, not yet understood, on the other side, myth (as alchemy) has the central mystery on which all creation is based as its object. The circumambulation of the mystery in myth and ritual helps to keep it alive and effective.

Let us also consider the aspect of change and fate which comes into being with the gods. From now on the gods coming forth from the original pair are ever more differentiated. They come as pairs of opposites which Babylonian man was familiar with from his daily experience with nature. The differentiation of the gods – psychologically of the archetypes, those forces which determine psychic life and experience – reflects the experience

---

[4]Latin: "(express) the unknown by the still more unknown."

of the contradictory aspects of which the concrete outer world is composed. Or, vice versa: the experience of a cut in the outer world follows the differentiation of the inner psychic images. Or, yet another option: both take place at the same pace and with a common meaning in the sense of synchronicity (Jung and Pauli 1955, Primas 1996); that is, the inner differentiation and the outer realization happen simultaneously.

The power is in the hands of the god who has the tablets of fate. The idea of destiny and fate is the original form and psychic experience of what was later to become causality in science. Following the myth, it comes into being long before consciousness appears, but presupposes the created gods. As we will see in Sect. 3, the respectful acceptance of fate is characteristic of the hero and knower; only this actually amounts to the coronation and redemption of consciousness with its one-sided attempts to establish an ultimate foundation of its existence.

In the ancient Babylonian mythical world, Lachnu and Lachamu are created and represent silt and slime (cf. Sproul 1979), that is, first differentiations and precipitations from the waters of chaos. Then the primordial parents bring forth Anshar and Kishar as horizons of the sky and the earth. Again, becoming conscious has intimately to do with – or is equivalent to – distinction, limits, definiteness, boundaries. Concerning this function of consciousness, Jung calls "the faculty of discrimination the *sine qua non of all consciousness*" (Jung 1968, par. 563) and urgently warns against "a regression of consciousness into unconsciousness" which occurs when consciousness looses this faculty as a consequence of taking "too many unconscious contents upon itself," i.e., having identified itself with them. The real distinction meant in the development of consciousness is that of consciousness and the unconscious; this distinction is the precondition for a mature relation of the conscious ego to the numinous background of the collective unconscious in the terminology of analytical psychology.

Now time began: "the days became long, the years multiplied", and more gods were created. The young gods began to disturb the primal gods, Apsu and Mummu, and to bewilder them by jumping around and dancing, "without dampening their yelling". The primal gods appear as entirely static, as just existing, without any purpose, with a tendency toward sleep and rest. By contrast, their younger offspring appear as troublemakers, as motion, dance, noise. The origin of motion in the motionless primal state reflects the old conflict of mankind: the opposition of the old and the new which disturbs the established state.

Apsu, the primal father and father of the Great Gods, calls Mummu who appears here as his messenger and as an independent deity. They go to Tiâmat to complain: "Unbearable is their [the young gods] behavior. During the day I cannot repose, during the night I cannot sleep. I will destroy them to put an end to their bustle. Peace shall prevail so that we will sleep again!" But Tiâmat takes the side of her offspring and wants to calm Apsu: "What? Shall we destroy what we have created? Certainly their behavior is embarrassing, however, let us be patient leniently!" Nevertheless Apsu and Mummu plan to destroy them. The young gods are desperate when they learn about this plan. "But the very clever, wise and powerful, the omniscient Ea recognized their intentions. He devised and made against him [Apsu] a tremendous spell, and his word made them stick to the waters." Then he strikes the sleeping Apsu dead, robs Mummu of his clothes and his insignia of power, locks him away and keeps him under control by keeping a tight rein on him. Over the abyss of Apsu, Ea is building his magnificent palace.

At the very beginning of creation, long before consciousness appears, the opposites become hostile to each other. We are faced with an unyielding dynamic of opposites, the old father god together with Mummu – who apparently evolved into something like a primitive form of a male or fatherly spirit after having mingled with Tiâmat, the great mother goddess and primordial dragon – on the one side, and the younger offspring on the other. The old state stands against the new dynamic, the dominating old "order" against the new forces of the future. This archaic conflict will be encountered again and again in various pictures in myths. Let us remember the sequence of the Greek father gods, Uranos, Kronos, and Zeus, each of them knowing that he will be deprived of power by one of his sons.

Now Ea together with Damkina, his wife, engenders a magnificent radiant hero Marduk, a sunlike child, whose appearance causes all the gods to rejoice. Anu creates the winds which bring unrest and disturb Tiâmat. Again the old gods plot the destruction of the new forces. This time the old forces succeed in convincing Tiâmat to put an end to the young, bright, restless gods. For this decisive fight, Tiâmat brings forth monster after monster, each of them more fearful than its predecessor, with terrible, primitive appearances and dreadful outfits, to make an end with the "gods in the heavens". She names Kingu, one of the first-borns, as the leader of her terrible horde and even provides him with the tablets of fate. When Ea and the young gods learn about Tiâmat's preparations, they are consumed with anxiety and panic. Different attempts to calm Tiâmat are in vain. Despair spreads out. At this perilous juncture, Marduk is asked to be the leader in the fight against Tiâmat and her horde. For this purpose he is provided by the young gods of the heavens with all the powers they possess. By his sparkling stature alone he confuses Kingu and his helpers. A bitter struggle ensues between Tiâmat and Marduk. He succeeds catching Tiâmat – who had been transformed from the life-giving primordial mother to a life-threatening dragon – with the help of a net, and lets the winds blow

Figure 2. The sun-god Marduk fighting with Tiâmat, Great Mother Goddess and dragon. (Reproduced from Eliot 1976, p. 19.)

it into two parts. From one part he makes the heaven where the waters are locked up, from the other he creates the earth. From the eyes he causes the Euphrat and Tigris to arise, her breasts become hills, and on her head he piles up mountains – then all the other familiar landscape is created from her dead body. Marduk is celebrated as the brilliant victor and a new order is consolidated. The city of Babylon is built so that the gods will have a homestead there. The defeated gods shall be slaves to the victorious forces.

With the birth of Marduk the opposition between new and old becomes even more accentuated; above and below, spirit and formless matter, shining hero and dark primordial matrix, consciousness and unconscious, progression and regression, purposeful directed action and unconscious proliferation of life. Also in the weaponry of the two enemies, the irreconcilable opposition is visible. Tiâmat's power relies on the primordial horror of monsters, abstruse shapes, poison, snakes, mixed figures. Marduk uses a net, the wind, and an arrow to fight. The net is an archetypal image of the delimiting, holding, selecting property of consciousness. The spirit is breath, wind which penetrates the unconscious. The arrow, like other pointed weapons (spear, lance, sword, knife, etc.), symbolizes the purposeful, fixing and differentiating activity of consciousness.

While the differentiation of the original One – the various pairs of gods arising at the beginning – is taking place in the unconscious and in harmony with it, there is now a merciless "either–or", which is typical of the

While the differentiation of the original One – the various pairs of gods arising at the beginning – is taking place in the unconscious and in harmony with it, there is now a merciless "either–or", which is typical of the adolescent state in the development of consciousness. During this phase of development it is of vital importance for the individual to leave the mother behind which represents the unconscious and its seductive effect on consciousness. After its victory over the unconscious (Fig. 2), it is now the heroic new consciousness which furthers the resulting separations and differentiations. "From now on instead of creation by the brute cosmic powers, which is lawless and just flowing by itself, order by spirit takes over the place" (Eliade 1980, p. 128). With it, outer reality arises as we know it in daily experience; the primal mother and matrix is taken as raw material. First the heavens are ordered, then the earth.

Now Marduk wants to accomplish still another great opus: "A tissue out of blood I shall make, bones I shall form, to let arise a being: man shall be its name. I shall create a being, man. He shall do the service of the [defeated] gods to make their fate easier. [...] One of their brothers [of the defeated gods] shall be selected. This one shall die so that mankind be created." Marduk wants to sacrifice the rebellious god who seduced Tiâmat into fighting. They name Kingu as the guilty one and hand him over: "They let him suffer his punishment, they cut his arteries. Of his blood he [Ea] created mankind. He imposed on them the service of the gods, to exempt those [of it]." The epic comes to its end with lengthy praises of Marduk.

Man, the conscious being, appears here as created from the blood of the slain Kingu, a brute power which opposed further development and differentiation as the outer reality was made of the dead body of Tiâmat. Insofar the *Enuma Elish* displays a state of consciousness fairly independent of nature and even in opposition to it. Marduk can be understood as the principle of a new consciousness, the force in the unconscious which pushes toward further differentiation and clearer order. This principle of consciousness is to a great extent emancipated from the unconscious and confronts it rather self-confidently and freely. The myth is full of violence and slaughter; there is a strong opposition and hostility between the origin and the new power.

A mature reconciliation of the unconscious and the conscious is not yet possible – not unlike the current situation in the sciences. Service of the gods is the duty of newly created man; from now on it is up to him to relieve the defeated gods of their punishment. On the other hand, mankind must also take over from them the menace of death. Mankind is punished for something it has not done consciously, the misdeed, rather, is something that happened long before reflecting consciousness was created, deep in the unconscious. Here we recognize the motif of primal sin, which can be found in many creation myths: consciousness as the sin of being separated from the original oneness.

In the course of the creation myth, various differentiations and cuts are made, quite gently at the beginning, such as the differentiation of salt water and fresh water, water and slit, heaven and ocean. Then the differences turn to hostilities and battles between the older and younger gods. Finally the dead body of Tiâmat is divided to form cosmos and earth, and man is created from the blood of the slain Kingu. At the end of the epic there is a separation of the upper and lower gods, and there is a cut between the gods and man, who is charged to make sacrifices to the gods and perform their duties. The process of becoming aware of reality is presented as a sequence of cuts. In the creation myth, violent confrontations take place behind each cut, each further separation. One side loses and remains as lifeless body serving as material for further developments. The other side withdraws more and more into heaven, that is, remains unconscious or is experienced as such.

The conflict between progressive and regressive powers runs through the epic and impresses us with its merciless violence. Of special importance for us is the motif of the creation of the concrete world from the dead body of Tiâmat. This mythologem of the sacrifice or self-sacrifice of a divine primordial being as the basis of the created world can be found in many cultures: in Germanic mythology the primordial giant Ymir sacrifices himself so that the world can come into existence, Purusha and Prajâpati are the first sacrifices in Indian mythology, P'an-Ku in China, and so forth. By sacrifice of the all-one, of the original oneness, polarity and eventually the plurality of existence emerges. The identity of creator and creation on one side, and their radical difference on the other indicates a fundamental paradox of human existence and could not be put more drastically. The concrete and familiar existence is opposed to something which cannot be grasped by the senses nor by reason; and yet the two are made of the same stuff.

Creation myths usually end at the point where the gods have more or less withdrawn and do no longer interfere with the world. The myth leads to the lived reality and makes it real just by showing its divine origin. The creation myth stops, so to say, at the actual state of human existence: so it is and not otherwise. The teleological question "what for?" is usually further dealt with in religious systems and gets its most sublime elaboration in the different mystical paths. In the modern secularized world, reason is charged with this question. Whether reason alone, in its present one-sidedness, is able to cope with this task, the reader may judge for himself.

The decay of the primordial totality and oneness in creation myths corresponds to the feeling of the human subject of being split and torn in pieces, which is the true companion of the growing differentiation of consciousness: "This is what fate means: to be opposite, to be opposite and nothing else, forever" – so laments Rilke in the eighth of the *Duino Elegies* (Rilke 1982). To overcome and heal the fundamental split, often represented by the image of a restoration of the original oneness, the primordial divine being, sometimes of cosmic size, is then the task of the individual – as an inner process – and is at the center of mysticism and religion (Franz 1994).

In Catholic theology this mythologem can be found, for example, in the idea of the church as the body of the mystical Christ (mystici corpus Christi), as a cosmic all-embracing Christ who is to be put together again. This mythical and mystical idea of the church has given rise to the claim of totalitarian outer power by misunderstanding and projecting the archetypal content of unity onto worldly things – a tragic, all-too human tendency that also reappears in the totalitarian claims of science, economy, modern state, etc. The archetypal tendency to overcome doubt and split, to re-establish the oneness and participate in an ultimate certainty is projected into something concrete and outer. In that respect we are like those among the alchemists who sought to produce concrete gold, the precious metal. There were also others, however, who knew that this treasure, no longer corruptible and so desperately desired and sought for, was an inner psychic process.

3.

Das Auge sieht das Auge nicht,
Das Schwert schneidet das Schwert nicht,
Das Wasser wäscht das Wasser nicht,
Das Feuer brennt das Feuer nicht.

(Keiji Nishitani 1982)

Mythological documents of the birth and growth of consciousness range from ancient images of the death of a primordial divine being whose dead body is the stuff of our (conscious) world to the Judaeo-Christian myth in which god drives the original parents of mankind, Adam and Eve, out of paradise as punishment for having trespassed the divine order not to eat the fruits of the tree of knowledge of good and evil. Including the results and outgrowths of the era of modern science, this is the story of a one-sidedness which is experienced as unavoidable and apparently originates in being expelled from a oneness or totality. The more primitive a creation myth, the deeper is the conflict rooted in the divine realm. With the growth of consciousness, splits and losses increasingly become the fault of man. Nevertheless, in the Christian liturgy of Easter this primal sin is celebrated as *felix culpa* since it also brought the necessity of redemption and initiated the process of healing the primal split. This, of course, enhances the responsibility and value of man's struggle tremendously. He is not just the victim of a drama in the "beyond", but is called upon to become a conscious collaborator and actor.

The loss of primordial (unconscious) totality is experienced as suffering, and consciousness, the product of the original split, attempts to heal the wound and restore the original oneness. Mysticism, religions, the arts, the sciences, ideologies, and certainly also the strivings of civilization are corresponding forms to deal with the split or to find meaning in it. In psychotherapy, there is no doubt that the search for meaning is experienced as healing. Psychologically it makes all the difference whether the totality is searched for in progression or regression, that is whether, after having lost paradise, the necessity to go through the suffering is accepted or one gets stuck in self-pity and a retrograde longing for the lost tranquility.

All human attempts to heal the split are accompanied by the tendency and the danger of projecting the lost oneness and totality into something too narrow, too one-sided. The aspired aim will always be an image of the totality we ourselves are capable of at that moment, i.e., it reflects our state of consciousness. In that respect many of our modern wishful notions are not unlike the aspirations of the disreputable goldmakers among the alchemists who really strove to produce the precious metal in its concrete chemical form. In our modern notions this kind of gold reappears, for example, in the search for an eternally flowing, non-polluting source of energy from nuclear fission or fusion, or in a "theory of everything" as a final picture of the world, or in genetic engineering as the key for a final victory over corporeal disease and decay, or in the many approaches toward joining together in monstrous economic trusts or political unions – the bigger the better – to guarantee economic prosperity and comfortable security. Giant size, finality of the advertised solution, a hastening to establish it, and absoluteness of the claims are symptoms of projections of totality into all-too-human and narrow realms. Those modern alchemical gold phantasies are wishful projections of a one-sided rational consciousness, and they are eventually deadly.

In complete contrast to such regressive attempts there is the *sacrifice* of the ego with all its certainties and wishes with which it is initially identical. This is the way mysticism, religion, or sometimes also art try to regain the lost totality. Instead of a stubborn attempt to establish the ego eternally by reducing and centering the entire existence to it, it enters consciously and becomes part of the cycle of life, which means permanent transformation. This sacrifice of the conscious ego's pretense to dominance, the *sacrificium intellectus* (Jung 1968, par. 59), is symmetrical, so to speak, to the sacrifice of the original divine oneness. It is the way the individual has to engage him- or herself and to suffer through alone. As conscious reality emerged from the (self-)sacrifice of the original oneness, the lost oneness can only be regained by sacrificing the one-sidedness of just this consciousness.

Let us try to illustrate this archetypal dynamics once again by means of a myth. The myth which lends itself for our theme is the myth of Oedipus. We must try to approach the story as openly as possible and without the cliché-ridden ideas which are especially associated with this myth. In her article "Ödipus ist keine Inzesttragödie", Denman (1995) tries such an approach (see also Jung 1976a); I am following her thoughts in their essential delineations.

As the beginning of the myth of Oedipus we can consider the foundation of the city of Thebes by Cadmus or, still before that, the kidnapping of Europa by Zeus disguised as a bull. All the different fates of heroines and heroes of the house of Cadmus – Europa, Cadmus, Harmonia, Labdacus, Laius, Jocasta, Oedipus, Antigone – follow each other and are interconnected in more than one way and form an all-embracing unit. They are self-representations of a coherent archetypal development. We should not loose sight of these connections; but neither should it prevent us from concentrating on the fate of Oedipus. A more comprehensive treatment would also make it necessary to take into account the different versions of the myth,[5] but this would exceed the frame of our scope of interpretation. Here we follow the version of Sophocles in his tragedy *Oedipus the King* (Sophocles 1991a).

Oedipus enters the stage as the reigning King of Thebes. After years of a florishing and prosperous rule, the city is then menaced by a plague. The people plead with the King to rescue the city from ruin, as he had already done once before when he liberated it from the stranglehold of the Sphinx. At that time, Oedipus succeeded in solving the riddle of the Sphinx, that monster with the head of a virgin, the body of a lioness, the wings of an eagle, and the tail of a snake (Fig. 3). She used to pose a riddle to everyone who passed by her column on the market place: "A double legged there is on earth, and a four legged called by the same word, and also triple legged. It changes its form alone of all living beings who move on earth, in the air, and in the sea. When it uses most of its legs for moving the speed of its limbs is the slowest" (Kerényi 1951, p. 83). Until then no one had been able to solve the riddle, and whoever was unable to provide the solution forfeited his life. It was Oedipus who finally defeated the monster. Thereupon the

---

[5] An excellent presentation of the subtle interconnections of the fates of the heroines and heroes of the house of Cadmus can be found in Calasso (1988). The classical works of Ranke-Graves (1955) and Kerényi (1951) are also recommended.

Figure 3.   Oedipus and the Sphinx, Attic bowl, Vatican Museum, Rome. (Reproduced from Guirand 1935, p. 117.)

Now facing a new menace in the form of a plague, Creon, the brother of Jocasta, had asked the oracle for help and just returned to Thebes. According to the oracle, the murderer of king Laius was still in the town. The unatoned murder was the cause of the plague. Oedipus only now learns that Laius was struck dead on his way to the oracle. "The riddling Sphinx induced us to neglect mysterious crimes and seek solution of troubles at our feet" (Sophocles 1991a, line 130), is Creon's explanation as to why at the time when the king was killed no one searched further for the murderer. Oedipus sees it as his task to bring light into the dark and thereby to rescue the town. "God will decide whether we prosper or remain in sorrow" (line 145). In order to find some trace of the culprit, Creon proposes asking the blind seer Teiresias. Aware, by the power of inner sight, of the terrible consequences, Teiresias tries to dissuade Oedipus from searching further for the truth. Oedipus, by nature being prone to violent outbursts of temper, becomes raving mad, because Teiresias hesitates to speak out. In his rage he even accuses Teiresias of having himself hatched the plan to murder Laius.

consequences, Teiresias tries to dissuade Oedipus from searching further for the truth. Oedipus, by nature being prone to violent outbursts of temper, becomes raving mad, because Teiresias hesitates to speak out. In his rage he even accuses Teiresias of having himself hatched the plan to murder Laius. Finally the seer reveals: "I say you are the murderer of the king!" (line 362). Outraged by this revelation, Oedipus counters with his own skillful salvation of the city of Thebes; by his cleverness he had seen through the plots of the Sphinx and had put her out of action: "But I came, Oedipus, who knew nothing, and I stopped her. I solved the riddle by my wit alone" (line 396).

Like Oedipus, Jocasta also rejects the art of prophecy and strengthens Oedipus in his attitude: "listen to me and learn that human beings have no part in the craft of prophecy" (line 709). And just then, immediately after the repudiation of any possibility of knowledge by inner sight, the tragic interconnections and fateful relations begin to unfold. At first the accounts of the sequence of events of the King's murder make it clear to Oedipus that it was Laius whom he had struck dead at the Cithairon, a narrow mountain pass. At that time he was just trying to get away from Corinth and his supposed parents, with the firm intention of escaping the oracle which he had consulted. He had asked the oracle to reveal who he was, after having been confronted with the reproach that he was not the son of Polybus, King of Corinth. The oracle of Delphi had revealed to him that he would be the murderer of his father and the husband of his mother. In order to escape this prediction he chose not to return to the court of Corinth, to his supposed parents, but turned his path toward Thebes. Coming through the narrow defile on the Cithairon, Oedipus confronted a carriage with an old man and his herald. When they demanded that he give way, tried to block his path, and struck him with a stick, he killed them in his rage.

Now a messenger from Corinth reports the death of king Polybus. Thus Oedipus' real origin is revealed: he is the son of Laius and Jocasta. It was just this messenger who had once handed over Oedipus, as a little foundling with pierced feet, to Polybus who was childless and adopted him. Laius, Oedipus' natural father, on the other hand, had decided to kill and abandon his newborn child in order to escape the oracle which had prophesied that he would be killed by his son, and that the son would marry the mother. When this course of events is gradually unraveled and all the fragmentary facts fit together like a mosaic, Jocasta begins to understand, even before Oedipus does, what really had happened and tries to prevent further revelations: "God keep you from the knowledge who you are!" (line 1068). But Oedipus insists on learning the whole truth. When it is obvious that the horrible predictions of the oracle have been fulfilled word by word, Jocasta hangs herself and Oedipus blinds himself.

In *Oedipus at Colonus*, Sophocles describes the further fate and the end of the tragic hero. Antigone, Oedipus' daughter, guides the blind man and beggar after he is expelled from the city of Thebes, in accordance with his own previous ruling. In the grove of the Eumenides in Colonus, Oedipus knows that he has reached his final place of peace. On the path down to the underworld, Hermes is his guide, and his daughters, Antigone and Ismene, follow him until they finally wash him and dress him in shroud. Then they have to remain behind and only Theseus, the hero and king of the nearby city of Athens, is allowed to witness Oedipus' last transition. Oedipus, following the goddess of the kingdom of the dead, disappears from his sight. The place is to remain a mystery as Oedipus describes it to Theseus: "The things within this ban, not to be uttered, yourself shall learn, when you come there alone, for I shall not declare them to anyone of these citizens, nor to my daughters, though I hold them dear. Keep them yourself always, and when you come to the end of life reveal them only to him that is nearest to you, and he in turn to his successor." (Sophocles 1991b, line 1741ff).

Denman calls the myth of Oedipus a "tragedy of knowing" (Denman 1995, p. 5), and indeed the inner climax of the course of events is the confrontation of Oedipus and the blind seer Teiresias, each of the two embodying one side of human knowledge. We all have both sources of knowledge within us, knowledge via conscious reason and immediate, spontaneous knowledge from the unconscious. Oedipus is proud of his victory over the Sphinx. He found the solution of her riddle, trusting only himself and his reason, and thus liberated Thebes from bleeding to death and losing the best of her youth. Let us remember Tiâmat and the monsters she brought forth in order to destroy the young gods, "the gods of the heavens". These gods of light and air represent the archetypal forces in man and nature which make man strive toward consciousness and are related to spirit, light, clarity, order, planning, reserved reflection – quite in opposition to Tiâmat and her horde which stand for primordial, primitive nature with its unrestricted, dark, formless, overpowering force, its excessive impulses, catastrophes, illnesses, scourges. The Sphinx comes from the underworld, offspring of Echidna, the snake goddess, and the dog Orthos, Echidna's own son. Here the motif of incest is touched upon already; we will have to deal with it later in more detail. The primordial matrix of nature, psychologically the collective unconscious, is confronting us here in its dark, destructive aspect, hostile to life and consciousness. It takes the hero, consciousness in all its capacity, to escape destruction.

Laius was regarded as the originator of pedophilia in ancient Greece. He had once kidnapped Chrysippus, the shining, golden boy, after having fallen in love with him. Since the King's pedophilic passion had besmirched the marriage bed, the marital order was put at risk. So Hera, the guardian of marital relationship, had sent the Sphinx as punishment. Just as Marduk as the sun hero had vanquished the threatening Tiâmat – not so much by brute force, but rather by purposeful, ordered skill – Oedipus succeeds in defeating the monster of the underworld by clear and sharp reasoning. But "little did he know that the riddle of the Sphinx can never be solved merely by the wit of man" (Jung 1976a, par. 264). The conflict between nature and man in their respective totalities is at stake. Nature, the primal mother, propounds herself as a riddle to the (male) spirit, which the hero is able to solve, thus depriving her of her menacing aspect and vanquishing her. Is this the final victory? The myth of Oedipus revolves around the answer to this question. The solution of the riddle of the Sphinx is man himself, as Oedipus recognizes, but apparently he misses the essential element of the confrontation with the devouring mother aspect of nature.

In analogy, it is nature which poses riddles to man in the sciences, and it is man himself who is to be found at the very end of the search in each science – no matter whether one wants this or not. Even in the exact sciences, where the human psyche – *res cogitans* – was thoroughly eliminated, it reappeared, e.g., in the problem of measurement in quantum physics or in Gödel's discoveries about the foundations of mathematics.[6] If a science advances far enough, it inevitably begins to show its self-referential character.[7] In the course of the tragedy it becomes clear that the victory over the Sphinx is not to Oedipus' advantage in the end, nor is it an ultimate solution. There is still another aspect of nature – psychologically speaking, of the unconscious – which demands to be recognized and lived. It is carved in the door of the Delphic temple of Apollo: γνωϑι σαυτον – "know thyself".

As important and inevitable as the successful solving of the riddle is, and as necessary as it is to face nature and the unconscious at this level, the

---

[6]See Breuer (1997) for a sound discussion of Gödelian self-reference in quantum theory.

[7]With respect to psychology Jung (1969, par. 429) writes: "The psychology of complex phenomena finds itself in an uncomfortable situation compared with other natural sciences because it lacks a base outside its object. It can only translate itself back into its own language, or fashion itself in its own image. The more it extends its field of research and the more complicated its objects become, the more it feels the lack of a point which is distinct from those objects. And once the complexity has reached that of the empirical man, his psychology inevitably merges with the psychic process itself. It can no longer be distinguished from the latter, and so turns into it. But the effect of this is that the process attains to consciousness. In this way, psychology actualizes the unconscious urge to consciousness. It is, in fact, the coming to consciousness of the psychic process, but it is not, in the deeper sense, an explanation of this process, for no explanation of the psychic can be anything other than the living process of the psyche itself. Psychology is doomed to cancel itself out as a science and therein precisely it reaches its scientific goal. Every other science has so to speak an outside; not so psychology, whose object is inside the subject of all science."

hero must not stand still there. The real confrontation, the γνωϑι σαυτον is still awaiting the hero. After having defined himself in the face of nature and the original matrix, it is up to him to find another way for dealing with and relating to nature and the unconscious. Mostly the sciences, but also many of the different psychologies, fall short with the development of a comprehensive and beautiful theory. As important as this may be and actually is, it will never represent a final solution. As long as one remains caught by the habit of solving riddles, the whole process of theory formation and improvement turns to a childlike and unrelated play in the long run. The fact that every rational theory is found to be limited indicates a desire to go further to find what really lies behind it.

The Sphinx was sent as punishment by Hera because Laius had dared replace his wife with the boy Chrysippus. What does pedophilia mean symbolically, what are its archetypal roots and implications? Man becomes fascinated with the beauty of the young body of an innocent boy; in a psychological sense it is the pleasure of consciousness in beauty and playful, unintentional ease. It is not the unconditional involvement of the true hero with the unknown Other which he is confronting as an inner necessity. Pedophilia in classical Hellas was certainly accompanied with a magnificent strengthening of the reflecting mind and an emancipation from the grip of an overpowering nature. By turning his back on femininity and its threats, and shirking its fascination and omnipotence, man was able to free himself of nature and turn to the realm of the spirit to explore it. The result was Greek philosophy in its unequalled magnificence. At the same time, this meant escape from the natural marital order which Hera was guarding carefully and jealously. According to Hera's order, the opposites – man and woman, masculine and feminine, consciousness and unconscious – have to relate to each other, whereas in the pedophilic relation equal relates to equal (see, e.g., Wittgenstein 1961, 6.12). The qualitatively new, however, will only be possible and can only emerge when the Ego relates to the Other. Without the genuine Other – the open, the dark, the abyss, the "apeiron" of Anaximander[8] – as opposite, archetypally and symbolically experienced by man as woman and by woman as man, nothing new can possibly emerge.

At the same time, the pedophilia of Laius symbolically touches on the theme of the old king who does not want to abdicate eventually and to be vanquished by the son. As king he symbolically represents the established collective consciousness; he is not willing to be drawn into the cycle of nature of coming into being and passing away. By his homoerotic passion

---

[8] *apeiron*, Greek: the unlimited, boundless. According to Anaximander the *apeiron* is the unchanging, indefinite primordial principle from which all things emerge and return to again.

Laius wants to avoid the father's fate of being murdered by the son. In all our compulsions to produce final solutions, this "pedophilic" aspect is touched upon, the anxiety of being swallowed and dissolved in an evolving and fading of nature.

The oracle prophesies that if Laius is going to procreate, it will result in destruction; therefore he refuses to unite with Jocasta. However, she lures him into drunkenness; so it happens anyway, and the destiny so anxiously avoided takes its course. Reality is always more complex than that which can be expressed by a system bound to the rigor of formal logic. His nature – curiosity, hybris, greed, etc. – always seduces man to apply his beautifully imagined intellectual games and theories to nature, and then fate takes its course: over-population, environmental problems, Chernobyl, ...

On the one hand, there is the eternal rhythm of becoming and dying in nature, on the other the deep longing of man toward something lasting, certain, incorruptible. Insofar as he lives consciously, man is destined to being torn apart, to homelessness and tension. In the Egyptian myth of Isis and Osiris, a first attempt to find a solution to these existential problems is recorded (Neumann 1954). It tells the story of the transformation of Osiris from a corn-god and a god of fertility, of the yearly growing and fading, to the eternal god of the dead who find final and incorruptible certainty in him. Similarly, the alchemical opus was an attempt to make possible the *coniunctio* of Sol and Luna (masculine and feminine, spirit and nature, conscious and unconscious), and thus to produce the philosophers' stone, the lapis, the gold, and so forth as a substance no longer corruptible – and to produce it not only in the "Beyond", but in the here and now (Jung 1968). It was the aspiration of the alchemists never to neglect the body and corporeal world in their attempts to find redemption. Sol and Luna were their symbols of the absolutely opposite, such as consciousness and the unconscious, Ego and Other.

Modern science functions with the Cartesian assumption that the dark as such does not exist, everything can be reduced to a rational consciousness. The abyss, the awe, the Other, the numinous is reduced to subjective fantasy. The dark is not admitted *per definitionem*. Wherever it is encountered, it is eradicated and subjected to the dictates of the "either–or", as in the narrow path of the Cithairon: either Laius or Oedipus. It is always the attempt to create oneself by oneself exclusively, not to confront the opposite, not to face an end, not to sacrifice oneself by fear to get dissolved in the eternal cycle. It is a regressive way of dealing with the human condition. Religions and mysticism try to find a solution *facing* the paradox as did the alchemists. The way in which they understood their gold shows with particular clarity whether the paradox was endured or reduced.

Oedipus is a foundling, not knowing his real origin. Only through the deepest suffering and humiliation does he eventually realize it and, furthermore, that right from the beginning he is and was enfolded in a fate exceeding his rational understanding. The twofold origin and parenthood – earthly and divine – is an attribute of the heroine and the hero often found in myth and religion: the many Greek heroes such as Hercules, Perseus, and the Dioscuri, Adam and Eve with their origin in dust and in divine breath, or Jesus in our own Judeo-Christian tradition. Symbolically, this expresses the (psychological) fact that all conscious existence – i.e., the hero in opposition to man whose only origin is in the collective with which he is identical – has a twofold origin: on the one hand the causal, perfectly comprehensible, on the other the being just-so and not otherwise which corresponds to the divine breath, the spontaneous creative act, the non-contingent.

Initially, Oedipus is conscious only of the one side, and correlated with this one-sidedness is the Sphinx-side of nature as an irreconcilable enemy, a threatening opponent. Seen this way, the story of Oedipus demonstrates the dogged and stubborn insistence not to become a hero or to remain a "hero" according to one's own limited comprehension, not to give up the one-sidedness of consciousness but to remain identical with rational consciousness. In what unfolds, like an initiation into the real mystery of human existence, Oedipus becomes conscious of his real, twofold origin. And facing the second origin, that of being embedded in a divine fate, the question of the apparent origin, that of earthly parents and causal relations, loses its exclusive character. What the hero has to live consciously – i.e., to be related to the Other – happens to Oedipus unconsciously at first; he does not unite with a beloved, desired woman, but with his mother. It seems as if the myth wanted to show that there are only the alternatives of being related to the Other consciously – symbolically, man as a mature being related to the woman – or being in an unconscious incestuous union with the Other. Reducing the Other to a manageable device is equivalent to incest, an unconsciously consummated union with one's own mother as in Oedipus' case. Consciousness can be symbolized as the mature hero who confronts the unconscious as the desired Other and celebrates the *coniunctio* with it, or as a "hero" by his own grace and comprehension who returns to the mother and is unconsciously contained in her.

This brings us to the important theme of incest, which has a central place in modern psychology. Besides the concrete sexual aspect which Freud (1953b) has dealt with extensively, incest also has immensely symbolic contents which are often overlooked. Jung has devoted a good part of his work to this aspect of incest (Jung 1966, 1968). In Jung's understanding, the growing kernel of an ego during mankind's early periods of development as well as in the growth of every individual consciousness, is constantly in danger of disappearing back into the unconscious, of dissolving in nature (as we have seen in the Babylonian story of creation in Sect. 2; see also Neumann 1954). What starts as a necessity for being contained is eventually experienced as being held back. This threat finally leads to culmination in the fight of a sun-like hero against the dissolving forces. During those early stages of development, it would be deadly for consciousness to relate to the overpowering nature and the unconscious in the sense of a *coniunctio*. The incest taboo is an expression of the danger of uniting with the unconscious during this early period of growth of the ego. However, what is an obligation and a strict order during the phase of a growing consciousness does not hold any longer for the grown-up, mature ego.

Strangely, incest was even prescribed and, as Jung points out, was a prerogative for the chief of the tribe, the king in early times, and the pharaoh in old Egypt. Jung's interpretation is that the mature, strong, grown-up ego, which was projected onto the king as the incarnated principle of consciousness in contrast to collective man, must relate back to this Other, its matrix and origin. At an early stage, such a relation would be incest and strictly forbidden; in the mature phase it is a prerogative and a holy duty. During the early phases of the sciences it was certainly necessary to depotentiate the Other, the dark, death, mystery, and to fight against its paralyzing effect. In refusing to take the Other into account and to relate to it in a mature conscious way, the scientific enterprise has turned into an unconsciously consummated, scandalous incest.[9] Oedipus believed he impregnated his partner, but in reality he was contained in his mother by doing this. Blurring the beloved woman with the mother, the son is sexually united with her in an incestuous way without knowing it. This seems to be an accurate picture of a good part of contemporary scientific activity – the Other is depotentiated, cut off, and declared irrelevant[10] rather than supported to become conscious and considered as an autonomous partner.

In the unfolding of the tragedy of Oedipus, the Other becomes ever more distinct and the real pole of reference. At the end of *Oedipus at Colonus*, a divine Other is the hero's guide. Oedipus enters straightforwardly and observantly, conscious of the depths. It is and will remain a mystery how the split between consciousness and the unconscious can be healed, how consciousness can relate and unite with the unconscious without one of them being diminished. Solving the riddle of the Sphinx, Oedipus solved only part of the mystery of nature and human existence, namely that one which can be and has to be dealt with by human reason. In Colonus he en-

[9]Pauli took on the task of confronting the Other and wrestling with it. In the inner image and dream figure of the Chinese woman, it became an intimate inner partner to him. It was she who told Pauli to think about a "Hintergrundsphysik". See Pauli's "Klavierstunde" (Pauli 1995) and Jung (1968).

[10]This is especially revolting and macabre in the medical sciences, whenever the confrontation with death and sickness becomes reduced to an engineering problem.

gages in knowing the Other, where both partners are equivalent and neither of them is reduced nor destroyed. It is the *coniunctio* that the alchemists, for example, were striving for, risking all their means and existence. Man has been circling around it ever since – unconsciously or consciously – as the tabooed incest or as the final truth. By contrast to the public place in Thebes, where the riddle of the Sphinx was solved to the benefit of all the people of the town, the second encounter is an individual mystery. Not even the faithful daughter Antigone, who was Oedipus' guide in his blindness, is allowed to accompany him. Only Theseus, the hero and King of Athens, is witness of the transition. The mystery is not to be disclosed to the public.

4.

Averroês: "Quelle sorte de solution as-tu trouvée par l'illumination et l'inspiration divine? Est-ce identique à ce que nous dispense à nous la réflexion spéculative?"
Je lui répondis: "Oui et non. Entre le oui et non les esprits prennent leur vol hors de leur matière, et les nuques se détachent de leur corps."[11]
(Henry Corbin 1958)

In addition to the knowledge he possessed and relied on exclusively at first, Oedipus encountered and had to acknowledge another kind of knowing and being known. Solving the riddle of the Sphinx – that devouring, threatening side of the mother archetype – and proud of his skill, he united unconsciously with the feminine in its mother aspect. The tabooed incestuous union of Oedipus with the mother is accompanied by not knowing who his real father is and murdering him unknowingly. We have seen that the theme of the twofold descent – natural and divine – is characteristic for the hero and indicates his knowing about the two sides of conscious reality. This knowledge and the courage to face it distinguish the hero. By contrast to collective man, who only knows and tries to stick to one of them, the supposedly concrete side of reality, it is the hero's very essence to confront them both. The replacement of one physical theory by a more complete subsequent theory and the irresistible urge toward an absolute and final one seems to point beyond the claim of reason to represent the foundation of concrete reality alone. It points toward searching for its other side.

The male line of the family of Cadmus ends with the sons (or brothers) of Oedipus, Polyneices, and Etiocles; after Oedipus left Thebes, they

[11]Dialog between the Arabic philosopher Averroes (Ibn Ruschd, 1126–1198) and the Arabic mystic Ibn'Arabî (1165–1240).

kill each other in a fight over the throne of the city. Thus the development which we began to follow with Laius and his pedophilic inclination, takes a dead end; there is no future this way. Equal confronts equal in the two brothers: there are no general commitments or obligations, nothing which exceeds the goal of the ego. Everybody just stands for his or her own interest. It is like the dead end of a philosophy of "anything goes". Such a conception of progress in the sciences might seem correct from a purely formal, rational, and retrospective point of view; but is hopelessly false regarding the responsibility and experience lived by the hero – i.e., the conscious ego – in the moment of confrontation and struggle. Then it is never just a detached choice, but rather the utmost possible confrontation: son versus father, conscious versus unconscious, the ego versus the Other. To avoid the fate of the father and of the hero means to try to escape the natural rhythm of becoming and dying, as Laius did. Nature and spirit aim for transformation. "It is not progress that is demanded of us, but to be at the service of transformation, since all life desires transformation" (Schweizer 1994, p. 32). Thus the future, as the myth shows, is in Oedipus and the transformation which accompanies his suffering. It is he who attains another level of consciousness by self-sacrifice.

As we have seen, the recognition of a twofold foundation of conscious reality is at stake in Oedipus. Today such a recognition would represent more profound and deeper capacities for science to reach its goal. The very essence of consciousness seems to aim at distinguishing between conscious and unconscious. Separating them and keeping them apart is a necessary achievement at first, in the growth of consciousness, but not a goal in itself. Distinction is the necessary condition for conscious relationships. Jung (1968, par. 562f.) writes about this fundamental aspect of consciousness:

"... ever since the Age of Enlightenment and in the era of scientific rationalism, what indeed was the psyche? It had become synonymous with consciousness. The psyche was 'what I know'. There was no psyche outside the ego. Inevitably, then, the ego identified itself with the contents accruing from the withdrawal of projections. ... The contents that were formerly projected were now bound to appear as personal possessions, as chimerical phantasms of the ego-consciousness. ... An inflated consciousness is always egocentric and conscious of nothing but its own presence. It is incapable of learning from the past, incapable of understanding contemporary events, and incapable of drawing right conclusions about the future. It is hypnotized by itself and therefore cannot be argued with. Paradoxically enough, inflation is a regression of consciousness into unconsciousness. This always happens when consciousness takes too many unconscious contents upon itself and loses the faculty of discrimination, the sine qua non of all consciousness. When

fate, for four whole years, played out a war of monumental frightfulness
on the stage of Europe – a war that nobody wanted – nobody dreamed
of asking exactly who or what had caused the war and its continuation. Nobody realized that European man was possessed by something
that robbed him of all free will. And this state of unconscious posses-
sion will continue undeterred until we Europeans become scared of our
'god-almightyness'. Such a change can begin only with individuals, for
the masses are blind brutes, as we know at our cost. It seems to me
of some importance, therefore, that a few individuals, or people individually, should begin to understand that there are contents which do
not belong to ego-personality, but must be ascribed to a psychic non-
ego. This mental operation has to be undertaken if we want to avoid a
threatening inflation."

The sacrifice of an exclusive pretense to dominance of reason, in the
myth of Oedipus the sacrifice of his eyes, finally makes the hero enter consciously and upright into the Beyond and celebrate the *coniunctio* with the
Great Mother Goddess – central mystery of the Eleusinian mysteries. Or,
as Jung has expressed it with the images rooting in our Judeo-Christian
heritage and with the experience of modern empirical psychology (Jung
1958, par. 740):

"The unconscious wants both: to divide and to unite. In this striving
for unity, therefore, man may always count on the help of a metaphysical advocate, as Job clearly recognized. The unconscious wants to flow
into consciousness in order to reach the light, but at the same time it
continually thwarts itself, because it would rather remain unconscious.
That is to say, God wants to become man, but not quite. The conflict
in his nature is so great that the incarnation can only be bought by an
expiatory sacrifice offered up to the wrath of God's dark side."

This sacrifice occurs essentially in the individual and in silence, not in
declaration, and yet it transforms the world.

References

Atmanspacher H. (1996): Erkenntnistheoretische Aspekte physikalischer Vorstellungen
  von Ganzheit. *Zeitschr. Grenzgeb. Psych.* **38**, 20–45.
Baudelaire C. (1857): Correspondances. In *Les Fleurs du Mal* (Flammarion, Paris 1991),
  pp. 62–63.
Breuer T. (1997): Quantenmechanik: Ein Fall für Gödel? (Spektrum, Heidelberg).
Calasso R. (1988): *Le nozze di Cadmo e Armonia* (Adelphi, Milano).
Corbin H. (1958): *L'imagination créatrice dans le soufisme d'Ibn'Arabî* (Flammarion,
  Paris), p. 39.
Denman C. (1995): Ödipus ist keine Inzesttragödie. *Gorgo. Zeitschrift für archetypische
  Psychologie und bildhaftes Denken* **29**, 12–16.
Deussen P. (1963): *Sechzig Upanishad's des Veda* (Wiss. Buchgesellschaft, Darmstadt).
Eliade M., ed. (1980): *Schöpfungsmythen* (Wiss. Buchgesellschaft, Darmstadt).

Eliot A., ed. (1976): *Mythen der Welt* (Bucher, Luzern).

Fechner G. T. (1861): *Über die Seelenfrage* (Amelang, Leipzig).

Franz M.-L. von (1994): *Archetypische Dimensionen der Seele* (Daimon, Einsiedeln).

Franz M.-L. von (1995): *Creation Myths. Patterns of Creativity Mirrored in Creation Myths* (Shambala, Boston).

Freud S. (1953a): *Interpretation of Dreams* (Hogarth, London).

Freud S. (1953b): *The Standard Edition of the Complete Psychological Works, Vol. 7.* (Hogarth, London).

Gebser J. (1977): *Ursprung und Gegenwart* (Novalis, Schaffhausen).

Guirand F. (1935): *Mythologie générale* (Librarie Larousse, Paris).

Hornung E. (1983): *Der Eine und die Vielen* (Wiss. Buchgesellschaft, Darmstadt).

Jung C.G. (1958): Psychology and religion. West and East. *Collected Works 11* (Princeton University Press, Princeton).

Jung C.G. (1966): The practice of psychotherapy. *Collected Works 16* (Princeton University Press, Princeton).

Jung, C.G. (1968): Psychology and alchemy. *Collected Works 12* (Princeton University Press, Princeton).

Jung, C.G. (1969): The structure and dynamics of the psyche. *Collected Works 8* (Princeton University Press, Princeton).

Jung, C.G. (1976a): Symbols of transformation. *Collected Works 5* (Princeton University Press, Princeton).

Jung, C.G. (1976b): Psychological types. *Collected Works 6* (Princeton University Press, Princeton).

Jung C.G. and Pauli W. (1955): *The Interpretation of Nature and the Psyche* (Pantheon, New York).

Kerényi K. (1951): *The Gods of the Greeks* (Thames & Hudson, London/New York).

Lévy-Bruhl L. (1923): *Primitive Mentality* (Macmillan, New York).

Lévy-Bruhl L. (1926): *How Natives Think* (Macmillan, New York).

Neumann E. (1954): *The Origins and History of Consciousness* (Harper, New York).

Nishitani K. (1982): *Was ist Religion?* (Insel, Frankfurt).

Pauli W. (1954): Naturwissenschaftliche und erkenntnistheoretische Aspekte der Ideen vom Unbewußten. *Dialectica* **8**, 283–301.

Pauli W. (1984): *Physik und Erkenntnistheorie* (Vieweg, Braunschweig).

Pauli W. (1995): Die Klavierstunde. In *Der Pauli–Jung Dialog und seine Bedeutung für die moderne Wissenschaft*, ed. by H. Atmanspacher, H. Primas, and E. Wertenschlag (Springer, Berlin), pp. 317–330.

Primas H. (1983): *Chemistry, Quantum Mechanics and Reductionism* (Springer, Berlin).

Primas H. (1992a): Umdenken in der Naturwissenschaft. *Gaia* **1**, 5–15.

Primas H. (1992b): Es gibt keine Einsicht ohne innere Bilder. *Gaia* **1**, 311–312.

Primas H. (1993): The Cartesian cut, the Heisenberg cut, and disentangled observers. In *Symposia on the Foundations of Modern Physics, Wolfgang Pauli as a Philosopher*, ed. by K.V. Laurikainen and C. Montonen (World Scientific, Singapore), pp. 245–269.

Primas H. (1995): Über dunkle Aspekte der Naturwissenschaft. In *Der Pauli-Jung Dialog und seine Bedeutung fur die moderne Wissenschaft*, ed. by H. Atmanspacher, H. Primas, and E. Wertenschlag (Springer, Berlin), pp. 205–238.

Primas H. (1996): Synchronizität und Zufall. *Zeitschr. Grenzgeb. Psych.* **38**, 61–91.

Ranke-Graves R. von(1955): *Griechische Mythologie – Quellen und Deutung (2 Bände)* (Rowohlt, Hamburg).

Rilke R.M. (1982): *The Selected Poetry of Rainer Maria Rilke* (Vintage, New York).

Schrödinger E. (1935): Die gegenwärtige Situation in der Quantenmechanik. *Die Naturwissenschaften* **23**, 807–812, 823–828, 844–849.

Schweizer A. (1994): *Seelenführer durch den verborgenen Raum. Das ägyptische Unterweltsbuch Amduat* (Kösel, München).

Sophocles (1991a): Oedipus the King. In *The Complete Greek Tragedies: Sophocles I*, ed. and transl. by D. Grene (University of Chicago Press, Chicago).

Sophocles (1991b): Oedipus at Colonus. In *The Complete Greek Tragedies: Sophocles I*, ed. and transl. by D. Grene (University of Chicago Press, Chicago).

Sproul B. (1979): *Primal Myths. Creating the World* (Harper, San Francisco).

Wittgenstein L. (1961): *Tractatus Logico-Philosophicus* (Routledge and Kegan Paul, London).

Wheeler J.A. and Zurek W.H., eds. (1983): *Quantum Theory and Measurement* (Princeton University Press, Princeton).

Carl Friedrich von Weizsäcker:
REFLECTIONS ON QUANTUM THEORY

There are two very fundamental questions in quantum mechanics:

– What are "elementary" quantum building blocks and how can one use them to compose more complex quantum systems?
– How can one define pure states, what is their interpretation, and which role do they play in quantum systems which are not completely isolated?

These questions puzzled both Hans Primas and Carl Friedrich von Weizsäcker for many years. They had a number of discussions on these matters and developed entirely different conceptual viewpoints.

Weizsäcker considered the simplest spin system (built up by three Pauli matrices and the unit matrix) as elementary and called it "ur" (or, more precisely, "ur-alternative"). Physical systems and the whole universe as such should be built up from a large (but finite) number of these urs and thereby give rise to complex quantum systems. Primas, on the other hand, considered an abstract quantum system – consisting only of an algebra of operators and a time evolution – as insufficient for a physical description and insisted on the use of additional structure such as different kinematical groups (Galilei group, Lorentz group, etc.) for different contexts.

As a consequence, "elementarity" has to be defined in a context-dependent way, i.e., depending on the relevant kinematical group. Starting from the elementarity concepts of irreducibility (in quantum mechanics) and transitivity (in classical mechanics), he and Anton Amann came to the conclusion that a quantum system should be called elementary if the respective kinematical group acts ergodically (Amann 1979, 1986). This means that every operator in the algebra of observables is nontrivially transformed by at least one symmetry (one group element) from the kinematical group. An ur is elementary from this point of view, when the kinematical group is given as $SU(2)$, but there are many other $SU(2)$-elementary systems, both *classical and quantum mechanical; there are even elementary systems with joint quantum and classical properties.*

Weizsäcker was skeptical about introducing classical properties at a fundamental level, whereas Primas' algebraic approach allows one to deal with classical and quantum properties in the same formal framework. Primas was puzzled by the fact that classical behavior can arise automatically in systems with infinitely many degrees of freedom. He argued that each physically relevant system actually is of this type insofar as it is coupled to the electromagnetic or gravitational field. Weizsäcker wanted to introduce classical behavior by stability considerations of systems with finitely many degrees of freedom, i.e., consisting of finitely many urs, by showing that certain states are unstable under "small perturbations," without reference to fields with infinitely many degrees of freedom.

What is the role of pure states in this debate? Pure states historically arose through the Hilbert space formalism of quantum mechanics together with the irreducibility postulate of von Neumann. Every state vector ψ in an irreducible Hilbert space representation gives rise to a pure state ϕ with expectation values $\phi_\psi(A)$,

$$\phi_\psi(A) = \langle \psi | A\psi \rangle, \tag{1}$$

for every observable A. The mathematical term "state" was then used for all (pure and non-pure) linear, normalized and positive expectation value functionals ϕ, and a pure state was shown to be extremal within the set of all states of a given algebra of observables.

Primas was not satisfied with "extremal normalized positive linear functionals" but wanted to have a more physical derivation of pure states ϕ_o. Together with Guido Raggio he worked out a solution based on truth-definite yes-no observables (Raggio 1981, Amann 1987). These are given by all projectors (i.e., observables p satisfying $p^2 = p = p^*$) whose expectation value $\phi_o(p)$ is either one (true) or zero (false), $\phi_o(p) \in \{0, 1\}$. They showed that a state ϕ is pure if and only if its set of truth-definite projections is maximal (among all projections in the algebra of observables). Hence a pure state ϕ_o represents nothing else than a maximal set of truth-definite projections. Though all expectation values of truth-indefinite projections and arbitrary observables can be reconstructed from the truth-definite ones, Primas considered such more general expectation values $\phi_o(p) \notin \{0, 1\}$ as physically irrelevant.

The truth-definiteness of some projector leaves two possible alternatives open: the corresponding proposition may be "true" ($\phi_o(p) = 1$) or "false" ($\phi_o(p) = 0$). In Weizsäcker's approach such alternatives are already built in as the "urs" or "ur-alternatives". Hence, Weizsäcker's elementary systems are also elementary alternatives or a kind of quantum mechanical yes-no *decisions. This is somewhat related to quantum bits or q-bits that play a significant role in contemporary discussions.*

For Weizsäcker, a pure state represents exhaustive knowledge about a physical system. In this scheme, every expectation value (even $\phi_0(p) \notin \{0,1\}$) can arise through a probability interpretation. This Copenhagen-inspired point of view is in contrast to Primas' view who does not attribute a probability interpretation to pure states. The question as to why our knowledge should follow the Schrödinger equation even led him to a skeptical view about probabilities at the level of non-pure states represented by density operators. Since a non-pure state can be decomposed into pure states in infinitely many different ways, the attribution of probabilities of pure states (arising in a "mixture") is disconcerting for Primas. What he accepts without hesitation is a (classical or central) decomposition of a non-pure state and an associated probability interpretation due to classical observables (in the center of the algebra of observables).

For both Weizsäcker and Primas pure states play a decisive role in quantum mechanics. Weizsäcker's ur-alternatives are "quantum yes-no decisions". He thinks in terms of a large but finite number of urs, i.e., in terms of systems which are entirely quantum mechanical, without any classical observables. Primas insists on classical observables, classical decompositions of non-pure states, and classical yes-no alternatives. One of his central ideas is that classical observables and classical decompositions of non-pure states enter automatically via infinitely many degrees of freedom.

References

Amann A. (1979): *Über eine Verallgemeinerung des quantenmechanischen Elementaritätsbegriffes* (Diploma Thesis, ETH Zürich).

Amann A. (1986): Observables in W* algebraic quantum mechanics. *Fortschr. Phys.* **34**, 167–215; here: pp. 171–177.

Amann A. (1987): Jauch-Piron states in W*-algebraic quantum mechanics. *J. Math. Phys.* **28**, 2384–2389.

Raggio G. (1981): *States and Composite Systems in W*-Algebraic Quantum Mechanics* (Thesis ETH Zürich No. 6824, ADAG AG).

REFLECTIONS ON THE FOUNDATIONS OF QUANTUM THEORY

CARL FRIEDRICH VON WEIZSÄCKER
Maximilianstr. 14c, D-82319 Starnberg, Germany

Concerning the foundations of quantum theory, I had discussions with Hans Primas for a number of years. I do not want to reconsider these discussions in detail here. What I rather want to do is to sketch some arguments which came to my mind recently.

Richard Feynman is often quoted with the statement that he himself does not understand quantum theory, and at present no one would be living who understands it (Feynman 1985, pp. 9f): "You're not going to be able to understand it ... my physics students don't understand it either. That is because I don't understand it. Nobody does. ... The theory ... describes Nature as absurd from the point of view of common sense."

What does "understanding" mean here? Quantum theory has a basic mathematical structure, the foundations of which can be summarized in compact form on a few pages. We know how to apply the mathematics for empirical results. There is a tremendous number of empirical results which agree with the predictions obtained from the mathematical formulation of quantum theory. In spite of careful search, no single empirical result has been found that would contradict quantum theory. On the other hand, each year new books are published on the interpretation of quantum theory. It was only recently that I once again witnessed a discussion about this interpretational issue among physicists who know each other for decades and nevertheless could not agree.

Why this difficulty? Contemporary physics still denotes its objects in the language of classical physics; space and time, fields and particles. In the terminology of Thomas Kuhn's view toward the history of science (Kuhn 1962), classical physics was a paradigm to solve scientific problems. In a certain sense, this classical paradigm has been superseded by a scientific revolution leading to a new paradigm provided by quantum theory. However, quantum theory still describes its objects using the language of the classical paradigm.

In terms of Heisenberg's view toward the history of science (Heisenberg 1948), formulated quite a time before Kuhn had developed his ideas, both paradigms are "closed theories", i.e., theories that cannot be further improved by any small modifications. The revolution that quantum theory brought along is a transition to a theory with radically novel concepts. Nonetheless, the success of its classical predecessor has to be explainable in terms of the later, more advanced theory. The concepts and structures of the older theory should be recovered as limiting cases of its successor.

The debate about the interpretation of quantum theory will remain unsolved as long as one tries to provide interpretations in terms of concepts that fit to its classical predecessor only. In my own interpretational attempts (Weizsäcker 1985), I have tried to understand quantum theory from the perspective of its non-classical preconditions and concepts, and to derive classical concepts in certain limits. The present contribution intends to contribute to a further elaboration of this program.

The main assumption is that quantum theory is a theory of the continuum. Let me explain this assumption by defending it against the obvious objection that the continuum is a concept of mathematics, quantum theory, however, is physics.

First defense: Mathematics and physics, both founded in ancient Greek history, share a number of common basic concepts. "Continuum" is a concept of geometry. "Geometry", however, means as much as topographic survey, and as such it represents a (partly) physical procedure. Mathematics then serves the purpose of analyzing a conceptual procedure (*pro-cedere* in the sense of a human activity). Physics applies mathematics for its description of natural experience (*experientia* in the sense of an encounter with reality).

Second, more detailed defense: The mathematical concept of a continuum remains insufficiently clarified until today. Aristotle defined continuum (*syneches*, that what holds together) as something which can be divided into parts of the same kind in an unlimited manner, but only in a finite number of steps (open finitism; cf. Weizsäcker 1971, Chap. IV.4). This view was maintained by leading mathematicians up to the late 19th century when Cantor introduced the continuum as an uncountably infinite point set. This produced the paradoxes of set theory from Russell to Gödel. And at the same time it led to the origins of quantum theory in physics – the infinite-dimensional state space of the electromagnetic field (Planck), and Rutherford's atom model as represented by Bohr in 1913. This demonstrates that the paradigm of classical physics failed due to an overly naive use of the set theoretical continuum. A scientific revolution often arises as the consequence of a sequence of internal, hidden self-contradictions within an existing paradigm.

For these reasons, I tried to construct an abstract quantum theory, re-nouncing a priori classical concepts such as space and fields and particles, but rather deriving them as secondary concepts by certain approximations (Weizsäcker 1985, Chaps. 8,9). This approach starts with the concept of a "finite, empirically decidable alternative" A_n ($n \in N$). Then states z can be introduced which are different from the disjoint outcomes x, y of an alternative, but can be characterized by conditional probabilities $p(z|x)$ and $p(z|y)$. From this one can derive an n-dimensional complex vector space as state space. Decomposing all A_n into two-dimensional "ur"-alternatives, a real, $(3+1)$-dimensional position (Minkowski) space can be obtained. This is possible since the two-dimensional complex Lie group $SL(2, C)$ is isomorphic to the real Lorentz group up to a trivial ambiguity. In the resulting position space concepts such as particles and fields can be defined by proper approximations.

This procedure does not yet solve the problem of the continuum though. The space of all possible probabilities and thus the resulting position space is still an uncountably infinite point set. This leads to the postulate of second quantization. However, this creates the same problem again and therefore seems to entail arbitrarily multiple quantizations. This result is not surprising since traditional quantum theory uses a classical set theo-retical concept of the continuum.

One can ask whether the geometry of the continuum can be constructed in an Aristotelian manner, i.e., without using the concept of a point set. First steps of a corresponding approach have been published elsewhere (Weizsäcker 1997). Let me give some brief philosophical comments on this approach.

The fundamental concept that we need here is the concept of time. Time is the only fundamental concept of classical physics which is indispensable in my approach, but it has to be interpreted more precisely than in classical physics. (A corresponding discussion can be found in Weizsäcker (1985), Chaps. 1–5.) This approach is motivated by Kant's idea that a theory which is confirmed by experience has to comprise a formulation of the necessary conditions for experience. Experience means to learn from the past for the future. Our theories comprise what we know about past and future events. What we know about the past is a discrete set of facts. What we know about the future are contingent possibilities. They are described as continua, e.g., with probabilities or ψ-functions. This has to be analyzed in more detail.

In our experience, time proceeds continuously. If the continuum is not assumed to be a point set then time isn't either. In a truly Aristotelian spirit, one can divide a period of time only into finitely many, perhaps very many, shorter periods of time. And the boundaries of such a period, its beginning as well as its ending, must not be conceived of as points. Which kind of mathematics do we have to choose in order to be able to describe this properly?

Following Brouwer, the natural numbers 1, 2, 3, 4, 5, 6, 7, ... are representations of the act of counting. The integer numbers result from operations within the natural numbers: addition and subtraction. In this way they form a group; as far as they are used to characterize temporal processes, they cover past and future. If we define the continuum of time by divisibility, in an Aristotelian manner, then we are led to the rational numbers, fractions with finite nominators and denominators.

Within this framework, quantum theory can be understood as a theory of temporal processes (Weizsäcker 1997). The simplest kind of such a process is a homogeneous process characterized by $x = vt$, where t is a measure of time. This may be considered as a basis of the theory. It is a representation of the law of inertia, free motion without additional forces.

If there are mutual interactions between arbitrary objects then the theory is more complicated. An ensemble of, say, n objects is described by a time-dependent real n-dimensional vector space, where the mutual interactions between the objects are described by $n \times n$-dimensional matrices. These vectors and matrices are to be constructed as rational numbers. Complex numbers arise as special nilpotent 2×2 matrices whose components are rational numbers. On this basis, abstract and, eventually, concrete quantum theory have to be established.

This approach offers a way to reconstruct in a fundamental manner the paradigm of classical physics as a limiting case within the paradigm of quantum theory. Classically, objects possess continuous properties. By quantization they are reduced to discrete values, but one needs to introduce the ψ-function as a continuous probability amplitude. By contrast, the procedure sketched here starts with introducing a ψ-function as an n-dimensional real vector for n interacting "ur"-objects. The possible values for the observables are then the eigenvalues of the corresponding $n \times n$ matrix. The eigenfunctions of the "ur"-objects are time-dependent real numbers. For the formalization of "urs" in a complex quantum theory, real four-vectors are needed (Weizsäcker 1997).

Due to an argument by Castell, one can thus construct the conformal group of relativity theory (Weizsäcker 1985, pp. 404ff). I did not succeed in deriving the interaction of particles when I worked at this problem. At present I suspect that it is necessary to include real vectors in order to achieve this. Real matrices are required to incorporate representations of the "ghosts" in the interactions.

Finally let me emphasize that all this is just a program – however, it is a program on which I place hope, and therefore I am grateful to discuss it with colleagues.

References

Feynman R.P. (1985): *QED - The Strange Theory of Light and Matter* (Princeton University Press, Princeton).

Heisenberg W. (1948): Der Begriff "abgeschlossene Theorie" in der modernen Naturwissenschaft. *Dialectica* **2**, 331–336.

Kuhn T.S. (1962): *The Structure of Scientific Revolutions* (University of Chicago Press, Chicago).

Weizsäcker C.F. von (1971): *Die Einheit der Natur* (Hanser, München).

Weizsäcker C.F. von (1985): *Aufbau der Physik* (Hanser, München).

Weizsäcker C.F. von (1997): Time – Empirical Mathematics – Quantum Theory. In *Time, Temporality, Now*, ed. by H. Atmanspacher and E. Ruhnau (Springer, Berlin), pp. 91–104.

PUBLICATIONS BY HANS PRIMAS (UNTIL 1998)

Osimitz F. and Primas H. (1950): Tüpfelreaktionen. *Schweiz. Laboranten-Zeitung* **7**, 2–7.

Primas H., Lasman H., and Osimitz F. (1950): Moderne Vorschriften zur qualitativen Kationenanalyse. *Schweiz. Laboranten-Zeitung* **7**, 98–114.

Primas H. and Günthard Hs.H. (1953): Die Infrarotspektren von Kettenmolekülen der Formel R'CO(CH"CH")$_n$COR". I. Rocking- und Twisting-Grundtöne. *Helv. Chim. Acta* **36**, 1659–1670.

Primas H. and Günthard Hs.H. (1953): Die Infrarotspektren von Kettenmolekülen der Formel R'CO(CH"CH")$_n$COR". II. Die Normalschwingungen des Symmetrietypus B_u. *Helv. Chim. Acta* **36**, 1791–1803.

Primas H. and Günthard Hs.H. (1954): Spectres infrarouges de derivés carbonyliques du type R'CO(CH"CH")$_n$COR" contenant plus de dix groupes méthyléniques. *J. de Physique et le Radium* **15**, 209–211.

Primas H. and Günthard Hs.H. (1954): Theorie der Form von Absorptionsbanden suspendierter Substanzen und deren Anwendung auf die Nujolmethode in der Infrarotspektroskopie. *Helv. Chim. Acta* **37**, 360–374.

Primas H. and Günthard Hs.H. (1955): Theorie der Intensitäten der Schwingungsspektren von Kettenmolekeln. I. Allgemeine Theorie der Berechnung von Intensitäten der Infrarotspektren von grossen Molekeln. *Helv. Chim. Acta* **38**, 1254–1262.

Primas H. and Günthard Hs.H. (1956): Theorie der Intensitäten der Schwingungsspektren von Kettenmolekeln. II. Zur Berechnung der Intensitäten der Infrarotspektren von freien Kettenmolekeln der Symmetrie C_{2h}. *Helv. Chim. Acta* **39**, 1182–1192.

Günthard Hs.H. and Primas H. (1956): Zusammenhang von Graphentheorie und MO-Theorie von Molekeln mit Systemen konjugierter Bindungen. *Helv. Chim. Acta* **39**, 1645–1653.

Primas H. (1957): Ein Kernresonanzspektrograph mit hoher Auflösung. I. Theorie der Liniendeformation in der hochauflösenden Kernresonanzspektroskopie. *Helv. Phys. Acta* **30**, 297–314.

Primas H. and Günthard Hs.H. (1957): Ein Kernresonanzspektrograph mit hoher Auflösung. II. Beschreibung der Apparatur. *Helv. Phys. Acta* **30**, 315–330.

Primas H. and Günthard Hs.H. (1957): Herstellung sehr homogener axialsymmetrischer Magnetfelder. *Helv. Phys. Acta* **30**, 331–346.

Primas H. and Günthard Hs.H. (1957): Field stabilizer for high resolution nuclear magnetic resonance. *Rev. Sci. Instrum.* **28**, 510–514.

Primas H. and Günthard Hs.H. (1957): Hochauflösender Kernresonanzspektrograph. *Chimia* **11**, 130–132.

Primas H., Frei K., and Günthard Hs.H. (1958): Protonenresonanzspektren einfacher cyclischer Aether und Ketone I. *Helv. Chim. Acta* **41**, 35–38.

Primas H. (1958): Ein Modulationsverfahren für die Kernresonanzspektroskopie hoher Auflösung. *Helv. Phys. Acta* **31**, 17–24.

Primas H. and Günthard Hs.H. (1958): Eine Methode zur direkten Berechnung des Spektrums der von quantenmechanischen Systemen absorbierten bzw. emittierten elektromagnetischen Strahlung. *Helv. Phys. Acta* **31**, 413–434.

Primas H. (1959): A new method for analyzing spectra in high resolution NMR spectroscopy. In *Proceedings of the Conference of Molecular Spectroscopy*, ed. by R. Thornton and H.W. Thompson (Pergamon, London), pp. 19–25.

Primas H. (1959): Anwendungen der magnetischen Kernresonanz in der Chemie. *Chimia* **13**, 15–23.

Primas H., Arndt R., and Ernst R. (1959): Die Konstruktion von Kernresonanz-Spektrographen hoher Auflösung Ia. *Z. für Instrumentenkunde* **67**, 293–300.

Primas H., Arndt R., and Ernst R. (1960): Die Konstruktion von Kernresonanz-Spektrographen hoher Auflösung Ib. *Z. für Instrumentenkunde* **68**, 8–13.

Primas H., Arndt R., and Ernst R. (1960): Die Konstruktion von Kernresonanz-Spektrographen hoher Auflösung. II. Die Konstruktion des Hochfrequenzteiles von Kernresonanz-Spektrographen hoher Auflösung. *Z. für Instrumentenkunde* **68**, 21–29.

Primas H., Arndt R., and Ernst R. (1960): Die Konstruktion von Kernresonanz-Spektrographen hoher Auflösung. III. Einige aktuelle Probleme der Kernresonanz-Instrumentierung. *Z. für Instrumentenkunde* **68**, 55–62.

Primas H. (1961): Über quantenmechanische Systeme mit einem stochastischen Hamiltonoperator. *Helv. Phys. Acta* **34**, 36–57.

Primas H. (1961): Eine verallgemeinerte Störungstheorie für quantenmechanische Mehrteilchenprobleme. *Helv. Phys. Acta* **34**, 331–351.

Primas H. (1962): 35 Jahre Quantenchemie. *Chimia* **16**, 281–289.

Primas H., Arndt R., and Ernst R. (1962): Group contributions to the chemical shift in proton magnetic resonance of organic compounds. In *Advances in Molecular Spectroscopy (Proceedings of the International Meeting of Molecular Spectroscopy, Bologna 1959)*, ed. by A. Mangini (Pergamon Press, Oxford), pp. 1246–1252.

Ernst R. and Primas H. (1962): High resolution NMR-instrumentation: recent advances and prospects. *Disc. Faraday Soc.* **34**, 43–51.

Ernst R. and Primas H. (1963): Nuclear magnetic resonance with stochastic high-frequency fields. *Helv. Phys. Acta* **36**, 583–600.

Primas H. (1963): Generalized perturabtion theory in operator form. *Rev. Mod. Phys.* **35**, 710–712.

Banwell C.N. and Primas H. (1963): On the analysis of high-resolution nuclear magnetic resonance spectra. I. Methods of calculating NMR spectra. *Mol. Phys.* **6**, 225–256.

Ernst R. and Primas H. (1963): Gegenwärtiger Stand und Entwicklungstendenzen in der Instrumentierung hochauflösender Kernresonanz-Spektrometer. *Ber. Bunsengesellschaft phys. Chemie* **67**, 261–267.

Primas H. (1964): Was sind Elektronen? *Helv. Chim. Acta* **47**, 1840–1851.

Huber A. and Primas H. (1965): On the design of wide range electromagnets of high homogeneity. *Nucl. Instruments and Methods* **33**, 125–130.

Primas H. (1965): Separability in many-electron systems. *Modern Quantum Chemistry, part 2*, ed. by O. Sinanoglu (Academic Press, New York), pp. 45–74.

Primas H. (1965): Was sind Elektronen? *Chimia* **19**, 399.

Primas H. und Riess J. (1966): Linear diamagnetic and paramagnetic response. In *Quantum Theory of Atoms, Molecules, and the Solid State. A Tribute to John C. Slater*, ed. by P.O. Löwdin (Academic Press, New York), pp. 319–333.

Günthard Hs. H. und Primas H. (1967): Prof. Dr. A. Stieger, 80jährig. *Titania Winterthur AHC*, 5–6.

Primas H. (1967): A density functional representation of quantum chemistry. I. Motivation and general formalism. *Int. J. Quantum Chem.* **1**, 493–519.

Primas H. (1968): Zur Theorie grosser Molekeln. I. Revision der Grundlagen der Quantenchemie. *Helv. Chim. Acta* **51**, 1037–1051.

Riess J. und Primas H. (1968): A variational principle for the phase of the wave function of molecular systems. *Chem. Phys. Lett.* **1**, 545–548.

Primas H. (1968): Lars Onsager – ein Meister der theoretischen Chemie. *Neue Züricher Zeitung*, Nr. 700, p. 5.

Primas H. (1973): Chemische Bindung. Ausgearbeitet von A. Wokaun (Verlag der Fachvereine, Zürich); 2. Auflage 1975.

Primas H. und Schleicher M. (1975): A density functional representation of quantum chemistry. II. Local quantum theories of molecular matter in terms of the charge density operator do not work. *Int. J. Quantum Chem.* **9**, 855–870.

Schleicher M. und Primas H. (1975): A density functional representation of quantum chemistry. III. Rigorous realization of the program in lattice space. *Int. J. Quantum Chem.* **9**, 871–886.

Primas H. (1975): Pattern recognition in molecular quantum mechanics. I. Background dependence of molecular states. *Theoret. Chim. Acta* **39**, 127–148.

Primas H. (1976): Gibt es eine theoretische Chemie? In *Studienführer Chemie*, hrsg. von der Vereinigung der Assistenten an den chemischen Laboratorien der ETH Zürich, 85–87.

Primas H. (1977): Theory reduction and non-Boolean theories. *J. Math. Biology* **4**, 281–301.

Primas H. and Müller-Herold U. (1978): Quantum mechanical system theory: A unifying framework for observations and stochastic processes in quantum mechanics. *Adv. Chem. Phys.* **38**, 1–107.

Saraswati D.K. and Primas H. (1978): A system theoretic representation of mechanical systems. I. Decomposition of a mechanical system into a hierarchy of orthogonal stationary linear dynamical systems. *J. Math. Phys.* **19**, 2646–2654.

Saraswati D.K. and Primas H. (1978): A system theoretic representation of mechanical systems. II. Stochastic interpretation. *J. Math. Phys.* **19**, 2655–2658.

Primas H. (1978): Kinematical symmetries in molecular quantum mechanics. In *Lecture Notes in Physics* **79** (Springer, Berlin), pp. 72–91.

Primas H. (1978): *Elemente der Gruppentheorie* (Verlag der Fachvereine, Zürich).

Primas H. (1979): Chemie, Reduktionismus und Quantenlogik. *Match* **7**, 217.

Primas H. and Gans W. (1979): Quantenmechanik, Biologie und Theoriereduktion. In *Materie–Leben–Geist. Zum Problem der Reduktion der Wissenschaften*, hrsg. von B. Kanitscheider (Duncker & Humblot, Berlin), pp. 15–42.

Primas H. (1980): Foundations of theoretical chemistry. In *Quantum Dynamics of Molecules. The New Experimental Challenge to Theorists*, ed. by R.G. Woolley (Plenum Press, New York), pp. 39–113.

Primas H. (1981): *Chemistry, Quantum Mechanics and Reductionism* (Springer, Berlin); 2. Auflage 1983.

Primas H. (1982): Chemistry and complementarity. *Chimia* **36**, 293–300.

Raggio G.A. and Primas H. (1982): Remarks on "'On completely positive maps in generalized quantum dynamics". *Found. Phys.* **12**, 433–435.

Primas H. (1983): Quantum mechanics and chemistry. In *Les Fondements de la Mécanique Quantique*, ed. by C. Gruber, C. Piron, T.M. Tâm (A. V. C. P., Lausanne), 255–270.

Primas H. (1984): Verschränkte Systeme und Komplementarität. In *Moderne Naturphilosophie*, hrsg. von B. Kanitscheider (Königshausen & Neumann, Würzburg), pp. 243–260.

Primas H. (1984): Concepts of quantum theory. Review of *Quantum Theory and Measurement* by J.A. Wheeler and W.H. Zurek (Princeton University Press, Princeton). *Nature* **308**, 782–783.

Primas H. and Müller-Herold U. (1984): *Elementare Quantenchemie* (Teubner, Stuttgart); 2. Auflage 1990.

Primas H. (1984): Was ist die Teilchenforschung wert? Beitrag zu einer Podiumsdiskussion. In *Bild der Wissenschaft* **21**(9), 126–141.

Primas H. (1985): Kann Chemie auf Physik reduziert werden? *Neue Züricher Zeitung*, Nr. 42, 67–68.

Primas H. (1985): Kann Chemie auf Physik reduziert werden? Erster Teil: das molekulare Programm, zweiter Teil: die Chemie der Makrowelt. *Chemie in unserer Zeit* **19**, 109–119, 160–166.

Primas H. (1986): Die Quantenmechanik als umfassende Theorie der Realität. In *Europäisches Forum Alpbach*, hrsg. von O. Molden, (Österreichisches College, Wien), pp. 140–155.

Primas H. (1987): Objekte in der Quantenmechanik. In *Grazer Gespräche 1986: Ganzheitsphysik*, hrsg. von M. Heindler und F. Moser (Technische Universität, Graz), pp. 163–201.

Primas H. (1987): Contextual quantum objects and their ontic interpretation. In *Symposium on the Foundations of Modern Physics 1987. The Copenhagen Interpretation 60 Years after the Como Lecture*, ed. by P. Lahti and P. Mittelstaedt (World Scientific, Singapore), pp. 251–275.

Primas H. (1988): Zum Theorienpluralismus in den Naturwissenschaften. In *Wozu Wissenschaftsphilosohie?* Hrsg. von P. Hoyningen-Huene und G. Hirsch (de Gruyter, Berlin), pp. 172–178.

Primas H. (1988): The essentials of the Copenhagen interpretation. Contribution to a panel discussion. In *Symposium on the Foundations of Modern Physics 1987. The Copenhagen Interpretation 60 Years after the Como Lecture.* Report Series TURKU-FTL-L45, Department of Physical Sciences, University of Turku, pp. 17–21 and p. 83.

Primas H. (1988): Visszavezethetö-e a kémia a fizikára? *Mérlwg* **88/3**, 247–266.

Primas H. (1988): Can we reduce chemistry to physics? In *Centripetal Forces in the Sciences, Vol. II*, ed. by G. Radnitzky (Paragon House, New York), pp. 119–133.

Primas H. (1988): Rebuttal to the comments by Marcelo Alonso. In *Centripetal Forces in the Sciences, Vol. II*, ed. by G. Radnitzky (Paragon House, New York), pp. 137–143.

Primas H. (1989): Great expectations. Review of *Beyond the Atom. The Philosophical Thought of Wolfgang Pauli* by K.V. Laurikainen (Springer, Berlin). *Nature* **338**, pp. 305–306.

Primas H. (1990): Biologie ist mehr als Molekularbiologie. In *Die Frage nach dem Leben*, hrsg. von E.P. Fischer und K. Mainzer (Piper, München), pp. 63–92. Partly reprinted in *Einzelmaterial 6, Sek. II, 6* RAAbits Biologie, November 1995.

Primas H. (1990): Zur Quantenmechanik makroskopischer Systeme. In *Wieviele Leben hat Schrödingers Katze?* Hrsg. von J. Audretsch und K. Mainzer (Wissenschaftsverlag, Mannheim), pp. 207–243.

Primas H. (1990): Realistic interpretation of the quantum theory for individual objects. *La Nuova Critica* **13–14**, pp. 41–72.

Primas H. (1990): Mathematical and philosophical questions in the theory of open and macroscopic quantum systems. In *Sixty-Two Years of Uncertainty. Historical and Physical Inquiries into the Foundations of Quantum Mechanics*, ed. by A.I. Miller (Plenum Press, New York), pp. 233–257.

Primas H. (1990): Induced nonlinear time evolution of open quantum objects. In *Sixty-Two Years of Uncertainty. Historical and Physical Inquiries into the Foundations of Quantum Mechanics*, ed. by A.I. Miller (Plenum Press, New York), pp. 259–280.

Primas H. (1990): The measurement process in the individual interpretation of quantum mechanics. In *Quantum Theory without Reduction*, ed. by M. Cini and J.M. Lévy-Leblond (Adam Hilger, Bristol), pp. 49–68.

Primas H. (1990): Beyond Baconian quantum physics. In *Kohti uutta todellisuuskäsitystä* (Yliopistopaino, Helsinki), pp. 100–112.

Primas H. (1991): Besprechung von *The Philosophy of Quantum Mechanics* von R. Healey (Cambridge University Press, Cambridge). *Physikalische Blätter* **47**, Nr. 1, 68.

Primas H. (1991): Necessary and sufficient conditions for an individual description of the measurement process. In *Symposium on the Foundations of Modern Physics 1990. Quantum Measurement Theory and its Philosophical Implications*, ed. by P. Lahti and P. Mittelstaedt (Singapore, World Scientific), pp. 332–346.

Primas H. (1991): Remarks on our conception of reality. In *Symposium on the Foundations of Modern Physics 1990. Quantum Measurement Theory and its Philosophical Implications*, ed. by P. Lahti and P. Mittelstaedt (Singapore, World Scientific), pp. 504–506.

Primas H. (1991): Vor-Urteile in den Naturwissenschaften. In *Wissenschaftstheorie und Wissenschaften*, hrsg. von H. Bouillon und G. Andersson (Duncker & Humblot, Berlin), pp. 49–63.

Primas H. (1991): Reductionism: palaver without precedent. In *The Problem of Reductionism in Science*, ed. by E. Agazzi (Kluwer, Dordrecht), pp. 161–172.

Primas H. (1992): Umdenken in der Naturwissenschaft. *Gaia* **1**, 5–15.

Primas H. (1992): Die Einheit der Wissenschaften: Ein gebrochener Mythos. In *Auf der Suche nach dem ganzheitlichen Augenblick. Der Aspekt Ganzheit in den Wissenschaften*, hrsg. von C. Thomas (Verlag der Fachvereine, Zürich), pp. 267–271.

Primas H. (1992): Mut zur Ganzheit. 400 Jahre einäugige Wissenschaft sind genug. *Kleine Schriften Nr. 21* (ETH Zürich).

Primas H. (1992): Time-asymmetric phenomena in biology. Complementary exophysical descriptions arising from deterministic quantum endophysics. *Open Systems & Information Dynamics* 1, 3–34.

Primas H. (1992): Umdenken in der Naturwissenschaft. *Vierteljahrschrift der Naturforschenden Gesellschaft in Zürich* 137, pp. 41–62.

Primas H. (1992): Warnung! Besprechung von *The Meaning of Quantum Theory* von J. Bagott (Oxford University Press, Oxford). *Nachr. Chem. Tech. Lab.* 40, 1152–1153.

Primas H. (1992): Ein Ganzes, das nicht aus Teilen besteht. Komplementarität in den exakten Naturwissenschaften. In *Mannheimer Forum* (Boehringer Mannheim, Mannheim), pp. 81–111. Reprinted in *Neue Horizonte 92/93* (Piper, München 1993), pp. 81–111.

Primas H. (1992): A propos de la mécanique quantique des systèmes macroscopiques. In *Erwin Schrödinger. Philosophie et Naissance de la Mécanique Quantique*, ed. par M. Bitbol et O. Darrigol (Editions Frontières, Gif-sur-Yvette Cedex), pp. 385–402.

Primas H. (1992): Es gibt keine Einsicht ohne innere Bilder. *Gaia* 1, 311–312.

Primas H. (1993): The Cartesian cut, the Heisenberg cut, and disentangled observers. In *Symposia on the Foundations of Modern Physics 1992. The Copenhagen Interpretation and Wolfgang Pauli*, ed. by K.V. Laurikainen and C. Montonen (Singapore, World Scientific), pp. 245–269.

Primas H. (1993): Mesoscopic quantum mechanics. In *Symposium on the Foundations of Modern Physics 1993. Quantum Measurement, Irreversibility and the Physics of Information*, ed. by P.J. Lahti, P. Busch and P. Mittelstaedt (World Scientific, Singapore), pp. 324–337.

Primas H. (1994): Vom sanften Umgang mit der Natur. In *Denkanstösse '95. Ein Lesebuch aus Philosophie, Natur- und Humanwissenschaften*, hrsg. von H. Bohnet-von der Thüsen (Piper, München), pp. 173–177.

Primas H. (1994): Umdenken in der Naturwissenschaft. *Rotary* 69(8), 48–51.

Primas H. (1994): Endo- and exo-theories of matter. In *Inside Versus Outside. Endo- and Exo-Concepts of Observation and Knowledge in Physics, Philosophy, and Cognitive Science*, ed. by H. Atmanspacher and G.J. Dalenoort (Springer, Berlin), pp. 163–193.

Primas H. (1994): Realism and quantum mechanics. In *Logic, Methodology and Philosophy of Science IX*, ed. by D. Prawitz, B. Skyrms and D. Westerståhl (Elsevier, Amsterdam), pp. 609–631.

Primas H. (1994): Hierarchic quantum descriptions and their associated ontologies. *Symposium on the Foundations of Quantum Mechanics 1994*, ed. by K.V. Laurikainen, C. Montonen and K. Sunnarborg (Editions Frontières, Gif-sur-Yvette), pp. 210–220.

Atmanspacher H., Primas H., Wertenschlag-Birkhäuser E., Hrsg. (1995): *Der Pauli-Jung-Dialog und seine Bedeutung für die moderne Wissenschaft* (Springer, Berlin).

Primas H. (1995): Über dunkle Aspekte der Naturwissenschaft. In *Der Pauli-Jung-Dialog und seine Bedeutung für die moderne Wissenschaft*, hrsg. von H. Atmanspacher, H. Primas, and E. Wertenschlag-Birkhäuser (Springer, Berlin), pp. 205–238.

Atmanspacher H. and Primas H. (1996): The hidden side of Wolfgang Pauli. An eminent physicist's extraordinary encounter with depth psychology. *Journal of Consciousness Studies* **3**, 112–126. Reprinted in *Journal of Scientific Exploration* **11**, 369–386 (1997).

Primas H. (1996): Synchronizität und Zufall. *Z. Grenzgeb. Psych.* **38**, 61–91.

Amann A. and Primas H. (1997): What is the referent of a non-pure quantum state? In *Potentiality, Entanglement, and Passion-at-a-Distance. Quantum Mechanical Studies in Honor of Abner Shimony*, ed. by R.S. Cohen, M. Horne, and J. Stachel (Kluwer, Dordrecht), pp. 9–29.

Primas H. (1997): The representation of facts in physical theories. In *Time, Temporality, Now*, ed. by H. Atmanspacher and E. Ruhnau (Springer, Berlin), pp. 243–266.

Primas H. (1998): Emergence in exact natural sciences. *Acta Polytechnica Scandinavica* **Ma 91**, 83–98.

LIST OF CONTRIBUTORS

Anton Amann
Universitätsklinik für Anästhesie
Leopold-Franzens-Universität
Anichstr. 35
A-6020 Innsbruck
Austria

Harald Atmanspacher
Institut für Grenzgebiete der Psychologie
Wilhelmstr. 3a
D-79098 Freiburg
Germany

and

Max-Planck-Institut
für extraterrestrische Physik
D-85740 Garching
Germany

Charles P. Enz
Département de Physique Théorique
Université de Genève
24 quai Ernest Ansermet
CH-1211 Genève 4
Switzerland

Richard R. Ernst
Laboratorium für Physikalische Chemie
ETH Zürich
CH-8092 Zürich
Switzerland

Bernard d'Espagnat
Laboratoire de Physique Théorique
et Hautes Energies
Université Paris-Sud
F-91405 Orsay Cedex
France

Karl Gustafson
Department of Mathematics
University of Colorado
Boulder, CO 80309
U.S.A.

and

International Solvay Institute
Université Libre de Bruxelles
CP 231, Campus Plaine
Boulevard du Triomphe
B-1050 Brussels
Belgium

Vittorio Hösle
Forschungsinstitut für Philosophie
Gerberstr. 26
D-30169 Hannover
Germany

Wilhelm Just
Donaulände 12
A-4100 Ottensheim
Austria

Frederick Kronz
Philosophy Department
University of Texas at Austin
Austin, TX 78712
USA

Günter Mahler
Institut für Theoretische Physik
Universität Stuttgart
Pfaffenwaldring 57
D–70550 Stuttgart
Germany

Ulrich Müller-Herold
Departement für
Umweltnaturwissenschaften
ETH Zürich
CH–8093 Zürich-Hönggerberg
Switzerland

H. Narnhofer
Institut für Theoretische Physik
Universität Wien
Boltzmanngasse 5
A-1090 Wien
Austria

Alfred Rieckers
Institut für Theoretische Physik
Universität Tübingen
Auf der Morgenstelle 14
D–72076 Tübingen
Germany

Dennis H. Rouvray
Department of Chemistry
University of Georgia
Athens, GA 30602–2556
USA

Abner Shimony
Departments of Philosophy and Physics
Boston University
Boston, MA 02215
USA

E.C.G. Sudarshan
Physics Department
and Center for Particle Physics
University of Texas
Austin, TX 78712-1081
USA

W. Thirring
Institut für Theoretische Physik
Universität Wien
Boltzmanngasse 5
A-1090 Wien
Austria

Carl Friedrich von Weizsäcker
Maximilianstr. 14c
D–82319 Starnberg
Germany

INDEX

The manufacturer's authorised representative in the EU is Springer
Nature Customer Service Centre GmbH, Europaplatz 3, 69115 Heidelberg,
Germany. If you have any concerns regarding our products, please
contact ProductSafety@springernature.com

Printed and bound by CPI Group (UK) Ltd, Croydon, CR0 4YY
24/04/2026
02096348-0009